# NUMBERPEDIA

EVERYTHING YOU EVER WANTED TO KNOW
(AND A FEW THINGS YOU DIDN'T)
ABOUT NUMBERS

**HERB REICH**

Skyhorse Publishing

Skyhorse Publishing books may be purchased in bulk at special discounts for sales promotion, corporate gifts, fund-raising, or educational purposes. Special editions can also be created to specifications. For details, contact the Special Sales Department, Skyhorse Publishing, 307 West 36th Street, 11th Floor, New York, NY 10018 or info@skyhorsepublishing.com.

www.skyhorsepublishing.com

10 9 8 7 6 5 4 3 2 1

Library of Congress Cataloging-in-Publication Data is available on file.
ISBN: 978-1-61608-084-6

Printed in Canada

# INTRODUCTION

Our lives are governed by numbers. Not the evanescent assumptions of the numerologist, but the widely applied codified numerical systems underlying our very existence.

Aside from among our family and friends, we are known not by name but identified by our social security number. We are located by the numbers in our home address. We are contacted via our telephone number. At birth, our readiness for joining the world at large is condensed into our Apgar score. In school we are graded one to one hundred and tested for IQ. Our entry to college is in part determined by our SATs, and once enrolled, our progress is monitored by our GPAs. We travel via flight numbers. We relax and/or seek our fortune by trying to pick winning Lotto numbers or attempting to control the numbers on a pair of dice or on a roulette wheel; sometimes we try to out-think the house in playing blackjack, aka twenty-one.

Our medical status report is replete with numerical estimates of our BP; our CBC, WBC, RBC, and PLT; our LYM, HDL, and LDL; for men, a PSA, and for women, an FSH or TSH. When we decide to redecorate our homes, we may refer to colors by name, but we order paint by number. In a neighborhood pub, those assembled will likely be discussing stats relating to the sport of choice. In an accident we exchange license numbers and insurance numbers. And so it is in just about every realm of human experience.

Society simply could not function in the absence of well-defined, mutually accepted number systems.

Numbers serve two functions: They may denote measurements (i.e., the weight of a box of cookies or the capacity of a gas tank) or designations (i.e., the address of a building or the number on a football player's jersey). Thus, they both quantify and identify. Some numbers take on meaning beyond their capacity to count or to name; for example, 7 is seen as lucky, 13 as unlucky, 666 as unholy.

Acknowledging that numbers describe us and rule us, this book examines the various meanings of numbers, and, more specifically, the associations each number conveys to us. Thus, 7 may refer to the number of dwarfs who accompany Snow White, the international calling code for Russia, the wonders of the ancient world, or even a winning roll of the dice. Or 40 could bring to mind the days of Lent, the atomic number of zirconium, the value of Roman XL, or even Mickey Mantle's record number of home runs in World Series games.

We limit our scope to the numbers 0 to 100. The range could obviously be extended upward, but the restraints of space demand prudence in setting the bounds of our coverage. Within our chosen parameters, we make no attempt to be exhaustive, but rather selective in choosing what seem to be the most interesting, most instructive, at times best known, at times most obscure, items.

This book is intended to serve as a Baedeker of meaningful numbers. Its 7,500 entries were extensively researched to include items from history, religion, mythology, current events, government, law, several sports, music, theater, movies, children's games, geography, literature, medicine, art, pop culture, and just about every field of endeavor one can mention. It should prove valuable in generating some interesting conversation, in provoking some arguments, and maybe even in settling a few. And it will no doubt be a boon to trivia enthusiasts by providing an unending supply of challenging minutiae, some trivial, some not so.

Items are accurate (at least we strove for such) as of the date the manuscript went into production, but some (such as sports records or the states' allotment of congressional seats) will possibly be outdated by the time the book comes off press. Given the mass of data, the inclusion of some errors is virtually inescapable, but we trust they are few. If you, the reader, find any, please convey them to the author through the publisher so corrections can be made in later printings.

Now, let us conclude this overview as all news stories once did: 30, 30, 30.

# 00

- retired basketball jersey number of:
  - ⇨ Johnny Moore, retired by the San Antonio Spurs.
  - ⇨ Robert Parish, retired by the Boston Celtics.
- shirt number worn by hockey player John Davidson, New York Rangers goalie, for one season in the 1970s.
- shirt number worn by hockey player Martin Biron in his rookie season as Buffalo Sabres goalie, 1995–96.

# 0

- in astrology, the number 0 is ruled by the dwarf planet Pluto.
- ISBN group identifier for books published in English-speaking areas: Australia, English-speaking Canada, Gibraltar, Ireland, New Zealand, Puerto Rico, South Africa, Swaziland, U.S., UK, and Zimbabwe (shared with 1).
- lowest season batting average of a major league pitcher (baseball record): Bob Buhl's BA was = .000 in 1962, at bat once for Milwaukee Braves, sixty-nine times for Chicago Cubs, all in thirty-five games. He struck out thirty-six times.
- nothing, nil, zero.
- number of actors in Samuel Beckett's play, *Breath*, which lasts thirty-five seconds and has no dialogue. First performed in Glasgow at the Close Theatre Club, October 1969.
- pH of concentrated hydrochloric acid, the most acidic substance.
- shirt number worn by hockey player Paul Bibeault, playing for the Montreal Canadiens as a backup goalie during the 1942–43 season.
- shirt number worn by hockey player Neil Sheehy, playing defense for the Hartford Whalers in 1988.

### Anton Bruckner's *Symphony No. 0*

- "Die Nullte" symphony (in D minor).

### ceiling zero

- a pilot's term for weather condition in which heavy cloud cover totally obscures the field of view.

### *Ceiling Zero*

- fast-paced 1935 film about commercial flyers, starring James Cagney and Pat O'Brien. Directed by Howard Hawks from a stage play by Frank "Spig" Wead, who also wrote the screenplay.

### goose egg

- slang for a score of zero; applicable to several sports and games.

### ground zero

- the exact point at which an explosion occurs.

### null set

- in mathematics, a set (i.e., a class of definable items) with no entries in it; written [0].

### zero

- a person or thing of no importance; a nonentity.

### Zero

- a Japanese fighter plane in World War II, produced by the Mitsubishi Corporation.

### zero degrees Celsius

- temperature at which water freezes (= 32 degrees Fahrenheit).

### zero hour

- the time when a military operation is scheduled to begin.

# *1/4*

- proportion of nickel in a U.S. five-cent piece (three-quarters is copper).

### farthing

- one-quarter of a penny (obsolete British coin).

### quarter

- U.S. coin worth twenty-five cents, a quarter-dollar, in circulation since 1796.

**quarter-bottle**

- champagne measure = 187 ml, called a split.

**quarter days**

- in the Middle Ages, traditional days in the British Isles when the quarterly rent was due.

**quarter-eagle**

- $2.50 gold U.S. coin, minted from 1796 to 1929.

# 1/3

- at his inaugural address on January 20, 1937, President Franklin D.Roosevelt, focusing on his objectives of social justice, said, "I see one-third of a nation ill-housed, ill-clad, ill-nourished . . ."

# 1/2

- the number of Haiuri, the half-god of the Xhosa people of South Africa, who has only one eye, one ear, one arm, and one leg.

**better half**

- a person's spouse.

**half**

- U.S. coin worth fifty cents, a half-dollar, in circulation since 1794.

**half a loaf is better than none**

- a phrase meaning: If you can't get it all, be satisfied with what you can get.

**half-and-half**

- (a) a coffee additive composed of half milk and half light cream.
- (b) in Britain, a mixed drink, usually half ale and half porter.

**half-assed**

- (adj.) insufficient, disorganized, inept; done with insufficient planning or forethought.

**half-baked**
- (adj.) ill-conceived, stupid.

**half-bottle**
- champagne measure = 375 ml.

**half-cent**
- U.S. coin in circulation from 1793 to 1857.

**half-cocked**
- (adj.) done with insufficient planning; done before fully ready to perform.

**half-dime**
- five-cent piece, U.S. coin minted from 1796 to 1860.

**half-dollar**
- a fifty-cent coin.

**half-eagle**
- five dollars gold piece, minted by the U.S. from 1795 to 1929.

***The Half-Naked Truth***
- Gregory LaCava's 1932 film comedy about a slick carnival pitchman (Lee Tracy) who promotes a sideshow exotic dancer (Lupe Velez) to instant celebrity.

**half-penny (ha'penny)**
- a former British coin, worth half a penny.

**half-pint**
- (a) (slang) a short person.
- (b) Pa's nickname for daughter Laura Ingalls in the TV series, *"Little House on the Prairie."*

**half-shot**
- (slang, adj.) Well on the way to being drunk; not in control of one's faculties.

## 3/4

**three-fourth**

- a type of book binding in which the leather (or cloth) back strip extends only one-third or halfway across the front and back boards, making the book bound mostly in leather or cloth, but not entirely.

**three-quarter**

- (adj.) describes the perspective of a portrait, midway between profile and front view.

**three-quarter time**

- a musical time signature indicating three quarter notes per measure; waltz time.

## 0.86

- lowest season ERA (baseball record), held by Tim O'Keefe, Troy Trojans, 1880.

## 0.96

- lowest "modern-era" (after 1900) season ERA (baseball record), held by Dutch Leonard, Boston Red Sox, 1914. He pitched 224.7 innings and won 19 of 36 games.

## 001

- medical classification code for cholera.

## 01

- number of the French Département of Ain.

## 1

- atomic number of the chemical element hydrogen, symbol H.
- badge number of Alexander Waverly, played by Leo G. Carroll, on the ABC-TV series, *The Man from U.N.C.L.E.* (1964–68).
- Beaufort number for wind force of 1-3 miles per hour, called "light air."

- dental chart designation for the right third molar (wisdom tooth).
- the first digit in the series 1, 2, 3, 4 . . .
- gauge (number) of a wood screw with shank diameter 5/64 of an inch.
- in astrology, the number 1 is ruled by the Sun.
- in baseball, 1 represents the pitcher's position.
- in Hebrew, the numerical value of the letter *aleph.*
- in mineralogy, Mohs Hardness Scale level characterized by talc; can be crushed by a fingernail.
- in the card game Casino, the value of an ace or the two of spades.
- ISBN group identifier for books published in English-speaking areas: Australia, English-speaking Canada, Gibraltar, Ireland, New Zealand, Puerto Rico, South Africa, Swaziland, U.S., UK, and Zimbabwe (shared with 0).
- number of decimal places in the inverse multiple designated by the International System prefix deci-, written "d-".
- number of seats in the House of Representatives allotted to each of the states of Alaska, Delaware, Montana, North Dakota, South Dakota, Vermont, and Wyoming (as of the 2000 Census).
- number of sides on a Möbius strip.
- number of 0s in the multiple designated by the International System prefix deca-, written "da-".
- only one satellite of the dwarf planet Pluto, named Charon, discovered in 1978.
- retired baseball uniform number of:
  - ⇨ outfielder Richie Ashburn, retired by the Philadelphia Phillies.
  - ⇨ second baseman Bobby Doerr, retired by the Boston Red Sox.
  - ⇨ manager Fred Hutchinson, retired by the Cincinnati Reds.
  - ⇨ second baseman/manager Billy Martin, retired by the New York Yankees.
  - ⇨ manager Billy Meyer, retired by the Pittsburgh Pirates.
  - ⇨ shortstop Pee Wee Reese, retired by the Los Angeles Dodgers.
  - ⇨ shortstop Ozzie Smith, retired by the St. Louis Cardinals.
- retired basketball jersey number of:
  - ⇨ Nate Archibald, retired by the Sacramento Kings.
  - ⇨ Walter A. Brown, retired by the Boston Celtics.

- ⇨ Frank Layden, retired by the Utah Jazz.
- ⇨ Oscar Robertson, retired by the Milwaukee Bucks. Robertson's number 14 with the Sacramento Kings has also been retired.
- ⇨ Larry Weinberg, retired by the Portland Trail Blazers.
- ⇨ Gus Williams, retired by the Seattle SuperSonics.
- retired football jersey number of:
  - ⇨ Ray Flaherty, retired by the New York Giants.
  - ⇨ Warren Moon, retired by the Tennessee Titans.
- retired hockey shirt number of:
  - ⇨ The Fans, retired by the Minnesota Wild.
  - ⇨ Ed Giacomin, retired by the New York Rangers.
  - ⇨ Glenn Hall, retired by the Chicago Blackhawks.
  - ⇨ Bernie Parent, retired by the Philadelphia Flyers.
  - ⇨ Jacques Plante, retired by the Montreal Canadiens.
  - ⇨ Terry Sawchuk, retired by the Detroit Red Wings.
  - ⇨ Wendell Young, retired by the Chicago Wolves.

- value of the medieval Roman numeral J.
- value of Roman numeral I.

### "All for one, one for all"

- motto of the Three Musketeers, in the novel by Alexandre Dumas (1844).

### back to square one

- a phrase meaning "having to start all over again."

### Bell X-1

- the plane, named *Glamorous Glennis*, in which Chuck Yeager broke the sound barrier on October 14, 1947. The first supersonic flight, at Mach 1.06, was achieved over Muroc Air Base in the Mojave Desert in southern California.

### Berlioz's *Symphony No. 1*

- Symphonie Fantastique

### Big Red One

- the U.S. Army's 1st Infantry Division, officially born in New York in June, 1917, as part of the American Expeditionary Force to France, to fight in World War I.

### *The Big Red One*

- World War II film, released in 1980, directed by Samuel Fuller and starring Lee Marvin, Mark Hamill, and David Carradine, that attempted to portray the war through the daily experience of the foot soldier. A good film, its title trading on the respected nickname of the Army's First Division, it was released when such movies were becoming passé.

### birdie

- a golf score of one stroke under par on a hole.

### bogey

- a golf score of one stroke over par on a hole.

### Canon 1 of the American Bar Association Code of Professional Responsibility

- A lawyer should assist in maintaining the integrity and competence of the legal profession.

### draw one

- (a) barroom code for a glass of tap beer.
- (b) lunch counter code for a cup of coffee.

### Dvorak's *Symphony No. 1*

- "The Bells of Zlonice" symphony (in C minor).

### FIFO

- first in, first out: inventory system ensuring that the oldest stock will be the first sold.

### first aid

- emergency medical help applied at the scene of an accident.

### First Amendment to the Constitution

- establishes freedom of religion, of speech, of press, of the right to assemble and to petition.

### First Arrondissement

- section of central Paris containing the Louvre museum.

**firstborn**

- the oldest child in the family.

**first-class**

- highest level of service or performance, usually the most expensive.

**first-class mail**

- class of mail in the U.S. postal system, including letters, postcards, and all mailable matter sealed against inspection.

**first come, first served**

- handling clients or issues in the order of their arrival.

**First Commandment**

- Thou shalt have no other gods before Me.

**First Crusade**

- called by Pope Urban II in 1095 to subdue the Turks. The crusader forces took Nicaea in 1097 and Antioch in 1098, then captured Jerusalem in 1099, where they massacred the Muslims and the Jews. As a result of the First Crusade, several small Crusader states were created, among them the Kingdom of Jerusalem. This was the only Crusade that ended in substantial victory.

**first-day cover**

- a stamped envelope postmarked on the day of the stamp's issue, posted from the city where it was issued.

**first-degree burn**

- a relatively undamaging burn in which the skin is reddened but not broken.

**first edition**

- the total first print run of a book. Within the first edition, however, minor corrections or changes may be made in the text, thus creating a sequence of printings (or impressions).

## First Empire

- the empire (1804 to 1814) established in France by Napoleon Bonaparte, replacing the First Republic.

## first estate

- the first of the three classes of society, the clergy.

## First Family

- the family with the highest social rank; (capitalized) the family of the President of the United States, or of a state governor or a city mayor.

## First Folio

- the first compilation of plays by Shakespeare, published in 1623, seven years after his death. Containing thirty-six plays, omitting only *Pericles*, it was collected by his fellow actors, John Heminges and Henry Condell.

## First French War of Religion

- the first of eight related wars (1562 to 1589) pitting the established Catholic monarchy and commoners ("one faith, one law, one state") against the new Reformation Protestants (in France, called Huguenots), among whom were many French nobles. Hostilities began in 1562, when servants of the Duc de Guise, a prominent Catholic, killed several Huguenot parishoners in the town of Vassy and set their church afire. The national Huguenot leadership mobilized under the Bourbon Prince de Condé and captured several strategic towns. The war ended with the Edict of Amboise in March 1563, which restricted some Protestant freedoms but permitted Huguenot nobles to worship as they wished on their own estates, a resolution that did not sit well with many Catholics.

## first-hand

- (adj.) new, of, or from the original source.

## First House

- in astrology, represents self-awareness and self-expression.

**first impression**
- the first run of printed material off press for the first edition of a work. Also called first issue.

**First International**
- International Workingmen's Association, organized in London in 1864 by Karl Marx, with the aim of uniting the trade unions around the world. Dissolved in 1974.

**first in war, first in peace, first in the hearts of his countrymen**
- an epithet for George Washington.

**First Lady**
- the wife of the President of the United States or of a state governor or city mayor.

**First Lateran Council**
- Roman Catholic council, called by Pope Callistus II in 1123, at which the Concordat of Worms was ratified, ending the Investiture Conflict. This council also forbade the marriage of clergymen and it annulled the ordinances of the antipope, Gregory VIII.

**First Law of Thermodynamics**
- (the law of conservation of energy): energy can be converted from one state to another, but cannot be created or destroyed.

**first light**
- dawn.

**first mate**
- a ship's ranking officer just below the captain.

**first-nighter**
- a person who attends a performance on opening night.

**first of Aries**
- the vernal equinox; the point at which the sun crosses the plane of the Earth's equator from south to north, making day and night approximately the same length.

**first offender**

- a person convicted of an offense for the first time.

**first-of-May**

- in circus slang, (n.) a novice; a new or inexperienced employee; (adj.) inexperienced.

**First Order of St. Francis**

- the Friars Minor, dating from 1209; priests and lay brothers who live in a community or friary, who have sworn to lead a celibate life of prayer, preaching, and penance. May be called friars, or simply Franciscans.

**first papers**

- in the U.S., the documents filed by an alien declaring the intention to become a citizen.

**first person**

- in grammar, the speaker.

**first position**

- in ballet, the position in which the heels are back to back and the toes point out to the sides, the feet forming a straight line.

**first president**

- George Washington, April 30, 1789 to March 3, 1797.

**first print**

- an early printed copy of a published work, usually at the proofing stage, before it is ready for general publication.

**first-rate**

- (adj.) of top quality, first-class, excellent.

**First Reich**

- The Holy Roman Empire until its dissolution in 1806, when Francis I, Emperor of Austria, gave up the old imperial crown.

**First Republic**

- the French Republic declared in September 1792 with the abolition of the monarchy following the French Revolution, replaced by the First Empire in 1804.

**first run**

- in its initial set of performances; not seen previously.

**first running**

- the first fraction collected from a fractional distillation, ordinarily carrying low boiling impurities.

**first state of the Union**

- Delaware ratified the Constitution on December 7, 1787.

**first string**

- the primary set of players, excluding replacements.

**First Triumvirate**

- in ancient Rome, the governing alliance of three men—Julius Caesar, Pompey, and Crassus—formed in 60 BC.

**first watch**

- (Navy use): the watch standing from 8:00 p.m. (2000 hours) to midnight (2400 hours).

**first water**

- highest degree of quality, with the best attributes.

**first wedding anniversary gift**

- paper is customary.

**first weight**

- in mining, the first measurement of roof pressure after removal of coal from a seam.

***First Wives Club***

- predictable 1996 film about three wives who have been dumped by their husbands, starring Goldie Hawn, Bette Midler, and Diane Keaton, with a first-rate supporting cast.

**first-word entry**

- a system of cataloging that utilizes the first word of the item name, ignoring any definite or indefinite article, such as "the" or "an," which may precede.

## First World

- collectively, the major industrialized nations of the world, including the U.S., Canada, those in Western Europe, and Japan.

## Group 1 elements (of the Periodic Table)

- the alkali metals: hydrogen (1), lithium (3), sodium, (11), potassium (19), rubidium (37), cesium (55), francium (87).

## hang one on

- (a) punch with the fist.
- (b) get drunk.

## *It Happened One Night*

- blockbuster film comedy of 1934, directed by Frank Capra, starring Claudette Colbert and Clark Gable. The first of the so-called "screwball comedies," it swept the Oscars, winning Best Picture, Best Director, Best Actor, Best Actress, and Best Screenplay (by Robert Riskin).

## a hot one

- a good joke.

## H1N1

- designation for the virus associated with swine flu.

## Leonard Bernstein's *Symphony No. 1*

- "Jeremiah" symphony.

## Mahler's *Symphony No. 1*

- "Titan" symphony (in D major).

## Mount Singleton

- name of a small mountain in the southwestern part of the Northern Territory, Australia, and one in Western Australia, near Lake Moor, about 200 miles north of Perth.

## Newton's First Law of Motion

- (the principle of inertia): a body at rest tends to stay at rest, and a body in motion tends to stay in motion at a uniform velocity in a straight line unless acted upon by an external force.

**number one**

- a child's euphemism for urination.

**number 1 billiard ball**

- the ball that is conventionally solid yellow.

*Once More, With Feeling*

- directed by Stanley Donen, from a screenplay by Harry Kurnitz (based on his play), this very funny 1960 film has Yul Brynner as a temperamental orchestral conductor and Kay Kendall, in her last film, as his protective wife. One high moment is supplied by Mervyn Johns, as the symphony's major supporter, when he comes in looking for Brynner and says to one of the musicians, "Take me to your leader."

**once-over**

- a visual inspection; a quick look.

*Once Upon a Time in America*

- 1984 gangster film from Sergio Leone traces the rise and fall of two New York City street kids, with outstanding performances by Robert De Niro and James Woods.

*Once Upon a Time in the West*

- Sergio Leone's monumental 1969 take on the film Western, with Henry Fonda as the epitome of the malevolent villain, and Charles Bronson, who has an unnamed score to settle. Ennio Morricone provides a memorable score.

**1 A**

- skylight photographic filter, of pale magenta tint, that reduces blue and ultraviolet light.

**one and only**

- one's sweetheart.

**1 April**

- April Fools' Day.

**one-arm bandit**

- slot machine.

## one-arm joint

- a cheap restaurant, the kind that once had chairs with a wide right armrest that served as a table for the diner's plate of food.

## one-at-one wheel

- in lace making, a small machine that winds a length of yarn on a bobbin, keeping a uniform tension.

## one-bagger

- in baseball, a one-base hit.

## one bell

- time check on-board ship; signifies 12:30 or 4:30 or 8:30, AM or PM.

## 1 Cranch 137

- citation for the Supreme Court decision in *Marbury v. Madison* (1803), in which the Court ruled that Congress exceeded its authority when it passed the Judiciary Act of 1789 and declared that legislation void, thereby establishing the Court's authority to invalidate acts of Congress that it finds inconsistent with the Constitution.

## *One Day at a Time*

- CBS-TV situation comedy featuring Bonnie Franklin, Mackenzie Phillips, Valerie Bertinelli, and Pat Harrington. First telecast December 15, 1975; ran through May 1984.

## 1 December, 1955

- in Montgomery, Alabama, Rosa Parks refuses to move to the back of the bus, thereby setting off an unstoppable drive for racial equality in the U.S.

## one dollar

- denomination of U.S. paper money that bears the portrait of George Washington.

## one dollar per hour

- minimum wage effective March 1, 1956.

**one down**

- lunch-counter code for an order of toast.

**one-end straight**

- in the game of poker, a sequence of four cards either ace-high or ace-low.

**one-eyed monster**

- TV set (CB slang).

**1 February**

- National Freedom Day, commemorating Lincoln's signing of the thirteenth Amendment to the Constitution on this day in 1865, abolishing slavery.

***One Flew Over the Cuckoo's Nest***

- Milos Forman directed this disturbing 1975 film about a rebellious inmate of a mental institution (Jack Nicholson) who challenges its restrictive dictates to the consternation of steely Nurse Ratched (Louise Fletcher). In addition to Best Picture and Best Director, Oscars went to Nicholson and Fletcher for Best Actor and Best Actress, and to the screenplay by Lawrence Hauben and Bo Goldman (based on the novel by Ken Kesey).

**one foot in the graveyard**

- lunch-counter code for an order of soup (probably derives from soup being a popular food for invalids).

**one for the road**

- a last drink before leaving.

**one from manhattan**

- lunch-counter code for an order of clam chowder.

**one-horse town**

- a small, insignificant town where nothing interesting happens.

***One in a Million***

- 1936 film introducing ice skater Sonja Henie, with Adolphe Menjou, Don Ameche, and Ned Sparks in supporting roles.

Humor by the Ritz Brothers; music by Borah Minevich and his Harmonica Rascals.

## 1-iron

- golf club called a driving iron, equivalent to the earlier cleek.

## 1 January

- New Year's Day; marked by the Orange Bowl and Rose Bowl football games.

## 1 July

- Canada Day, which commemorates the creation of the Dominion of Canada in 1867 by terms of the British North America Act.

## *One Life to Live*

- soap opera on ABC-TV since July 15, 1968. Originally a half-hour, it went to forty-five minutes in 1976, then to a full hour in 1978. The cast has included over 250 different performers. At this writing, the show is still on the air.

## *One Man's Family*

- TV soap opera on nighttime NBC from November 4, 1949 to June 21, 1952, switching to daytime in March 1954, and finally cancelled in April 1955. Originally a radio serial, it debuted on a local NBC affiliate in California on April 29, 1932, and went to network NBC on May 17, 1933. Aired 3,256 episodes, in various formats, through May 8, 1959, the longest running serial drama in American radio.

## 1 May

- (a) May Day; observed as a holiday by workers in several countries.
- (b) Lei Day in Hawaii.
- (c) Bird Day in Oklahoma.

## *One More Spring*

- Robert Nathan's 1933 parable of Christian charity, set in New York City during the Great Depression.

## 1 November

- All Saints' Day, Christian feast day.

### One of Our Aircraft Is Missing

- RAF plane is forced down in the Netherlands and its crew works its way back to England in this 1942 British film with Godfrey Tearle, Eric Portman, Hugh Williams, and Bernard Miles.

### One of Our Girls

- 1885 play by Bronson Howard; a social comedy contrasting American and French standards.

### One of Ours

- Willa Cather's 1922 Pulitzer Prize–winning novel on the short life of Claude Wheeler, a boy from the West who is killed in World War I.

### one of the boys

- a popular fellow who fits in well.

### one on

- lunch-counter code for a hamburger on a bun.

### one on the city

- lunch-counter code for a glass of water.

### one on the country

- lunch-counter code for a glass of buttermilk.

### one on the house

- lunch-counter code for a glass of water.

### one penny

- one cent; U.S. coin in circulation since 1793.

### 1 Place de la Concorde

- Paris address of the Jeu de Paume, once the Impressionist museum, now primarily a venue for photography exhibits.

### One Sided Lake

- a natural lake in southwestern Ontario Province, Canada.

### one silver star

- insignia of a brigadier general of the U.S. Army.

### "One singular sensation . . ."

- first line from the song "One," a hit tune from the 1975 smash Broadway show, *A Chorus Line.*

### one-striper

- in the navy, an ensign; in the army, a private first class, whose rank is shown by a single stripe.

### one too many

- describes having had enough whiskey to be drunk.

### one-track mind

- an attribute of focusing on one idea or task to the exclusion of all others.

### One Tree Hill

- a section of the city of Auckland, New Zealand.

### one-two

- a boxing strategy in which the first blow is followed immediately by a second with the other hand.

### one-two punch

- a decisive blow.

### *One, Two, Three*

- a frantic comedy with James Cagney as a Coca-Cola company manager in Germany charged with squiring the ditzy daughter (Pamela Tiffin) of a home-base exec when she visits West Berlin. But she marries an East Berlin Communist hippie (Horst Buchholz) just before her blustering father (Howard St. John) is due to arrive, the situation threatening Cagney's expected promotion. Billy Wilder produced and directed this 1961 rapid-fire farce.

### one up

- (a) bartender code for a glass of beer.
- (b) lunch-counter code for a cup of hot chocolate.

### *One Upmanship*

- a humorous book explaining how to stay one-up on (i.e., one foot ahead of) neighbors or competitors, by Stephen Potter (1952).

**one with**

- lunch-counter code for a hamburger with onion.

***One Woman's Life***

- Robert Herrick's 1913 novel follows the business and romantic life of Milly Ridge, an ambitious, self-centered woman intent on propelling herself into a social career.

**1-wood**

- golf club called a driver.

**"a partridge in a pear tree"**

- the gift on the first day of Christmas in the carol "The Twelve Days of Christmas."

**Prokofiev's *Symphony No. 1***

- "Classical" symphony (in D major).

**a quick one**

- (a) a single shot of whiskey.
- (b) an abbreviated sexual encounter.

**Schumann's *Symphony No. 1***

- "Spring" symphony (in B-flat major).

**single**

- (a) in baseball, a one-base hit.
- (b) a one-dollar bill.
- (c) (adj.) unmarried.

**single-action**

- (adj.) describes a type of gun for which the hammer must be cocked before each shot is fired.

**single-blind**

- refers to a type of experiment in which the researchers, but not the subjects, know who is receiving the active treatment and who is receiving the placebo. The purpose is to eliminate the anticipatory reaction, known as the placebo effect, among the subjects.

**single-breasted**
- (adj.) describes a jacket or coat that closes with only one row of buttons.

**single entry**
- a simple accounting system which records only amounts due to be received and amounts due to be paid by a business.

**single-measure**
- describes a wooden joint which is square on both sides.

**single-minded**
- (adj.) sincere, honest; having only one all-consuming purpose.

**single-name paper**
- a promissory note with no cosigning guarantor other than the borrower.

**single-phase**
- (adj.) relating to a circuit operated by a single alternating electric current.

**singles**
- (a) a sports match with only one player on each side.
- (b) small roofing tiles, about 8 x 12 inches.

**single-spaced**
- (adj.) typed or printed with no blank spaces between lines.

**single-sticker**
- a vessel with only one mast, as a sloop or a cutter.

**single tax**
- a tax that provides the one and only source of public revenue.

**singleton**
- in several card games, a card that is the only one of a suit in a hand.

## Singleton

- name of towns in: Texas (northwest of Houston); Lancashire, England; Sussex, England (few miles north of Chichester); New South Wales, Australia (about 80 miles north of Sydney); and in the central part of the Northern Territory, Australia.

## square one

- the beginning; e.g., "back to square one," start over.

## *The Taking of Pelham One Two Three*

- an exciting 1974 film about a subway and its passengers held for ransom with Walter Matthau as the good transit cop, and Robert Shaw as the cold-blooded leader of the hijackers. Remade in 2009 with Denzel Washington in the Matthau role and John Travolta in the Shaw role; directed by Tony Scott, with a modified title, *The Taking of Pelham 1 2 3*. Also a made-for-TV version in 1998 with Edward James Olmos and Vincent D'Onofrio.

## Tchaikovsky's *Symphony No. 1*

- "Winter Daydreams" symphony (in G minor).

## Tiger 1

- The main tank of German forces in World War II.

## Vaughan Williams's *Symphony No. 1*

- A "Sea Symphony" (in D major), a choral symphony set to poems by Walt Whitman.

## "The Wonderful 'One-Hoss Shay' "

- also called "The Deacon's Masterpiece"; this Oliver Wendell Holmes poem with its odd rhythmic structure, first published in 1858, is a parable of the breakdown of Calvinism.

## *You Only Live Once*

- Fritz Lang's 1937 filmed tragedy of a Depression-era couple, portrayed by Sylvia Sidney and Henry Fonda, who were turned into criminals by bad luck and the hardships of the times.

## *1.5*

- 1.5 inches, record one-minute rainfall, at Barot, Guadeloupe, November 26, 1970.

### One and a Half Degree Channel

- a passage between Haddummati Atoll and Suvadiva Atoll at the southern end of the Maldive Islands in the Indian Ocean.

### One and a Half Mile Opening

- a strait north of Cook's Passage, in the Great Barrier Reef, Australia.

### one-and-a-half-striper

- a naval lieutenant, junior grade (U.S. Navy usage).

## *1.82*

- lowest career ERA (baseball record), earned by Ed Walsh, with Chicago White Sox, 1904–16.

## *002*

- decimal ASCII code for [start of text].
- medical classification code for typhoid and paratyphoid diseases.

## *02*

- number of the French Département of Aisne.

## *2*

- Article of the U.S. Uniform Commercial Code that deals with sales.
- atomic number of the chemical element helium, symbol He.
- badge number of Agent Illya Kuryakin, played by David McCallum, on the NBC-TV spy spoof, *The Man from U.N.C.L.E.* (1964–68).
- base of the binary number system.
- Beaufort number for wind force of 4 to 7 miles per hour, called "light breeze."

- bekas in a shekel (ancient Hebrew measure of weight).
- considered a favorable number in Chinese culture because in Cantonese, the word for "two" sounds the same as the word for "easy." The Chinese say good things come in pairs.
- cups in a pint.
- dental chart designation for the right second molar.
- duality, an important concept in philosophy and many religions, shown in such pairings as good-evil, yin-yang, right-wrong, man-woman, night-day, etc.
- gauge (number) of a wood screw with a shank diameter of 3/32 of an inch.
- in astrology, the number 2 is ruled by the Moon.
- in baseball, 2 represents the catcher's position.
- in Hebrew, the numerical value of the letter *beth.*
- in mineralogy, Mohs Hardness Scale level characterized by gypsum; can be scratched by a fingernail.
- in rugby union, the number of the hooker.
- ISBN group identifier for books published in French-speaking areas: France, Luxembourg, and French-speaking Belgium, Canada, and Switzerland.
- lethechs in a homer (ancient Hebrew measure of dry capacity).
- number of decimal places in the inverse multiple designated by the International System prefix centi-, written "c-".
- number of "every living thing" Noah is told to bring to the ark (Genesis 6:19).
- number of faces on the ancient Roman god Janus, enabling him to look backward at the past and forward to the future at the same time.
- number of participants it takes to tango.
- number of presidents of France under the Fourth Republic: Vincent Auriol, 1947; and René Coty, 1954–59. (The Fifth Republic followed.)
- number of presidents of France under the Second Republic: Louis-Eugène Cavaignac, 1848; and Louis-Napoléon Bonaparte, 1848–52. (The Second Empire followed.)
- number of satellites circling the planet Mars: Phobos, which circles Mars in about seven hours, and Deimos, which revolves around the planet in about thirty-one hours.

- number of seats in the House of Representatives allotted to each of the states of Hawaii, Idaho, Maine, New Hampshire, and Rhode Island (as of the 2000 Census).
- number of 0s in the multiple designated by the International System prefix hecto, written "h-".
- pints in a quart.
- quarts in a magnum (spirits measure).
- retired baseball uniform number of:
  - ➪ second baseman Nellie Fox, retired by the Chicago White Sox.
  - ➪ second baseman Charlie Gehringer, retired by the Detroit Tigers.
  - ➪ manager Tommy Lasorda, retired by the Los Angeles Dodgers.
  - ➪ second baseman Red Schoendienst, retired by the St. Louis Cardinals.

- retired basketball jersey number of:
  - ➪ Red Auerbach, retired by the Boston Celtics.
  - ➪ Junior Bridgeman, retired by the Milwaukee Bucks.
  - ➪ Chuck Daly, retired by the Detroit Pistons.
  - ➪ Alex English, retired by the Denver Nuggets.
  - ➪ Moses Malone, retired by the Philadelphia 76ers.
  - ➪ Mitch Richmond, retired by the Sacramento Kings.
  - ➪ Malik Sealy, retired by the Minnesota Timberwolves.

- retired hockey shirt number of:
  - ➪ Doug Harvey, retired by the Montreal Canadiens.
  - ➪ Tim Horton, retired by the Buffalo Sabres.
  - ➪ Rick Ley, retired by the Hartford Whalers.
  - ➪ Brian Leetch, retired by the New York Rangers.
  - ➪ Al MacInnis, retired by the St. Louis Blues.
  - ➪ Eddie Shore, retired by the Boston Bruins.
- the smallest and the only even prime number.
- spans in a cubit (ancient Hebrew measure of length).
- sum of the first three Fibonacci numbers $(0 + 1 + 1)$.
- symbolic of independence in Finland, where two candles are displayed on Independence Day, signifying division, and thus, independence.
- tablespoons in a fluid ounce.
- Title of the United States Code that deals with the Congress.

- two years, average life of a five dollar bill.
- value of Roman numeral II.
- value of the ten of diamonds in the card game Casino.

**Berlioz's *Symphony No. 2***
- "Harold in Italy" symphony.

**Big Twin Lake**
- a geographic feature about 5 miles south of Winthrop, Washington State.

**Canon 2 of the American Bar Association Code of Professional Responsibility**
- a lawyer should assist the legal profession in fulfilling its duty to make legal counsel.

**Copland's *Symphony No. 2***
- the "Short" symphony.

**CV-2**
- registry of the U.S. Navy aircraft carrier, USS *Lexington*. Fatally damaged in the Battle of the Coral Sea.

**The Dioscuri**
- Castor and Pollux, sons of Zeus.

**double**
- in baseball, a two-base hit.

**double-action**
- (adj.) describes the type of gun for which each pull of the trigger both cocks and fires it.

**double agent**
- a spy who spies on the country to which he nominally has allegiance.

**double bass**
- the largest of the violin family of stringed instruments, usually played standing up.

**Double Bayou**
- a town in eastern Texas, about 30 miles northeast of Galveston.

**double-beat valve**

- a hollow valve used to control high-pressure fluids.

**double-bellied**

- (adj.) describes a baluster which has been turned alike at both ends.

**double-bill**

- to charge twice, either the same or different accounts, for one transaction.

**double bind**

- a psychological term describing a situation in which someone is given conflicting demands, at different levels of communication, putting the person in a "no-win" situation.

**double-blind**

- refers to a type of study in which neither the researchers nor the subjects know who is in the experimental group and who is in the control group. The arrangement is designed to eliminate the possibility of unintended bias by the experimenters.

**double bogey**

- a golf score of two strokes over par on a hole.

**double bond**

- a chemical connection by two covalent bonds between two atoms of a molecule.

**double-breasted**

- (adj.) describes a jacket or coat having two parallel vertical rows of buttons on the front.

**double-check**

- to examine again in order to ensure accuracy or effective functioning.

**double-clutch**

- in driving an automobile, to shift gears twice, once into neutral and then into the desired gear, releasing the clutch twice, once for each gear change.

**double cream**

- a French soft cheese mostly of cow's milk with cream added to comprise at least 60 percent fat.

**double cross**

- betrayal of colleague or co-worker.

**double crown**

- a standard size of printing paper, 20 x 30 inches.

**double dagger**

- the diesis (‡), a reference mark frequently used in footnotes.

**double date**

- a date in which two couples go together.

**double-dealing**

- being duplicitous, deceitful, treacherous.

**double demy**

- a standard size of printing paper, 22-½ x 35 inches.

**double-dip**

- to earn a salary from one position while drawing either a pension or a salary in another role from the same employer, particularly as a government employee.

**double Dutch**

- a form of jumping rope where two ropes are used concurrently, synchronized and swung inward in opposite directions.

**double exposure**

- in photography, having two images on the same piece of film or plate.

**double eagle**

- a gold coin with a face value of twenty dollars, minted by the U.S. from 1849 to 1933.

**double entendre**

- a word or a phrase of ambiguous meaning that is open to two interpretations, one of which is usually risqué.

## double-entry

- an accounting method in which each transaction is entered twice in a ledger once as a debit in one account, and once as a credit in another.

## double fault

- in tennis, squash, and some other sports, two faults in succession which results in the loss of the point.

## double feature

- two films shown together for the price of a single admission to the theater.

## double flat

- in music, a symbol [♭♭] that lowers the following note two semitones.

## double foolscap

- a standard size of printing paper, 17 x 27 inches.

## Double Headed Shot Cays

- small islands on Cay Sal Bank, in the western Bahamas, about 80 miles due south of Key Largo, Florida.

## doubleheader

- two separate sport matches, usually between the same teams, one immediately following the other on the same day; most common in baseball.

## Double Hill

- a land feature outside Dunedin on South Island, New Zealand.

## double-hung

- (adj.) describes a window which has top and bottom sashes, each balanced by a separate sash weight so as to be movable vertically in its own grooves.

## double imperial

- a size of brown paper used in England, 29 x 45 inches.

## *Double Indemnity*

- 1944 film noir classic in which insurance salesman Fred MacMurray is lured into a murder plot by seductive Barbara

Stanwyck in order to cash in on a double indemnity insurance policy. Edward G. Robinson plays the investigator who begins to unravel the plot. Directed by Billy Wilder, who also wrote the screenplay with Raymond Chandler, from a James N. Cain novel.

**Double Island Point**
- a cape into the Pacific Ocean about 100 miles north of Brisbane in Queensland, Australia.

**double jeopardy**
- the process of trying a person a second time for the same crime.

*A Double Life*
- directed by George Cukor, this 1947 film won Ronald Coleman a Best Actor Oscar for his riveting portrayal of an actor whose onstage Othello takes over his offstage life.

**Double Mountain**
- a land feature in the Tehachapi Mountains in California, about 40 miles southeast of Bakersfield.

**Double Mountain Fork**
- a 64-mile bend in the Brazos River, from Clairmont to Old Glory, in Kent Stonewall County, Texas, attractive for its white-water paddling and kayaking.

**double nickel**
- slang for 55, especially when referring to a speed limit of 55 miles per hour.

**Double Peak**
- a geographic feature in the Chigmit Mountains, west of Cook Inlet, in southern Alaska.

**double pair royal**
- in cribbage, holding four cards of the same denomination; scores twelve points.

**double parting**
- double set of tracks in a mine to allow tram traffic in two directions.

**double play**

- in baseball, a play in which two outs are made.

**Double Point**

- a spit of land facing the Great Barrier Reef about 15 miles south of Innisfail in the northern part of Queensland, Australia.

**double reed**

- relates to a woodwind instrument having a mouthpiece of two reeds vibrating one against the other, such as an oboe or a bassoon.

**double royal**

- a standard size of printing paper, 24 x 38 inches; in England, 25 x 40 inches.

**double sharp**

- in music, a symbol [##] that raises the preceding note two semitones.

**double small**

- a standard size of cut card, 3-½ x 5 inches; in England, 3-5/8 x 4-¾ inches.

**double-spaced**

- (adj.) typed or printed with each pair of typed or printed lines separated by a blank line.

**double spread**

- two facing pages occupied by an advertisement, photo, or story.

**Double Springs**

- a town west of Sipsey Creek in northwestern Alabama.

**doubletalk**

- gobbledygook; either nonsense or circumlocution used as an evasion tactic or to cause intentional confusion.

**Dual Alliance**

- an October 1879 agreement between Germany and Austria-Hungary intended to assure Russia of Germany's peaceful intentions in the Balkans, but also to guarantee military assistance if either were attacked by Russia. Duration was five years, but it was renewed repeatedly until 1918.

**eagle**

- a golf score of two strokes under par on a hole.

***The Eagle with Two Heads***

- 1948 French film by Jean Cocteau (based on his stage play) about an anarchist poet intent on assassinating the Queen who develops a passion for her instead, love overcoming violence.

**Group 2 elements (of the Periodic Table)**

- the alkaline earth metals: beryllium (4), magnesium (12), calcium (20), strontium (38), barium (56), radium (88).

**hero twins of Mayan myth**

- Xbalanque and Hunahpu, sons of Hun-Hunahpu.

***A Kid for Two Farthings***

- a delightful comedy-fantasy about a poor boy from London's East End who searches for a magical unicorn but finds a goat instead. Directed by Carol Reed, this 1956 charmer is from a screenplay by Wolf Mankowitz, who wrote the novel on which it is based.

**Kingdom of the Two Sicilies**

- name given by Bourbon King Ferdinand IV of Naples to his domain, newly merged in 1916 after the end of the Napoleonic era. (Before the French Revolution he had been King of Sicily and King of Naples, separate states.) The kingdom lasted until 1860, when King Francis II was overthrown by Garibaldi.

**K2**

- second-highest mountain in the world at 28,250 feet, in the Karakoram Range in northern Kashmir; also called Godwin Austen.

**Leonard Bernstein's *Symphony No. 2, for Piano and Orchestra***

- "The Age of Anxiety" symphony.

## Mahler's Second Symphony

- "Resurrection" symphony (in C minor).

## Mendelssohn's Second Symphony

- "Lobgesang" symphony (in B-flat major).

## Newton's Second Law of Motion

- a force exerted upon a body will change the body's velocity in the direction of the force, the acceleration directly proportional to the force and inversely proportional to the mass of the body.

## not worth two cents

- said of something perceived to have no value.

## number two

- a child's euphemism for a bowel movement.

## number 2 billiard ball

- the ball that is conventionally solid blue.

## PDD-2

- Presidential Decision Directive 2 deals with Organization of the National Security Council, issued by President Bill Clinton on January 20, 1993.

## *The Postman Always Rings Twice*

- classic 1946 film adaptation of James M. Cain's steamy novel has John Garfield and Lana Turner killing her husband (Cecil Kellaway) so they can use his savings to finance their torrid love affair. Directed by Tay Garnett, who managed to keep the undertone of sex without offending the watchful movie censors of the period. Filmed twice before, in France (1939) and in Italy (1942), and once later, in 1981, with Jack Nicholson and Jessica Lange, but Garnett's version is by far the best.

## *Prisoner of Second Avenue*

- stage comedy-drama by Neil Simon about a successful business executive who gets close to a nervous breakdown after he loses his job. Opened on Broadway in November 1971, with Peter Falk and Lee Grant. Made into a film in

1975, directed by Melvin Frank, with Jack Lemmon and Anne Bancroft in the leading roles.

## Rimsky-Korsakoff's Second Symphony

- "Antar" symphony (in C minor).

## Second Amendment to the Constitution

- guarantees the right to keep and bear arms.

## secondary color

- a color formed by mixing roughly equivalent parts of two primary colors.

## second banana

- the straight man in a comedy routine who feeds lines to the comedian.

## Second Cataract

- a section of white water in the Nile River, near the border between Egypt and Sudan.

## second childhood

- state of dotage or senility.

## *Second Chorus*

- 1940 musical film featuring Fred Astaire and Burgess Meredith, both trumpet players, vying for the hand of Paulette Goddard. Songs from several composers, mostly Johnny Mercer.

## *Second City TV*

- a satirical TV comedy series that parodied TV shows. The Second City group had begun as an improvisational-comedy troupe in 1959 Chicago, but it was a Toronto branch that began the TV series in 1977. In 1981 the series name changed to *SCTV Network 90*, and then to *SCTV Channel* from 1983 to 1984.

## second-class

- (adj.) inferior, mediocre.

## second-class citizen

- a person considered inferior in status and rights.

## second-class mail

- class of mail in the U.S. postal system, including newspapers and magazines.

## second coming

- Christ's return on the last day of the world to judge all humankind.

## Second Commandment

- Thou shalt make no graven images.

## second cousin

- a child of one's first cousin.

## Second Crusade

- following the Turkish conquest of the town of Edessa in 1144, armies of France, led by Louis VII, and of Germany, led by Holy Roman Emperor Conrad III, independently set out for Asia Minor in 1147. Both suffered defeats by Turkish forces until they finally merged in 1148. Their joint attack on Damascus also failed, and the leaders returned home in 1148 and 1149, the Second Crusade having ended in woeful failure.

## second-degree burn

- in which the outer layer of skin is damaged, most often with blistering and liquid collected in the tissue below the burn.

## Second Empire

- the empire (1852–70) established in France by Louis-Napoléon Bonaparte; replaced the Second Republic.

## second estate

- the nobility as a class of society.

## second fiddle

- one of inferior rank or lesser quality; second best.

## Second French War of Religion

- the Catholic Duc de Guise had been killed in the First War, and his brother, the Cardinal de Lorraine, now argued for more suppression of the Huguenots. When Spanish troops appeared along the eastern French border in 1567, heading to support the Spanish subjugation of the Netherlands, the Huguenots feared a Catholic plot to exterminate them and they attempted a coup at Meaux to seize the king. The plot failed and led to the Second War, mostly a repeat of the First, ending in 1568 with the Peace of Longjumeau, which reaffirmed the earlier Peace of Amboise.

### second-guess

- to correct or criticize after the result is known.

### second hand

- (adj.) previously owned or used by another person.

### Second House

- in astrology, deals with material resources and possessions.

### Second International

- (Socialist International), set up headquarters in Brussels in 1889 to promote the unity of Socialist parties in several countries. Fractured by World War I.

### Second Lateran Council

- held in 1139, convoked by Pope Innocent II, to reunify the Church after the schism of the antipope Anacletus II, who died in 1138. This Council also condemned usury and forbade monks from studying medicine and civil law.

### Second Law of Thermodynamics

- (increase in entropy): the quality of energy is degraded irreversibly because there is always an increase in disorder; the principle of degradation of energy.

### Second Order of Saint Francis

- the Poor Ladies or Poor Clares (after Clare of Assisi), founded in 1212; women who live in a community or nunnery, and who have sworn to lead a celibate life of poverty, prayer, and penance. Also known simply as Clares, they divide into two branches: Colettines, who live in rigorous cloisters with strict rules of obedience, and Urbanists, less austere, who sometimes work outside their convents.

## Second Peace of Paris

- the final French settlement (November 20, 1815) following Napoleon's disastrous defeat at Waterloo, by which France had to cede several land areas to victorious nations, to give up fortresses on her frontiers, to return art treasures ransacked from all over Europe, and to pay 700 million francs to offset the cost of the war.

## second person

- in grammar, the person who the speaker addresses.

## second position

- in ballet, position in which the feet are spread slightly apart and are at right angles to the direction of the body, the toes pointing out in opposite directions.

## second president

- John Adams, March 3, 1797 to March 3, 1801.

## second-rate

- (adj.) inferior; of low quality.

## Second Reich

- the German Empire, from 1871 to 1919.

## second sight

- the faculty of seeing future events; clairvoyance.

## second state of the Union

- Pennsylvania ratified the Constitution on December 12, 1787.

## second-story man

- a burglar adept at entering premises through an upstairs window.

## second string

- a player or group of players that substitute for the starters.

## Second Triumvirate

- in ancient Rome, the governing alliance of three men—Augustus, Antony, and Lepidus—formed in 43 BC.

**second wedding anniversary gift**
- cotton is customary.

**second wind**
- restored energy or strength after having tired earlier.

**Shostakovich's Second Symphony**
- "To October" symphony (in B-flat major).

**snake eyes**
- in craps, when both dice turn up, each with one spot showing.

*A Tale of Two Cities*
- Charles Dickens's intricate 1859 novel that interweaves several personal stories into the events of the French Revolution. The novel opens with one of the most quoted lines in all of literature: "It was the best of times, it was the worst of times." Several attempts to film the story have been made, the most successful in 1935, with Ronald Colman as Sidney Carton, the doomed martyr, and Blanche Yurka as the knitting Madame Defarge. Others have included an impressive 1917 silent film, with William Farnum playing both Carton and the titled Charles Darnay, a 1958 version with Dirk Bogarde, and one in 1980, made for TV.

**Tchaikovsky's Second Symphony**
- "Little Russian" symphony (in C minor).

*Tea for Two*
- 1950 musical comedy film with Doris Day and Gordon MacRae, an updated rewrite of the 1924 Broadway hit, *No, No, Nanette*, using the technique of flashbacks to capture the period feeling of the original.

**twice cooked pork**
- a spicy Chinese dish that really is cooked twice. The pork is first simmered, then stir-fried in hot and sweet bean sauce with scallions or leeks and bell peppers.

*Twice-Told Tales*
- a collection of thirty-nine stories by Nathaniel Hawthorne, originally published in 1837, and later expanded in 1842.

## Twin

- a town on the south side of the Strait of Juan de Fuca in Washington State.

## Twin Beds

- 1914 stage play by Salisbury Field and Margaret Mayo, the granddaddy of all bedroom farces, noted at the time for being "naughty." Several film adaptations were made; the first, a 1920 silent film, with Carter and Flora Parker DeHaven in their film debut; then, two sound versions, the first in 1929, the script only tangentially related to the original, and the second in 1942, a limp rendition starring George Brent and Joan Bennett.

## Twin Bridges

- name of towns in southwestern Montana on Beaverhead River, and in northern California, about 15 miles south of Lake Tahoe.

## Twin Brooks

- a town in northeastern South Dakota.

## Twin Buttes Reservoir

- a water catchment area south of San Angelo, Texas.

## Twin Cities

- Minneapolis and St. Paul, both in Minnesota.

## Twin City

- (a) a town west of Fort William in the southwestern part of Ontario Province, Canada.
- (b) a city in south-central Idaho.

## Twin Heads Mountain

- a peak in the northern part of Western Australia.

## Twin Hills

- a town in southwest Egypt.

## Twin Humps Peak

- a mountain in the McIlwraith Range on Cape York Peninsula in Queensland, Australia.

**Twin Lake**

- (a) a town in Michigan, about 60 miles northwest of Grand Rapids.
- (b) a lake in Michigan, about 100 miles east of Traverse City.

**Twin Lakes**

- (a) pairs of lakes in: north-central Arkansas (Lake Norfolk and Bull Shoals Lake); west central Colorado, southeast of Mount Elbert; central Oregon, south and west of Lookout Mountain; Cook County, Illinois; western Kenosha County, Wisconsin (Lakes Mary and Elizabeth); the Toiyabe National Forest, north of Yosemite National Park in California; the Tongass National Forest, Alaska; and in north-central Newfoundland, Canada.
- (b) name of towns in: Colorado, at the foot of Mount Elbert; southeastern Wisconsin, about 20 miles west of Kenosha; Broward County, Florida, north of Fort Lauderdale; Lowndes County, Georgia; Litchfield County, Connecticut; Mono County, California; on the Keweenaw Peninsula, north of Muskegon, Michigan; and in northeast Pennsylvania, about ten miles west of Port Jervis.
- (c) reservoirs in central Colorado, south of Leadville, and in southern Idaho, north of Twin Falls.
- (d) a state park southwest of Richmond, Virginia.
- (e) a state beach in Santa Cruz, California.

**Twin Mountain**

- a geographic feature in the White Mountains in north-central New Hampshire.

**Twin Peaks**

- highland pairs in central San Francisco, California; in the Salmon River Mountains in south-central Idaho; in the Cascade Range east of Everett in Washington State; and in the southwestern part of Western Australia.

*Twin Peaks*

- David Lynch's surreal TV series decribed by reviewers as "quirky." On ABC from April 1990 to August 1991, featuring

Kyle MacLachlan as FBI special agent Dale Cooper and Michael Ontkean as sheriff Harry S. Truman. Ranked number 20 in the *TV Guide* list of the 25 Top Cult Shows Ever (May 30, 2004). Reborn in 1992 as a prequel movie, titled *Twin Peaks: Fire Walk with Me.*

## Twin Rivers

- a town in central New Jersey, about 10 miles east of Princeton.

## Twin Rocks

- (a) islands in the Archipelago of the Recherche off the southern coast of Western Australia.
- (b) islands in Aden Harbor in the southwestern part of South Arabia.

## The Twins

- (a) a town in central South Australia.
- (b) a peak on the northern end of South Island, New Zealand.
- (c) Minnesota Twins, major league baseball team with home field in Minneapolis.
- (d) familiar name for Romulus and Remus, legendary founders of the city of Rome.

## Twins Creek

- a waterway in the eastern part of South Australia.

## Twin Valley

- a town in western Minnesota, about 35 miles northeast of Moorhead.

## 2A

- number of the French Département of Corse-du-Sud.

## two-a-day

- vaudeville patois for big-time (derives from playing big cities where vaudeville theaters did two full shows per day).

## *Two Admirals*

- James Fenimore Cooper's 1842 novel about the British navy before the Revolutionary War. Cooper's realistic portrayal of a great naval battle was intended to press the need for an effective American fleet.

### Two Against Nature

- Steely Dan's first studio album in twenty years, Grammy–winning Album of the Year for 2000.

### 2 April

- Pascua Florida Day, a state holiday commemorating the discovery of Florida by Ponce de Leon in 1513. He called the area Pascua Florida, likely because his discovery occurred near Easter.

### two aspects of language

- the metonymy (similarity) and the metaphor (continuity). (R. Jakobson, *The Two Aspects of Language*, 1956)

### 2B

- number of the French Département of Haute-Corse.

### two-bagger

- in baseball, a two-base hit.

### two bells

- time check on-board ship, signifies 1:00 or 5:00 or 9:00, a.m. or p.m.

### two-bit

- (adj.) worth twenty-five cents; small-time, inferior.

### two bits

- slang for twenty-five cents.

### Two Bridges

- a town in Devon, England, about 25 miles southwest of Exeter.

### Two Butte Creek

- a waterway in the southeastern corner of Colorado.

### Two Buttes

- a town in the southeastern corner of Colorado.

## two-by-four

- a piece of lumber measuring 1-½ x 3-½, of varying length.

## two-cent piece

- U.S. bronze coin in circulation from 1864 to 1873.

## two cents' worth

- one's opinion or remark interjected in a conversation.

## Two Creeks

- a town in the southwest corner of Manitoba Province, Canada.

## two dollars

- denomination of U.S. paper money that bears the portrait of Thomas Jefferson.

## Twodot

- a town in Montana, about 100 miles east of Helena.

## two-faced

- (adj.) deceitful, deceptive.

## 2 February

- (a) Candlemas Day; Christian feast day commemorating the Presentation of Jesus.
- (b) Groundhog Day.

## twofer

- a ticket, usually to some performance, that admits two people for the price of one.

## Twofold Bay

- an inlet off the Tasman Sea in New South Wales, Australia, about 30 miles north of Cape Howe.

## *Two for the Money*

- (a) TV quiz show, emceed first by Herb Shriner, later by Sam Levenson. First telecast on NBC on September 30, 1952, it switched to CBS in August 1953, and was last aired on September 7, 1957.
- (b) 2005 film about sports betting, with Al Pacino, Matthew McConaughey, and Rene Russo, directed by D. J. Caruso.

### *Two for the Seesaw*

- playwright William Gibson's first Broadway hit, opening in January 1958, starring Henry Fonda and a then-unknown actress named Anne Bancroft (brought in to replace Lainie Kazan, who bowed out during rehearsals), who won that year's Tony for Best Actress. The play became a film in 1962, directed by Robert Wise, with Robert Mitchum and Shirley MacLaine in the lead roles. It then metamorphosed into a 1973 stage musical with a score by Cy Coleman and Dorothy Fields, starring Michele Lee and Ken Howard, with Tommy Tune in his first Broadway role.

### 2,4,5-T

- 2,4,5-trichlorophenoxyacetic acid, known as Agent Orange.

### *Two Gentlemen of Verona*

- one of Shakespeare's earliest comedic plays, about friends Valentine and Proteus, rivals for the hand of Julia, daughter of the Duke of Milan.

### Two Harbors

- a town on Lake Superior in northeastern Minnesota.

### Two Headed Island

- piece of land in the Gulf of Alaska, off the southeast corner of Kodiak Island, Alaska.

### two heads are better than one

- a saying meaning that two people can complete a task or solve a problem more expeditiously than one.

### Two Hills

- a town in the eastern part of Alberta, Canada.

### 2-iron

- golf club equivalent to the earlier mid-iron.

### two kilograms

- weight of men's discus.

### two-light frame

- a window containing one mullion that divides the window space into two glass areas.

### Two Little Confederates

- an 1888 tale by Thomas Nelson Page of two Southern boys aiding the Confederate cause during the Civil War; a very popular work in the late nineteenth century.

### The Two Magics

- Henry James's 1898 volume of supernatural stories.

### 2 March

- Texas Independence Day, celebrating the anniversary of the 1836 declaration of Texas's independence from Mexico.

### Two Men of Sandy Bar

- 1876 play by Bret Harte, dramatized from one of his short stories.

### Twomileborris

- a town in County Tipperary, Ireland, about 20 miles west of Kilkenny.

### two-minute warning

- in football, a time-out called by the officials to inform the players that only two minutes of playing time remain; occurs twice: at the end of the first half, and at the end of the game.

### Two Natures

- an 1894 sculpture by George Grey Barnard, probably his best known, showing a pair of opponents, one standing over the prostrate form of the other, the two symbolizing "good" and "evil," but not clearly defining which is which.

### Two Noble Kinsmen

- romantic play by John Fletcher, published in 1634, about the love of two men for one woman. A tournament is planned in which the winner will win the hand of the fair Emilia, and the loser will be executed. The loser is about to be beheaded when the winner falls from his horse and, with his last breath, gives up his right to the lady in favor of the other.

### 2 November

- All Souls' Day, Christian feast day.

**Two Ocean Pass**
- a pass through the Absaroka Range in northwest Wyoming.

**"Two Old Crows"**
- a 1915 poem by Vachel Lindsay (1897–1931).

*Two Orphans*
- an 1874 French play by Adolphe d'Ennery and Eugène Cormon (adapted for the American stage by N. Hart Jackson) about two sisters who search for one another for three acts, only to discover they are not sisters at all.

**Two-Penny Act**
- a law passed by the Virginia Assembly in 1758 in response to a devastated tobacco crop. Some taxes and debts had been payable in tobacco, but the failed crop reduced the ability of planters to meet their obligations. A crisis measure, the Two-Penny Act allowed debts in tobacco to be paid in cash, fixing an equivalency rate of two pence per pound, even though the market price had tripled. The clergy objected because, by a 1748 law, their salary had been set at 16,000 pounds of tobacco a year, but the Two-Penny Act deprived them of the additional income they would have enjoyed if they had received payment in resellable high-priced tobacco rather than currency.

**two-penny nail**
- a carpenter's common nail, 1 inch long (coded 2d).

**two-phase**
- (adj.) alternating current electrical system using two phases, whose voltages are displaced one from the other by 90 electrical degrees (or diphase).

**Two River**
- a town in Wisconsin on Lake Michigan, about 80 miles north of Milwaukee.

**The Two Rivers**
- a town on Reindeer River in eastern Saskatchewan, Canada.

**"Two Rivers of Greed and Anger"**

- (Japanese) Shan-tao's famous parable which illustrates how one awakens faith in the midst of evil passions.

**Two Rivers Reservoir**

- an installation near Roswell, New Mexico.

*Two Rivulets*

- a collection of mixed prose and poetry pieces by Walt Whitman; appeared in 1876 as the second volume of a set whose first volume was a later edition of *Leaves of Grass*.

**2 rue de Richelieu**

- Paris address of the Comédie Française, founded by the Molière company in 1680; now the state theater specializing in productions of classical French plays.

**2 rue Louis Boilly**

- Paris address of the Marmottan Museum, the world's largest collection of Monet paintings.

**two's company, three's a crowd**

- a phrase suggesting that the presence of a third person can interfere with the amatory intentions of a pair.

**2 September 1945**

- V-J Day (Victory over Japan), marking the end of the Pacific phase of World War II.

**two shakes (of a lamb's tail)**

- slang for a few seconds; a very short time.

**two sides to every question**

- a phrase suggesting that one should consider alternatives when attempting to understand an issue, and before reaching a conclusion.

**two silver stars**

- insignia of a major general of the U.S. Army.

**two-spot**

- a two-dollar bill.

**two-step**

- a ballroom dance in three-quarter time characterized by long, gliding steps.

**two-stepper**

- hobo slang for a chicken.

**two-striper**

- a navy lieutenant, senior grade.

**Two Thumbs Range**

- mountains in central South Island, New Zealand.

**two-time**

- (v.) to be unfaithful; to double-cross.

**two-time loser**

- someone who has been convicted twice; a person who is seen as having no chance of success in any endeavor.

**Two-Ton Tony**

- nickname of Tony Galento, a heavyweight boxer of the late 1930s and early 1940s, alluding to his being obviously overweight. Reputed to train on beer, and often described as an ambulatory beer barrel, Galento was known as a dirty fighter. He fought champion Joe Louis at Yankee Stadium on June 28, 1939, and after knocking the champ down in the third round, he proceeded to take a beating from Louis until referee Arthur Donovan stopped the fight two minutes and twenty-nine seconds into the fourth round.

**"two turtledoves"**

- the gift on the second day of Christmas in the carol, "The Twelve Days of Christmas."

***Two Way Stretch***

- a very funny 1961 British film comedy starring Peter Sellers, about a jewelry robbery planned in prison.

***Two Weeks in Another Town***

- Irwin Shaw novel dealing with the problems of movie people shooting a film in Rome, turned into a film by Vincente

Minnelli in 1962, featuring Kirk Douglas, Edward G. Robinson, Cyd Charisse, and George Hamilton.

### *Two Women*

- a 1961 film set in wartime Italy, adapted from an Alberto Moravia novel and directed by Vittorio De Sica. This film won Sophia Loren the Best Actress Oscar, the first awarded to a non-English-speaking woman in a film.

### 2-wood

- golf club, sometimes called a brassie.

### *Two Years Before the Mast*

- 1840 novel by Richard Henry Dana Jr. realistically depicting the hardships and injustices that American sailors experienced at that time. Filmed in 1946, with Alan Ladd as the spoiled young man who is shanghaied, Brian Donlevy as Dana, and Howard da Silva in a memorable performance as the tyrannical Captain Thompson.

### "Under the Double Eagle"

- a march by Josef Franz Wagner, "the Austrian march king," composed in 1893, as "Unter dem Doppeladler," when Wagner was bandmaster for the 47th Austrian Regiment. The title is taken from the double eagle in the Austro-Hungarian monarchy's coat of arms.

### *Under Two Flags*

- a romance novel of 1867, at once sentimental and adventurous, about the Honorable Bertie Cecil of the First Life Guards, by the widely read English novelist Ouida (Marie Louise de la Ramée). Adapted to a film in 1932, with Ronald Colman as the French foreign legionnaire and Claudette Colbert as the camp follower.

### U2

- Irish rock band formed in 1976 by students at Mount Temple High School in Dublin, Ireland. Features vocals by Bono, backed by Adam Clayton on bass, Dave Evans ("the Edge") on guitar, and Larry Mullin Jr. on drums. Their music features social and religious messages.

## U-2

- CIA high-altitude reconnaissance plane during the Cold War between the U.S. and the Soviet Union. Became known when one, flown by pilot Gary Powers, was downed near Svedlovsk on May 1, 1960. Powers was convicted of espionage by the Soviets and was sentenced to ten years, but was traded for Soviet spy Rudolph Abel in February 1962.

## Vaughan Williams's *Symphony No. 2*

- a "London" symphony.

## V-2 Missile

- the *Vergeltungswaffe,* "vengeance weapon," Nazi rockets fired at London from Peenemünde, on the Island of Usedom on the Baltic Coast, in an attempt to weaken British resolve to continue the war against Germany. This forty-six-foot-long terror weapon was the world's first ballistic missile. Flying at five times the speed of sound, hundreds were directed at London in 1943.

## 2-1/2

- lunch-counter code for a small glass of milk.

## Proposition 2-½

- a Massachusetts act passed in 1980 that put a 2.5 percent cap on annual property-tax increases.

## two-and-a-half-striper

- a naval lieutenant commander.

## *You Only Live Twice*

- the fifth of the James Bond flicks, set mostly in Japan, has 007 (Sean Connery) saving the world from SPECTRE bad guy Donald Pleasence. Released in 1967, this was not one of the better films in the series.

## Y2K or Y2K Problem

- Year 2000 problem: the anticipated computer dysfunction when the year changed from 1999 to 2000. Date-related processing was expected to develop problems because the year in many programs was represented by two digits, and when 99 turned to 00, it was feared that computers would interpret this as the year 1900.

## 003

- Decimal ASCII code for [end of text].
- medical classification code for salmonella infections.

## 03

- number of the French Département of Allier.

## 3

- Article of the U.S. Uniform Commercial Code that deals with commercial paper.
- atomic number of the chemical element lithium, symbol Li.
- badge number of Agent Napoleon Solo, played by Robert Vaughn, on the NBC-TV spy spoof, *The Man from U.N.C.L.E.* (1964–68).
- badge number of W. C. Fields's character, Egbert Sousé, in the 1940 comedy film, *The Bank Dick.*
- Beaufort number for wind force of 8 to 12 miles per hour, called "gentle breeze."
- considered a favorable number in Chinese culture because the word for "three" sounds like the word for "alive."
- dental chart designation for the right first molar.
- dimensions in pre-Einsteinian space: length, height, and width.
- feet in a yard.
- gauge (number) of a wood screw with the shank diameter of 3/32 of an inch.
- the great river goddesses of India: Ganga, Yamuna, and Saraswati.
- a group of three is called a triad.
- in astrology, the number 3 is ruled by the planet Jupiter.
- in baseball, 3 represents the first baseman's position.
- inches in a palm (linear measure).

- in football, the number of points scored for a field goal.
- in Hebrew, the numerical value of the letter *gimel.*
- in mineralogy, Mohs Hardness Scale level characterized by calcite; can be scratched by a copper coin.
- in numerology, a number associated with sexual potency and procreation.
- ISBN group identifier for books published in German-speaking areas: Germany, Austria, and German-speaking Switzerland.
- kabs in a hin (ancient Hebrew measure of liquid capacity).
- magnums in a Rehoboam (champagne measure).
- number of counties in the state of Delaware.
- number of decimal places in the inverse multiple designated by the International System prefix milli-, written "m-".
- number of Earth orbits by Lieutenant Colonel John H. Glenn Jr. in the Mercury capsule *Friendship 7,* on February 20, 1962, the first U.S. astronaut to achieve orbit.
- number of emperors of France. First empire: Napoleon I, 1894, and Louis XVIII, 1814–15. Second empire: Napoleon III 1852–70. (The Third Republic followed.)
- number of eyes on Shiva, the Hindu god of ascetics and the Great Yogi.
- number of films that have won all four major Academy Awards (Best Film, Best Director, Best Actor, and Best Actress): *It Happened One Night* (1934), *One Flew Over the Cuckoo's Nest* (1975), and *The Silence of the Lambs (1991).*
- number of Job's daughters (Job 1:2).
- number of Lancastrian rulers of England: Henry IV, 1399; Henry V, 1413; Henry VI, 1422 and 1470–71. (The Yorkist Dynasty followed.)
- number of presidential electoral votes apportioned to each of the states of Alaska, Delaware, Montana, North Dakota, South Dakota, Vermont, Wyoming, and the District of Columbia. (May be reapportioned for the 2012 election.)
- number of rings around the planet Jupiter, found by *Voyager 1.*
- number of seats in the House of Representatives allotted to each of the states of Nebraska, Nevada, New Mexico, Utah, and West Virginia (as of the 2000 Census).
- number of Yorkist rulers of England: Edward IV, 1461 and 1471; Edward V, 1483; Richard III, 1483–85. (The Tudor Dynasty followed.)

- number of 0s in the multiple designated by the International System prefix kilo-, written "K-".
- retired baseball uniform number of:
  - ⇨ outfielder Earl Averill, retired by the Cleveland Indians.
  - ⇨ outfielder Harold Baines, retired by the Chicago White Sox.
  - ⇨ infielder Harmon Killebrew, retired by the Minnesota Twins.
  - ⇨ outfielder Dale Murphy, retired by the Atlanta Braves.
  - ⇨ outfielder Babe Ruth, retired by the New York Yankees.
  - ⇨ first baseman Bill Terry, retired by the San Francisco Giants.

- retired basketball jersey number of:
  - ⇨ Dennis Johnson, retired by the Boston Celtics.
  - ⇨ Drazen Petrovic, retired by the New Jersey Nets.

- retired football jersey number of:
  - ⇨ Tony Canadeo, retired by the Green Bay Packers.
  - ⇨ Bronko Nagurski, retired by the Chicago Bears.
  - ⇨ Jan Stenerud, retired by the Kansas City Chiefs.
- retired hockey shirt number of:
  - ⇨ Ken Daneyko, retired by the New Jersey Devils.
  - ⇨ Bob Gassoff, retired by the St. Louis Blues.
  - ⇨ Al Hamilton, retired by the Edmonton Oilers.
  - ⇨ Lionel Hitchman, retired by the Boston Bruins.
  - ⇨ Harry Howell, retired by the New York Rangers.
  - ⇨ Keith Magnuson and Pierre Pilote, both retired by the Chicago Blackhawks.
  - ⇨ J. C. Tremblay, retired by the Quebec Nordiques.

- scruples in a dram (apothecary's measure).
- seahs in an ephah (ancient Hebrew measure of dry capacity).
- sides on a triangle.
- the smallest odd prime number.
- statute miles in a league.
- teaspoons in a tablespoon.
- three central figures of Indian philosophy: Shankara, Ramanuja, and Madhva.
- three days and three nights that Jonah spent in the belly of the whale (Jonah 1:17).
- three feet; height of a tennis net in the center, rising to three-and-a-half feet at the supporting posts.

- three-headed, fire-breathing son of Vulcan, named Cacus.
- three inches: diameter of a regulation hockey puck (by one inch thick, weighing between five-and-a-half and six ounces).
- three percent of the world's water is fresh water; the remainder is saline.
- three years:
  - ⇨ the age of all horses that run in the Kentucky Derby.
  - ⇨ the average life of a ten dollar bill.

- Title of the California Penal Code, Part I, that deals with "offenses against the sovereignty of the state."
- Title of the United States Code dealing with the President.
- the TV channel usually used for connecting VCRs or cable systems.
- value of Roman numeral III.
- wheels on a tricycle.

**anapest**

- a metrical foot of three syllables, two short (unstressed) followed by one long (stressed).

**apartment 3B**

- at 1030 East Tremont Avenue in the Bronx, home of Molly Goldberg and her family, on the TV sitcom *The Goldbergs*, starring Gertrude Berg.

**Avasthas**

- in yoga, the three stages of consciousness all humans experience: the waking state *(jagrat)*, the dreaming state *(swapna)*, and the deep sleep state (*sushupti).*

**barleycorn**

- an old measure of length equal to one-third of an inch.

**Beethoven's Third Symphony**

- "Eroica" symphony (in E-flat major).

**Berlioz's Third Symphony**

- "Romeo and Juliet" symphony (in E-flat major).

## The Big Three

- so named powers that attended the Yalta Conference in February 1945 to plan for dealing with post-World War II Europe. Roosevelt (for the U.S.), Churchill (for Great Britain), and Stalin (for Russia) agreed to partition Germany into four occupation zones, agreed to aid postwar governments in liberated nations, and created a voting procedure for the Security Council of the soon-to-be United Nations. In return for concessions in Asia, Russia also agreed to declare war on Japan within ninety days of Germany's defeat. Most of these arrangements remained secret until after the war.

## *Camera Three*

- a Sunday-morning local TV show on WCBS New York that offered innovative programming on literature, music, and the arts. The highly acclaimed show, which lasted from January 1956 to January 1979, won an Emmy in 1966.

## Canon 3 of the American Bar Association Code of Professional Responsibility

- a lawyer should assist in preventing the unauthorized practice of law.

## Cape Three Points

- a knob of land in southwest Ghana on the Gulf of Guinea.

## China's Three Sage Kings

- from prehistory: Yao, Shun, and Yu the Great.

## the Christian Trinity

- the Father, the Son, and the Holy Spirit.

## *Close Encounters of the Third Kind*

- director Steven Spielberg's intelligent 1977 sci-fi film about Earth's first contact with aliens, featuring Richard Dreyfuss, Françcois Truffaut, and Teri Garr. Notable for its superb special effects and memorable score by John Williams.

## Copland's Third Symphony

- "Quiet City" symphony, for organ and orchestra. Based on music written for Irwin Shaw's play of the same name. This symphony uses Copland's "Fanfare for the Common Man" as its climax.

## CV-3

- number of the U.S. Navy aircraft carrier, USS *Saratoga*. Sunk in atomic bomb tests at Bikini Atoll, June 25, 1946.

## dactyl

- a metrical foot of three syllables, one long (stressed) followed by two short (unstressed).

## DC-3

- called the most successful passenger plane ever flown, the Douglas Commercial 3 entered service with American Airlines in 1936. Designed to fly above most bad weather, at an altitude ceiling over 20,000 feet, the twin-engine DC-3 exhibited a number of innovative features, including retractable landing gear, wing flaps, and variable-pitch propellers. Production ended in 1944.

## double eagle

- a golf score of three strokes under par on a hole.

## "fiddlers three"

- a line from the nursery rhyme, "Old King Cole":

  Old King Cole was a merry old soul,
  A merry old soul was he.
  He called for his pipe,
  He called for his bowl,
  And he called for his fiddlers three.

## "Goldilocks and the Three Bears"

- children's story about a young girl who wanders into the home of the three bears in the woods, tastes their porridge, sits in their chairs, and finally falls asleep in one of their beds, the papa bear's bed being too hard, the mama bear's, too soft, but the baby bear's bed being just right.

## Group 3 elements (of the Periodic Table)

- scandium (21), yttrium (39), lutetium (71), lawren-cium (103).

## hat trick

- in ice hockey, scoring three goals in one game.

## the Horae

- in Greek mythology, goddesses of the seasons, of death and rebirth. Hesiod called them Eunomia (order), Dike (justice), and Eirene (peace). The Athenians knew them as Thallo, Auxo, and Carpo.

## *I Led Three Lives*

- a McCarthy-era propaganda TV series reflecting the Red scare of the period, syndicated from May 1952 to mid-1956. Purported to recount the experiences of Herbert A. Philbrick (played by Richard Carlson), a Boston ad exec who infiltrated the Communist Party as a counterspy for the FBI.

## Later Three Years' War

- Minamoto no Yoshiie, governor of the Mutsu province, put down a revolt by the Kiyowara family in northern Japan, 1083–87, adding to the military prestige of the Minamoto clan.

## Law of the Three

- George Ivanovich Gurdjieff's rules for the workings of the universe (active, passive, and neutral) and the human body (carnal, emotional, and spiritual).

## Leonard Bernstein's *Symphony No. 3*

- "Kaddish" symphony.

## Lesser Three Gorges

- A scenic waterway off the Yangtze River in China noted for its 2,000-year-old hanging coffins of the Ba people and an ancient plank road cut into the side of the cliff.

## Mendelssohn's Third Symphony

- "Scottish" symphony (in A minor).

### My Three Sons

- long-lived TV situation comedy of a widowed father and his male brood, starring Fred MacMurray. Ran on ABC from September 1960 to September 1965, then on CBS until August 1972. In the last few years, the three sons were played by three real-life brothers: Robbie, Charley, and Steve Douglas Jr.

### Newton's Third Law of Motion

- to every action there is an equal and opposing reaction.

### the Norns

- in Norse mythology, three goddesses of fate who tend the Yggdrasil, the Tree of Life. They are the past, present, and future, controlling a person's destiny: Urd (who manages Fate), Verdandi (who does the present), and Skuld (the future).

### number 3 billiard ball

- the ball that is conventionally solid red.

### Prokofiev's Third Symphony

- "Lieutenant Kijé" symphony (in C minor).

### Richard III

- Shakespeare's take on the much maligned monarch, who is best known to high school students for his cry at the 1485 Battle of Bosworth Field: "A horse. A horse! My kingdom for a horse!" The great Bard's play was adapted for a film in 1956, with Laurence Olivier doing double duty, both directing and playing a humanized Richard. The British Film Academy awarded it Best British Film, Best Film, and Best Actor; the Oscar that year went to Yul Brynner for his role in *The King and I*. An updated and somewhat sensationalized version (set in the 1930s) was filmed in 1995, with Ian McKellen as Richard, reprising his performance in his very successful stage presentation.

### Schumann's Third Symphony

- "Rhenish" symphony (in E-flat major).

### Sergeants 3

- Frank Sinatra and the Rat Pack in a lukewarm 1962 rewrite of Gunga Din as a Western. John Sturges directed, Sinatra produced, and Sammy Davis Jr. played the Sam Jaffe role.

### Shostakovich's Third Symphony

- "First of May" symphony (in E-flat major).

### The Stranger on the Third Floor

- 1940 film in which Peter Lorre (the Stranger) may or may not have committed murder. Notable is the weird dream sequence that occupies a large part of the film. All but forgotten, this gripping film is credited with being the first film noir.

### Tchaikovsky's Third Symphony

- "Polish" symphony (in D major).

### These Three

- William Wyler directed this first film version of Lillian Hellman's *The Children's Hour*, which substitutes a heterosexual triangle for the original play's lesbian story. This 1936 Oscar-nominated film starred Miriam Hopkins, Merle Oberon, and Joel McCrea.

### Third Amendment to the Constitution

- establishes conditions for the quartering of soldiers.

### Third Cataract

- a section of white water in the Nile River, north of Kerma in northern Sudan.

### third-class mail

- class of mail in the U.S. postal system, including all printed matter, except newspapers and magazines, that weighs less than 16 ounces and is unsealed.

### Third Commandment

- Thou shalt not take the name of the Lord in vain.

## Third Crusade

- Pope Gregory VIII preached the crusade following the 1187 recapture of Jerusalem by Saladin, and in 1189 the third began, directed by several of Europe's leading figures: Richard I of England, Philip II of France, and Frederick I, Holy Roman Emperor. Frederick died in Cilicia in 1190, and after the strained alliance of English and French, retook Acre from the Muslims in 1191, Philip returned home. Richard fashioned a truce with Saladin in 1192 and left the Holy Land that year.

## third degree

- rough or prolonged mental or physical coercion used to compel a prisoner to confess or to release information.

## third-degree burn

- in which skin and its underlying tissue are destroyed and sensitive nerve endings are exposed, sometimes accompanied by shock.

## third estate

- the common people as a class of society.

## Third French War of Religion

- undeterred by the 1658 Peace of Longjumeau, the Catholic Cardinal de Lorraine devised a plot to capture the Huguenot leaders, Duc de Condé and Admiral de Coligny, and thus devalue Huguenot influence in France. The plot failed, and Condé and Coligny raised another army to begin a third war, in which Huguenot forces suffered several defeats. But with coffers severely depleted by the war, the crown negotiated yet another peace, finalized at St. Germain in 1570. The fighting in central and southern France had extended the suffering to new areas and further increased tensions between Catholics and Protestants.

## Third House

- in astrology, deals with thinking and communicating.

## Third International

- also called the Comintern; organized by the Bolsheviks in 1919 to coordinate Communist activities worldwide, advocating revolution to achieve their ends. Dissolved in 1943.

## Third Lateran Council

- convened in 1179 by Pope Alexander III in order to settle political conflicts between the Vatican and Frederick I of Germany. Also known as the eleventh ecumenical council, it prescribed a two-thirds majority of cardinals to elect a pope.

## Third Law of Thermodynamics

- the entropy of a substance approaches zero as its temperature approaches absolute zero, but absolute zero can never be reached.

## third lieutenant

- one who has completed his/her officer's training, but has as yet not been commissioned.

## *The Third Man*

- Graham Greene's novel of intrigue set in Vienna after World War II, which became the basis for a tense 1950 film starring Joseph Cotten and Orson Welles, directed by Carol Reed.

## Third Order

- in Roman Catholicism, a branch of a religious order whose members are laypeople leading secular lives.

## Third Order of Saint Francis

- the Brothers and Sisters of Penance, founded in 1221; for men and women who live in a state between the world and the cloister, who engage in ordinary professions of secular life but are dedicated to Franciscan service through vow and discipline. Called tertiaries, they may be men or women, single or married, may or may not be celibate, may or may not live in a community, but have made a lifetime commitment to the rule of penance.

## third party

- (a) any new American political group offering an alternative to Republicans and Democrats.

- (b) a person who is involved only incidentally in a legal proceeding.

**third person**

- in grammar, the person referred to, other than the speaker or the person spoken to.

**third position**

- in ballet, position in which the feet are at right angles to the direction of the body, toes pointing out in opposite directions, one foot directly in front of the other, with the heel of each foot touching the middle of the other foot.

**third president**

- Thomas Jefferson, March 4, 1801 to March 3, 1809.

**third rail**

- the rail that carries the high voltage powering an electric railway.

**Third Reich**

- designates the German State from 1933 to 1945, the Nazi period, under the leadership of Adolf Hitler.

**Third Republic**

- the French Republic from the fall of the Commune (1871) until the German occupation in 1940.

**3rd Rock from the Sun**

- TV comedy series about an alien family learning the mores of living on Earth, with John Lithgow, Kristen Johnston, Joseph Gordon-Levitt, French Stewart, and Jane Curtin. Debuted on January 9, 1996, with the last show airing on May 22, 2001.

**third state of the Union**

- New Jersey ratified the Constitution on December 18, 1787.

**The Third Violet**

- 1897 novel by Stephen Crane about a young artist's infatuation with a rich girl, reflecting Crane's hopeless love affair with actress Helen Trent when he was nineteen.

## third wedding anniversary gift

- leather is customary.

## third wheel

- a superfluous person who contributes nothing to a social event or a task that must be completed.

## Third World

- collectively, the developing countries of Africa, Asia, and Latin America.

## Three-Age System

- a system for classifying prehistoric artifacts according to stages of technological development, divided into Stone, Bronze, and Iron Ages.

## three attributes of discourse

- (Roman): the composito, the elegantia, and the dignitas.

## three B's

- the great composers of classical music: Johann Sebastian Bach (1685–1750), Ludwig von Beethoven (1770–1827), and Johannes Brahms (1833–97).

## three-bagger

- in baseball, a triple; a three-base hit.

## "three bags full"

- line from the nursery rhyme:

    Baa, baa, black sheep,
    Have you any wool?
    Yes, sir, yes, sir,
    Three bags full.
    One for my master,
    One for my dame,
    And one for the little boy
    Who lives down the lane.

## three basic texts of Vedanta

- the chief Hindu philosophy: the Upanishads, Bhagavad Gita, the Brahma-Sutras. Together, they are called *Prasthana-traya*—the triple crown of the Vedanta.

**"Three Baskets of the Law"**
- holy writings of the Buddhists, the Tripitaka: "Abhidhamma Pitaka," metaphysical and philosophical doctrines; "Vinaya Pitaka," the law and regulations of discipline; and "Sutta Pitaka," the parables and sermons of Gautama.

**three bells**
- time check on-board ship, signifies 1:30 or 5:30 or 9:30, A.M. or P.M.

***The Three Black Pennys***
- 1917 three-part novel by Joseph Hergesheimer following the rise and fall of a family involved with the Pennsylvania iron industry.

**"Three Blind Mice"**
- the nursery rhyme:

> Three blind mice. Three blind mice.
> See how they run. See how they run.
> They all ran after the farmer's wife,
> Who cut off their tails with a carving knife.
> You never saw such a sight in your life
> As three blind mice.

**three brass balls**
- (a) the conventional symbol of a pawn shop.
- (b) the family crest of the Medici family of Florence, Italy.

**Three Bridges**
- a town in Somerset County, New Jersey.

***The Three Caballeros***
- 1944 Disney film inventively combining live and animated sequences, with music from several Latino composers. Sequences move from Mexico to Brazil and back to Mexico.

***Three Came Home***
- 1954 film with Claudette Colbert in a fine performance as American writer Agnes Keith, imprisoned during World War II in a Japanese internment camp, administered by Colonel

Suga (played by Sessue Hayakawa). Based on the book by Agnes Keith.

## three-card monte

- a gambling game in which the dealer shows three cards to a player, then turns them face down, shuffles them, and the player tries to identify the position of a particular card. Much favored by street hustlers.

## three-cent piece

- U.S. coin minted in silver from 1851 to 1873, and in nickel from 1865 to 1889.

## Three Charities

- (See *Three Graces*).

## three cheers

- the customary commendation in recognition of a job well done.

## three children of Llyr, God of Wales

- Bendigeidfran, Branwen, and Manawydan.

## *Three Coins in the Fountain*

- story about the romantic complications for three women in Rome living out the legend that anyone who throws a coin into the Trevi Fountain will return to the Eternal City. Directed by Jean Negulesco, the three women in this 1954 film are Dorothy McGuire, Jean Peters, and Maggie McNamara. The complications are supplied, respectively, by Clifton Webb, Rossano Brazzi, and Louis Jourdan.

## three-color

- printing or photographic process in which three primary colors, on three separate plates, are combined in different intensities to produce a vast array of other colors.

## *Three Comrades*

- with a screenplay coauthored by F. Scott Fitzgerald (based on the novel by Erich Maria Remarque), this bleak 1938 film follows three German soldiers Franchot Tone, Robert Taylor, and Robert Young as they return to their war-ravaged homeland following World War I. Margaret Sullavan, as their female compatriot, received an Oscar nomination and was named best actress of the year by the New York Film Critics.

### The Three-Cornered Hat

• an 1874 Spanish comic novel by Antonio de Alarcón, in which the miller Lucas suspects he is being cuckolded, and the various misunderstandings that ensue.

### Three Creek

• a town in southwestern Idaho.

### three Cyclopes

• Brontes, Steropes, and Arges; in Greek mythology, creatures who each had only one eye in the middle of the forehead, known for their strength and manual dexterity. They helped the Olympians defeat the Titans.

### 3-D

• (a) abbreviation for three-dimensional, in general usage.
• (b) specifically, the three-dimensional movie process or the film made using it.

### Three Days' Battle

• Battle of Gettysburg, the decisive battle of the Civil War, June 1–3, 1863. Confederate forces in several attacks, including Pickett's disastrous charge, were unable to dislodge the entrenched Union positions, and Lee fell back, with severe losses to both sides. The Union troops lost over 23,000 men, and the Confederates, 38,000. Five months later, on the battlefield, President Lincoln delivered his memorable Gettysburg Address on November 19, 1863.

### Three Days of the Condor

• 1975 CIA thriller directed by Sydney Pollack, with Robert Redford as an intelligence reader dropped into the middle of an internecine plot.

### three-decker

• a novel published in three separate volumes, usually in cloth binding; popular in Victorian England at the end of the nineteenth century.

## three-dimensional

- (adj.) defines a solid, which has length, and width, and height.

## three divisions of the Buddhist scriptures

*1.* Sutras, the teachings of the Buddha;
*2.* Precepts; and
*3.* Commentaries, or discourses on the Buddha's teachings.

## Three Dog Night

- popular singing group performing since 1969 (with a hiatus from 1975 to 1981), with twenty-one consecutive Top 40 hits, eighteen straight Top 20 hits, eleven Top 10 hits, three number-1 hits, and twelve straight RIAA Certified Gold LPs.

## three-dollar bill

- an odd person, or something that seems out of place at the moment.

## three-dollar gold piece

- U.S. coin minted from 1854 to 1889.

## three doshas

- in Ayurveda, the ancient Vedic system of medicine from India, health is balanced by three subtle forces: vata, pitta, and kapha. According to Sushruta, an ancient contributor to Indian medicine, the doshas represent three humors of the body: vata = air, pitta = fire, kapha = mucus.

## Three Emperors' League

- (a) informal alliance among Germany, Austria-Hungary, and Russia, activated in 1873 by a conference attended by the emperors William I, Francis Joseph, and Alexander II. The intent was to assert monarchical solidarity and to maintain peace between Austria-Hungary and Russia. Renewed in 1881, the alliance remained in force until replaced by the Austro-German Alliance of 1879.
- (b) also known as the Three Emperors' Alliance, initiated by Bismarck in June 1881 to reassure Russia following the Dual Alliance between Germany and Austria.

**"The Three Enemies"**

- an 1862 poem by Christina Rosetti (1830–94), the title alluding to the Flesh, the World, and the Devil.

**the three evil passions**

- in Buddhism, greed, anger, and stupidity.

**3 F's**

- fair rent, fixity of tenure, and freedom from sale; the three demands of the Irish Land League, founded in 1879 by Charles Stewart Parnell and Michael Davitt.

***The Three Faces of Eve***

- 1957 film starring Joanne Woodward in an Oscar-winning performance as an emotionally disturbed woman with three distinct personalities. Produced, written, and directed by Nunnally Johnson, the film was based on a true story.

**Three Fates**

- in classical mythology, goddesses of destiny: Clotho, who spins the thread of life; Lachesis, who determines its length; and Atropos, who cuts it off. Known to the Greeks as Moerae, to the Romans as Parcae. Daughters of Zeus and Themis.

**Three Fingers**

- a mountain in western Washington State, about 30 miles northeast of Everett.

**"The Three Fishers"**

- an 1849 poem by Charles Kingsley (1819–75) about three fishermen who die at sea, containing the classic line "For men must work, and women must weep."

**threefold ministry**

- hierarchy of the ministry in both Catholic and Episcopal churches: bishop, presbyter, deacon.

**threefold refuge**

- three affirmations made by Buddhist monks and laypersons: "We take refuge in the Buddha. We take refuge in the Dharma. We take refuge in the Sangha."

## Three Forks

- a town in western Montana, about 30 miles northwest of Bozeman.

## "three French hens"

- the gift on the third day of Christmas in the carol "The Twelve Days of Christmas."

## Three Furies

- in classical mythology, goddesses of vengeance: Alecto (the unresting), Megaera (the jealous), and Tisiphone (the avenger). Known to the Greeks as the Erinyes and Eumenides, and to the Romans as Furiae.

## *Three Godfathers*

- John Ford's 1948 Western, his first film in color, has John Wayne as an outlaw who finds redemption in caring for an infant left by a dying woman.

## Three Gorges

- a scenic passage along the Yangtze River in China: Xiling Gorge, from Yichang to Badong (106 miles, encompassing the Three Gorges Dam at Sandouping); Wu Gorge, from Badong to Wushan (72 miles); and Gutang Gorge, from Wushan to Wanxian (13 miles).

## three Gorgons

- daughters of Porcys and Ceto, hideous creatures with serpents for hair and huge boarlike tusks; they were named Sthenno, Euryale, and Medusa. Medusa was the only one of the three who was mortal; the other two immortal.

## Three Graces

- (a) in Christian theology, faith, hope, and charity. Also called theological virtues or cardinal virtues (1 Corinthians 13:13).
- (b) in Greek mythology, Aglaia, Thalia, and Euphrosyne: goddesses of grace, charm, and beauty. They are daughters of Zeus and Eurynome. Also known as the Three Charities.

**three Graeae**

- hideous misshapen sisters and guardians of the Gorgons, named Pemphredo, Enyo, and Dino. They lived in the land where the Sun never shone, and they shared one eye and one tooth. They are the daughters of Phorcys and Ceto.

**the three great men of Laotian and Northern Thailand myth**

- Pu Lang Seung, Khun K'an, and Khun K'et, who broke open three gourds and released the people who settled Southeast Asia.

**three great river goddesses of Hinduism**

- in India: Ganga, Yamuna, and Saraswati.

**three gunas**

- in the philosophy of yoga, three qualities of nature that account for the diversified objects of experience: *tamas* (for thought and stability), *rajas* (for activity and restlessness), *sattva* (for orderliness and restraint).

**three harpies**

- the predatory winged demons in Greek mythology: Aello, Celaeno, and Ocypete.

**three-headed giant**

- Geryones, from whom Heracles (Hercules) stole the cattle as one of his twelve labors.

**three heavens of Chinese Buddhism**

- the Kama Heaven, full of earthly pleasures; above that, the Rupa Heaven, of forms but no earthly pleasures; and, above all others, Arupa Heaven, with neither forms nor human conceptions.

**three Hesperides**

- in Greek mythology, the Nymphs of the Setting Sun: Aegle (Brightness), Erythia ("Scarlet"), and Hesperarethusa ("Sunset Glow"). Daughters of Atlas, they guarded the tree that produced golden apples.

## three high gods: the Sumerian pantheon

- An, heaven-god; Enlil, air-god; and a goddess, Ninhursag, the Lady of the Mountain.

## Three Hills

- a town in southeastern Alberta Province, Canada.

## Three Holy Children

- Shadrach, Meshach, and Abednego; three Jews who were thrown into the fiery furnace by Nebuchadnezzar and saved by an angel (Daniel 3:19–26).

## Three Hours

- a Roman Catholic religious observance practiced on Good Friday between noon and 3:00 P.M., in memory of the three hours Jesus hung on the cross.

## Three Hummock Island

- an island in the Bass Strait, northwest of Tasmania.

## 3-iron

- golf club equivalent to the earlier mid-mashie.

## Three Jewels of Taoism

- compassion, moderation, and modesty or humility.

## three judges of Hades

- in Greek mythology, Minos, Rhadamanthys, and Aeacus.

## 3 June

- birthday of Jefferson Davis, observed in the Southern states.

## three kinds of Buddhist sages

- Shravakas, pratyekabuddhas, and bodhisattvas.

## three kinds of pain

1. physical and emotional pain caused by illness, hunger, thirst, etc.;
2. pain of losing something or someone that one is attached to;
3. pain caused by vicissitudes of the world.

**three kingdoms**

- forms of matter: animal, mineral, vegetable.

**Three Kingdoms**

- period in Chinese history from 220 to 265, following the collapse of the Han Dynasty. The name reflects the division of China into three realms: Wei, Shu, and Wu.

**Three Kingdoms Period**

- after eliminating the Chinese in 313, Korea remained for several centuries fractured into Koguryo in the north, Paekche in the southwest, and Silla in the southeast.

**Three Kings Island**

- a small island off the northern tip of the North Island, New Zealand.

**Three Knobs**

- a mountain in the Northern Territory of Australia.

**Three Lakes**

- a town in northern Wisconsin, about 60 miles north of Wausau.

**three-legged mare**

- the gallows.

**three-legged race**

- in which each pair of contestants run as one with their adjacent legs tied together.

**three levels of existence in Tibetan Buddhism**

- world of the senses (*kama-loka*), world of fine materials (*rupa-loka*), and immaterial world (*arupa-loka*).

**three-light window**

- a window in which two mullions separate the window space into three glass areas.

**Three Links**

- direct telephone, postal, and airline links between mainland China (Fujian Province) and Taiwan (islands of Quemoy and Matsu) since 2001. Previously, all such contacts were made through Hong Kong. Now, of course, e-mail is also available.

## Three Little Maids

- Yum-Yum, Peep-Bo, and Pitti-Sing, schoolgirl wards of Ko-Ko in the Gilbert and Sullivan operetta, *The Mikado.*

## "Three Little Pigs"

- (a) children's tale of three little pigs who are threatened by a hungry wolf. The first two build their insubstantial houses of straw and branches, and the wolf, who warns them, "I'll huff and I'll puff, and I'll blow your house down," succeeds in doing just that. They flee to the house of the third, which is built of bricks and is too sturdy for the wolf to destroy.
- (b) an 8:41-minute film short that won a 1933 Oscar for Disney in the category Short Subjects / Cartoon.

## *Three Little Words*

- 1944 MGM film biography of the famed songwriters Bert Kalmar (Fred Astaire) and Harry Ruby (Red Skelton), directed by Richard Thorpe.

## *Three Lives*

- a 1909 collection of three stories by Gertrude Stein, each a character study of a different woman: "The Good Anna," "Melanctha," and "The Gentle Lena."

## 3M

- familiar name for the Minnesota Mining and Manufacturing Company, based in St. Paul, Minnesota, which produces over 60,000 products in a variety of industries, including chemicals, automotives, electronics, medical, pharmaceuticals, etc.

## three meanings of Tao

- Tao is the way of ultimate reality; Tao is the way of the universe; Tao suggests the way a man should order his life.

## *Three Men and a Baby*

- directed by Leonard Nimoy (Dr. Spock on *Star Trek*), a 1987 comedy film about three swinging bachelor roommates, Tom

Selleck, Steve Guttenberg, and Ted Danson, who become responsible for caring for an infant. English-language version of a 1985 French film *Three Men and a Cradle*.

### Three Men and a Little Lady

- 1990 sequel to *Three Men and a Baby;* same principals, with the setting moved to England, and the "baby" now six years old.

### "three men in a tub"

- from the nursery rhyme:

  Rub-a-dub-dub
  Three men in a tub,
  And who do you think they be?
  The butcher, the baker,
  The candlestick maker,
  They all jumped out of a rotten potato!
  Turn 'em out, knaves all three.

### Three Men on a Horse

- 1936 comedy film directed by Mervyn LeRoy, with Frank McHugh as a writer of greeting cards who has an uncanny talent for picking the winners of horse races. Based on a stage play by John Cecil Holm and George Abbott.

### Three Mile Island

- an island in the Susquehanna River, near Harrisburg, Pennsylvania, where, on March 28, 1979, nuclear reactor TMI-2 suffered a near disastrous meltdown that focused national attention on the potential dangers of nuclear energy.

### three-mile limit

- three miles off the shore, sometimes defining the country's territorial waters.

### The Three Musketeers

- rousing 1844 novel by Alexandre Dumas recounting the exploits of, actually, four friends: D'Artagnan (a real historical character) and three of Louis III's guardsmen, Athos, Porthos, and Aramis. Brought to the screen several times, first as a silent film in 1921, with Douglas Fairbanks Jr., who was perfect as the swashbuckling D'Artagnan. Five sound films followed, the first in 1935; in 1939, as a comic musical, with Don Ameche as a singing D'Artagnan and the Ritz Brothers as his comrades; in 1948, with Gene Kelly as D'Artagnan, fancying Lana Turner, with Vincent Price as Cardinal Richelieu, and Frank Morgan as King Louis XIII; in 1974, with Michael York's D'Artagnan befriended by Oliver Reed, Richard Chamberlain, and Frank Finley as the three swordsmen, Raquel Welch as the female attraction, and Charlton Heston as the evil Richelieu; and once again in 1993, heavy on the action, with Chris O'Donnell as D'Artagnan teaming up with Charlie Sheen, Kiefer Sutherland, and Oliver Platt to battle the forces of the offbeat Tim Curry as Richelieu.

**Three Oaks**
- a town outside South Bend, Indiana.

*Three O'Clock Dinner*
- 1945 novel by Josephine Pinckney, exploring a family scandal in Charleston, South Carolina.

**three of a kind**
- a poker hand with three cards of the same denomination in different suits; it beats a two pair but loses to a straight.

*Three on a Match*
- (a) quiz show on daytime NBC-TV, hosted by Bill Cullen. Debuted August 2, 1971, and ran until June 28, 1974.
- (b) 1932 gangster film, directed by Mervyn LeRoy, with Humphrey Bogart in his first gangster role. Joan Blondell, Ann Dvorak, and Bette Davis all tempt fate by lighting their cigarettes from the same match.

**threepence**
- an obsolete British three-penny coin.

**three-penny nail**
- a carpenter's common nail, 1 1/4 inches long (coded 3d).

### The Threepenny Opera

- (*Die Dreigroschenoper*, 1928) a work of social criticism by Bertolt Brecht, with music by Kurt Weill, based on John Gay's *The Beggar's Opera*. This ballad opera focuses on the social conditions that produce crime and criminals. First U.S. stage production in April 1933, with several revivals following. The longest running production debuted in September 1955 at the Theatre de Lys in New York City, closing in December 1961, after six years and 2,611 performances. Cast featured Scott Merrill as MacHeath and Lotte Lenya in the Jenny role that Brecht wrote especially for her. The first film adaptation was done in Germany in 1931, titled *Die 3 Groschen-Oper*, directed by G. W. Pabst and starring Rudolf Forster and Lotte Lenya. Again in 1962, with Curt Jurgens and Hildegarde Neff, and another in 1990, titled *Mack the Knife*, with Raul Julia playing MacHeath and Julia Migenes as Jenny.

### three-piece suit

- a jacket, trousers, and vest.

### three-point landing

- in which a plane touches down with its two main wheels at the same time as its nose wheel or its tail wheel or tail skid; frequently called a perfect landing.

### 3-point type

- typographer's measure, known as "excelsior."

### Three-Power Naval Conference

- a meeting of the U.S., Great Britain, and Japan, in the summer of 1927, in an attempt to set quotas for cruisers, destroyers, and submarines. The conference failed.

### Three-Power Pact

- an agreement between Germany, Italy, and Japan, ostensibly "to promote the prosperity of their peoples," concluded in Berlin on September 27, 1940. The pact pledged mutual support for a period of ten years.

### three-pricker

- RAF term for a three-point landing.

### Three Principles of the People

- political philosophy eschewed by Sun Yat-sen as part of a design to make China a free and powerful nation:
1. Nationalism
2. Democracy
3. Socialism

### three primary colors

- red, yellow, and blue.

### Three Pure Gods (San Qing)

- the highest dieties in the Taoist pantheon: Yu Qing (Jade Pure), Shang Qing (Upper Pure), and Tai Qing (Great Pure).

### three Rs of President Roosevelt's New Deal

- relief, recovery, and reform.

### three R's of schoolwork

- reading, 'riting, and 'rithmetic: the basics of schoolwork, and thus of learning. The phrase is attributed to Sir William Curtis, an illiterate British alderman, later Lord Mayor of London.

### "The Three Ravens"

- a popular English ballad that starts:

    There were three ravens sat on a tree,
    They were as black as they may be.

### three religious centers of the Old Kingdom of Egypt

- Heliopolis, Memphis, and Hermopolis.

### three religious vows

- those taken by novitiates to membership in a Catholic religious order: poverty, chastity, and obedience.

### three-ring circus

- (a) in which performances are occurring simultaneously in three separate ring stages.

- (b) slang for a confused event in which several activities are occurring simultaneously.

**Three Rivers**

- (a) name of towns in: central California, about 50 miles southeast of Fresno; south-central New Mexico, about 30 miles north of Alamagordo; south Texas, about 70 miles south of San Antonio; and (Trois Rivières) on Basse-Terre Island in Guadeloupe.
- (b) name of cities in southwest Michigan, on the St. Joseph River, about 20 miles south of Kalamazoo, and in Quebec Province, on the St. Lawrence River, about midway between Montreal and Quebec City.

**Three Rivers Stadium**

- former Pittsburgh sports venue serving as the home field for the baseball Pirates and the football Steelers teams. Opened in 1970, demolished in 2001. The stadium was named for its location, where the Monongahela joins the Allegheny to form the Ohio River.

**Three Rock Cove**

- a town on Port-au-Prince Peninsula in western Newfoundland, Canada.

**three roots of the Yggdrasil**

- in Norse mythology, the Yggdrasi, the giant ash tree that supported all creation, had three roots: one reached Niflheim, the northern region of cold and darkness ruled by Hel; one grew to Asgard, the heavenly home of the Norse gods; and one extended to Jotunheim, the land of the giants.

**"3s & 8s"**

- sign-off, meaning "best wishes" (CB slang).

*Three's Company*

- titillating sitcom about two single girls sharing an apartment with a single guy, an innocent arrangement but open to all kinds of double entendre. Debuted on ABC-TV in March, 1977, starring John Ritter, Suzanne Somers, and Joyce DeWitt, and ran, with several changes of cast, for 172 episodes, until September 1984.

**three secondary colors**
- green, orange, and purple.

**three sheets to the wind**
- slang for being drunk.

**Three Signs of Being**
- in Buddhism, impermanence, suffering, and absence of permanent soul.

**three silver stars**
- insignia of a lieutenant general of the U.S. Army.

**three Sirens**
- singers, in Greek mythology, whose lovely voices lured men away from their duties and their families; they were named Ligeia, Leucosia, and Parthenope.

**Three Sisters**
- (a) a group of mountains in west-central Oregon, about 65 miles east of Eugene.
- (b) a town in the Union of South Africa, about 30 miles south of Victoria West.
- (c) a small island in the Torres Strait north of Cape York, the northern tip of Queensland, Australia.

***Three Sisters***
- Anton Chekhov's 1901 four-act play examines the dreary existence of three sisters who hope for a better life. Regarded by some critics as possibly the best play in modern drama. Adapted to film in 1966, the sisters were played by Geraldine Page, Kim Stanley, and Sandy Dennis. And a second film, in 1970, directed by Laurence Olivier, starred Jeanne Watts, Joan Plowright, and Louise Purnell.

***Three Soldiers***
- a 1921 novel by John Dos Passos debunking the glory and glamour of war and exposing its brutal reality.

**threesome**
- any group of three, especially of golfers playing a round together.

**three sons of Adam and Eve**

- Cain, Abel, and Seth.

**three sons of Apollo**

- in Greek mythology, Laodocus, Dorus, and Polypoetes.

**three sorts of pain**

- those from which yoga aims to free humans: our own infirmities and wrong conduct; our relation with other living things; and our relation with external nature.

**Three Springs**

- a town in the southwestern part of the state of Western Australia.

**three-step drop**

- in football, a short, quick drop-back by the quarterback after the snap, intended to protect the quarterback when the offensive line is weak. After the drop, the quarterback usually throws a short pass to a nearby receiver and hopes he doesn't get swamped.

***The Three Stooges***

- slapstick comedy trio of several movies. Originally Larry (Fine) and (the brothers) Moe and Curly (Howard). Curly was replaced by Shemp Howard, another brother, in 1947, who was replaced by Joe Besser in 1955, who was replaced in turn by Curly Joe (DeRita) in 1958.

***Three Stories and Ten Poems***

- the first published volume by Ernest Hemingway, released in Paris in 1923, dealing with issues of "the Lost Generation," they who had lost hope and faith and had turned to cynicism. The three stories are "Up in Michigan," "Out of Season," and "My Old Man."

***Three Strange Loves***

- Ingmar Bergman directed this 1949 Swedish film about a three-way love relationship.

***Three Strangers***

- odd 1946 film about three strangers who share in a sweepstakes ticket that brings bad luck to each of them.

Directed by Jean Negulesco. The three are played by Geraldine Fitzgerald, Sydney Greenstreet, and Peter Lorre.

## three strikes

- number of pitches needed to retire a batter in the game of baseball.

## three-striper

- a navy commander, described by his rank insignia.

## the three superior powers of Black Magic

- Lucifer, Beelzebub, and Astaroth, according to the "Grimorium Verum" of 1517.

## *Three Taverns*

- 1920 book of poems by Edwin Arlington Robinson about people who live out their passions.

## 3-to-an-em

- an obsolete measure of spacing (equal to 1/3 em) used to justify a typographical line (i.e., spread out the words to fill the line to the margins); known also as a "thick."

## three tragic poets of ancient Greece

- Aeschylus (525–426 BC), Sophocles (496–406 BC), and Euripides (485–406 BC).

## three-up

- a gambling game played with three coins, all three falling heads or all three tails in order to win.

## three vowels

- obsolete British slang for an I.O.U.

## three-way switch

- one of a pair of switches that control lights or outlets from two points, typically for a stairway light that can be turned on or off from either the top or bottom, or a hallway fixture that can be operated from a switch at either end.

## *Three Weeks*

- Elinor Glyn's 1907 romantic novel, one of the first suppressed in Boston for its focus on illicit sex, which of course made it extremely popular.

**three-wheeler**

- a tricycle.

**"Three Wise Men of Gotham"**

- a Mother Goose nursery rhyme:

  > Three wise men of Gotham
  > They went to sea in a bowl.
  > If the bowl had been stronger
  > My poem would be longer.

**Three Wise Men of the East**

- the Magi who came to see baby Jesus in Bethlehem: Gaspar, Melchior, and Balthazar. According to Matthew 2, they had followed the Star of Bethlehem to find and adore the baby Jesus.

**three wise monkeys**

- see-no-evil, hear-no-evil, speak-no-evil.

**three wishes**

- reward customarily granted by the genie to the lucky finder of the magic lamp.

**three witches**

- characters that open the play, *Macbeth*.

**3-wood**

- golf club sometimes called a spoon.

**the three worlds**

- earth, heaven, and hell.
- in Hindu mythology, the three worlds refer to the Earth, the atmosphere or middle space, and the sky.

**"Three Years She Grew"**

- 1800 poem by William Wordsworth (1770–1850), about the loss of a child and the future that will never be.

**tricolor**

- a three-color flag (i.e., those of France, Italy, and Luxembourg).

**Trimurti**

- the Hindu trinity: Brahma, Vishnu, and Shiva, corresponding respectively to the creative, conservative, and destructive aspects of human activity. Usually portrayed as three faces on the one head of the god, Iswara.

**triple**

- in baseball, a three-base hit.

**Triple Alliance**

- (a) the alliance of Great Britain, Sweden, and the Netherlands against France (1668).
- (b) the alliance of France, Great Britain, and the Netherlands against Spain (1717).
- (c) the alliance of Great Britain, Holland, and Prussia against Catherine II of Russia, in defense of Turkey (1789).
- (d) the alliance of Germany, Austria-Hungary, and Italy for mutual security (1882–1915). Intended to protect Germany against Russian and French aggression; to ensure Italy's neutrality in the event of a war between Austria-Hungary and Russia; and to protect Italy in the event of a war with France.

**triple bogey**

- golf score of three strokes over par on a hole.

**triple bond**

- a chemical connection by three covalent bonds between two atoms of a molecule.

**triple cream**

- a French soft cheese mostly of cow's milk, with cream added to comprise at least 72 percent fat.

**Triple Crown**

- horse racing's most coveted honor: winning all three of racing's most-famous Thoroughbred events in the same year (the Kentucky Derby, Preakness Stakes, and Belmont Stakes). The long races, for three-year-olds, are held within a five-week period, at three different tracks, in three different states. Only eleven horses have captured this elusive prize;

the first was Sir Barton in 1919, and the most recent, Affirmed, in 1978. (See full list under 11.)

**triple-decker**

- a sandwich made of three slices of bread with two layers of filling, which are usually different.

**Triple Entente**

- an informal alignment among Great Britain, France, and Russia drawing on the Franco-Russian military alliance of 1894, an Anglo-French accord of 1904, and an Anglo-Russian entente of 1907 seen as a counterpoise to the Triple Alliance of Germany, Austria-Hungary, and Italy.

**triple play**

- in baseball, a play in which three putouts are made.

**triple point**

- the temperature and pressure at which the solid, liquid, and gaseous states of a particular substance are each in equilibrium with one another.

**triple sec**

- a liqueur flavored with sour orange.

**triplet**

- a chemical bond between two atoms in which three electrons are shared.

**triple threat**

- (a) a person who is expert in three different skills or in three related activities.
- (b) a football player who is skilled at running, passing, and kicking.
- (c) a baseball player who is skilled at playing the infield, playing the outfield, and hitting.

**The Triumvirate in English history**

- the Duke of Marlborough controlling foreign affairs, Lord Godolphin controlling Council and Parliament, and the

Duchess of Marlborough controlling the court and the Queen.

## The Triumvirate of Italian poets

- Dante, Boccaccio, and Petrarch.

## trivium

- the three studies of the lower division of the seven liberal arts in the Middle Ages: grammar (i.e., literature), rhetoric, and logic.

## *The Unholy Three*

- directed by Tod Browning, a 1925 silent film with Lon Chaney as a sideshow ventriloquist who cooks up a moneymaking scheme with friends Victor McLaglen, a strongman; Mae Busch, a pickpocket; and Harry Earles, a midget.

## Vaughan Williams's *Symphony No. 3*

- a "Pastoral" symphony.

## War of the Three Henrys

- the final phase (1584–89) of a series of religio-political struggles to define the leadership and character of sixteenth-century France. The dramatis personae were Henry III (King of France), Henry of Navarre (Protestant heir presumptive to the throne), and Henry of Guise (leader of the Catholic League).

## *3.1416*

- pi, the ratio of the circumference of a circle to its diameter.

## *3.3*

- percentage of the Earth's land surface occupied by Europe.

## *004*

- Decimal ASCII code for [end of transmission].
- Dewey Decimal System designation for books on data processing and computer science.

# 04

- number of the French Département of Alpes-de-Haute-Provence.

# 4

- Article of the New York State Labor Law that deals with employment of minors.
- Article of the U.S. Uniform Commercial Code that deals with bank deposits.
- atomic number of the chemical element beryllium, symbol Be.
- Beaufort number for wind force of 13 to 18 miles per hour, called "moderate wind–moderate breeze."
- autterflies and dragonflies have four wings.
- dental chart designation for the right second premolar (second bicuspid).
- farthings in a penny (obsolete British coin).
- firkins in a barrel (obsolete British measure of capacity).
- fluid drams in a tablespoon.
- fluid ounces in a gill.
- gills in a pint.
- four feet:
  ⇨ height of the goal cage on a regulation ice hockey rink.
  ⇨ width of baseball batter's box (by six feet).
- four most important AOC (Appellation contrôlée) designations of Bordeaux wines: Pomerol, Médoc, Graves/Pessac-Léognan, and Saint-Emilion.
- four years: average life of a twenty dollar bill.
- gauge (number) of a wood screw with the shank diameter of 7/64 of an inch.
- the Hindu god Vishnu is portrayed as having four hands: one holding a shell (*shankha*); the second, a quoit (*chakra*); the third, a lotus; and the fourth, a club.
- hogsheads in a tun (large cask).
- in astrology, the number 4 is ruled by the planet Mars.
- inches in a hand (height measure for horses, from ground to top of shoulders).
- in Hebrew, the numerical value of the letter *daleth*.
- in Mesopotamian myth, number of gods who planned the flood that was to have destroyed all life on Earth. Anu (the

father), Enlil (the counselor), Ninurta (the throne carrier, son of Enlil), and Ennugi (the canal inspector).

- in mineralogy, Mohs Hardness Scale level characterized by fluorite; can be scratched by glass.
- in yoga, the four means to attaining knowledge of Brahman:
  1. Viveka
  2. Vairagya
  3. Shatsampatti
  4. Mumukshutva

- ISBN group identifier for books published in Japan.
- logs in a kab (ancient Hebrew measure of liquid capacity).
- magnums in a Methuselah (champagne measure).
- mahi in a neut (ancient Egyptian unit of length).
- most field goals kicked in a Super Bowl game (football record): twice, by Don Chandler for Green Bay Packers against Oakland Raiders in SB-II (1968); and by Ray Wersching for San Francisco 49ers against Cincinnati Bengals in SB-XVI (1982).
- most Super Bowl wins by a coach (football record): by Chuck Noll for Pittsburgh Steelers, SB-IX, 1975; SB-X, 1976; SB-XII, 1978; SB-XIV, 1980.
- most times named MVP in the NFL (football record): by Peyton Manning, 2003, 2004, 2008, and 2009.
- most U.S. Open golf tournaments won: four times, by Willie Anderson (1901, 1903–05), by Bobby Jones (1923, 1926, 1929–30), by Ben Hogan (1948, 1950–51, 1953), and by Jack Nicklaus (1962, 1967, 1872, 1980).
- noggins in a U.S. fluid pint.
- number of Danish kings of England: first, Sweyn Forkbeard in 1013, for a few months interrupting the reign of the Saxon Ethelred II; then, Canute from 1016; Harold I (Harefoot), 1035; Harthacanute, 1040–42, when Edward the Confessor restored the Saxon monarchy of independent England.
- number of gold medals won by the legendary Jesse Owens at the 1936 "Hitler Olympics" in Berlin, Germany: 100-m (Olympic record), 200-m (world record), long jump, and 4 x 100-m relay. The outstanding performance by the African-American athlete shamed Hitler's expectation that the games would serve to display white supremacy.
- number of House of Windsor monarchs of Great Britain, starting with George V, 1910; Edward VIII, 1936; George VI, 1936; to the present Queen Elizabeth II, since 1952.

- number of lawful wives allowed at a time to a Muhammadan, according to Surah 4 of the Quran.
- number of lovebirds attending the Celtic love god, Aonghus.
- number of nations in the United Kingdom: England, Scotland, Northern Ireland, and Wales.
- number of Norman rulers of England, from William the Conqueror in 1066; William II (Rufus), 1087; Henry I (Beauclerc), 1100; Stephen 1135–54. (The Plantagenet Dynasty followed.)
- number of presidential electoral votes apportioned to each of the states of Hawaii, Idaho, Maine, New Hampshire, and Rhode Island. (May be reapportioned for the 2012 election.)
- number of presidents in the Provisional Government of France following World War II: Charles de Gaulle, 1944; Félix Gouin, 1946; Georges Bidault, 1946; Léon Blum, 1946. (After Vichy.)
- number of seats in the House of Representatives allotted to each of the states of Arkansas, Kansas, and Mississippi (as of the 2000 Census).
- number of strings on a violin, a viola, a cello, and a double bass.
- number of U.S. presidents named William: Henry Harrison, McKinley, Taft, and Clinton.
- number thought to be unlucky in both Chinese and Japanese cultures. In both languages the word for "four" (in pinyin, "si") sounds like the word for "death."
- the only number whose value is equal to the number of letters in its English name.
- pecks in a bushel (dry measure).
- percentage of the labor force in agriculture in 1970, compared with 52 percent in 1870. In 2009, it was less than 2 percent.
- pice in an anna (former bronze coin of British India).
- players on a polo team.
- quarts in a gallon.
- record number of home runs hit in a nine-inning baseball game: held by several players, the first in the "modern" period by New York Yankee Lou Gehrig, on June 3, 1932, the most recent on September 5, 2003, by Carlos Delgado, with the Toronto Blue Jays.
- retired baseball uniform number of:
    ⇨ shortstop Luke Appling, retired by the Chicago White Sox.

⇨ shortstop Joe Cronin, retired by the Boston Red Sox.
⇨ first baseman Lou Gehrig, retired by the New York Yankees on July 4, 1939, ending his 2,130 consecutive game streak, the first uniform number ever to be retired.
⇨ outfielder Ralph Kiner, retired by the Pittsburgh Pirates.
⇨ third baseman Paul Molitor, retired by the Milwaukee Brewers.
⇨ outfielder Mel Ott, retired by the San Francisco Giants.
⇨ outfielder Duke Snider, retired by the Los Angeles Dodgers.
⇨ manager Earl Weaver, retired by the Baltimore Orioles.

- retired basketball jersey number of:
  ⇨ Adrian Dantley, retired by the Utah Jazz.
  ⇨ Joe Dumars, retired by the Detroit Pistons.
  ⇨ Wendell Ladner, retired by the New Jersey Nets.
  ⇨ Sidney Moncrief, retired by the Milwaukee Bucks.
  ⇨ Jerry Sloan, retired by the Chicago Bulls.
  ⇨ Chris Webber, retired by the Sacramento Kings.
- retired football jersey number of Tuffy Leemans, retired by the New York Giants.
- retired hockey shirt number of:
  ⇨ Barry Ashbee, retired by the Philadelphia Flyers.
  ⇨ Jean Beliveau, retired by the Montreal Canadiens.
  ⇨ Bobby Orr, retired by the Boston Bruins.
  ⇨ Scott Stevens, retired by the New Jersey Devils.

- rods in a chain (surveyor's measure).
- sides on a square, a rectangle, or a parallelogram, of any quadrilateral.
- the smallest non-Fibonacci number.
- the square of 2.
- subchapter IV of the Uniform Code of Military Justice deals with court-martial jurisdiction.
- sum of the first four Fibonacci numbers $(0 + 1 + 1 + 2)$.
- Title of the United States Code dealing with the flag and seal, the seat of government, and the states.
- value of Roman numeral IV.
- years in an Olympiad.

## Anton Bruckner's *Symphony No. 4*

- "Romantic" symphony (in E-flat major).

**au four**

- (in French, not a number): oven-baked.

**The Big Four**

- U.S., Great Britain, France, and the Soviet Union; The countries that were most influential in international affairs after World War II.

*Born on the Fourth of July*

- Ron Kovic's 1976 best-selling memoir of his experience in the Vietnam War, and how it changed him from a gung ho marine into a peace activist. Returning paralyzed from the chest down, and suffering further from poor treatment in VA hospitals, Kovic became the country's most ardent anti-Vietnam spokesman. The book was converted into a powerful film by Oliver Stone in 1989, with an intense performance by Tom Cruise as Kovic, which won Stone the Best Director Oscar. Kovic co-wrote the screenplay.

**Canon 4 of the American Bar Association Code of Professional Responsibility**

- a lawyer should preserve the confidences and secrets of a client.

**F-4**

- U.S. Navy aircraft, the Phantom II, prized fighter of the Vietnam War.

**Fearsome Foursome**

- nickname of the great defensive line of the Los Angeles Rams in the mid-1960s, made up of Lamar Lundy and David (Deacon) Jones, tackles, and Merlin Olsen and Rosey Grier, ends.

**Final Four**

- colloquial name for the national semifinals of the NCAA Men's Division I Basketball Championship tournament held each spring; one of the nation's most-watched sports events.

**"Four Affirmations" in Shinto**

*1.* tradition and the family
*2.* love of nature
*3.* physical cleanliness
*4.* "Matsuri," the honoring and worship of the Kami and ancestral spirits.

**four ages of history**

- according to Indian tradition: the Satya Yuga (golden age); the Treta Yuga, in which righteousness (*dharma*) decreased by one-fourth; the Dvapara Yuga, in which righteousness decreased by one-half; and the present Kali Yuga, in which righteousness has decreased by three-fourths, considered the most evil of all.

**four ages of poetry**

- (Thomas Love Peacock, 1820): iron, gold, silver, and brass.

**four aims of life (*Purushartha*)**

- according to Indian tradition: virtuous living (*dharma*), means of life (*artha*), pleasure (*kama*), and self-liberation (*moksha*).

**four Alexandrian poets**

- (c. 250 BC): Apollonius, Theocritus, Bion, and Moschus.

**four animals at the mouth of Ga-Oh's cave**

- in Iroquoian myth, the Iroquoian Wind Giant had at the mouth of his cave a bear, a panther, a moose, and a fawn.

**Four Archers**

- a mountain in the northeastern part of the Northern Territory of Australia.

**Four Articles of Prague**

- demands for church reform in Bohemia, made in 1420 by followers of Jan Hus, including:
  *1.* freedom to preach the word of God by others than adherents to the Church of Rome;
  *2.* sacrament of the Eucharist in both kinds (bread and wine), to both priests and the laity alike;
  *3.* exclusion of the clergy from secular activity
  *4.* the clergy not to be exempted from civil law.

## four Bacabs

- the sons of Itzamna and Ixchel in Mayan mythology; the four Gods of the Cardinal Points of the Compass who hold up the four corners of the sky: Mulac points north, Cauac points south, Kan points east, and Ix points west.

## four-bagger

- in baseball, a home run.

## four bases of virtue in Buddhism

1. seeking truth
2. giving gifts
3. destroying karmic evils
4. cultivating wisdom.

## four basic operations of arithmetic

- addition, subtraction, multiplication, and division.

## four bells

- time check on-board ship, signifies 2:00 or 6:00 or 10:00, A.M. or P.M.

## four bits

- slang for fifty cents.

## four blood groups in humans

- A, B, AB, and O.

## Four Books

- the four principal works in the Confucian canon: *Analects of Confucius* (Lun Yu), *The Great Learning* (Ta Hsueh), *Book of Mencius* (Meng Tzu), and *Doctrine of the Mean* (Chung Yung), until recent times the primary textbooks in China's education system. These four and the Five Classics (q.v.) comprise the modern Confucian canon. (Yuan).

## four books of the Veda

- the sacred writings of Hinduism: *Rig-Veda* (hymns to the gods), *Sama-Veda* (priests' chants), *Athara-Veda* (magical chants), *Yajur-Veda* (sacrificial formulae in prose).

## 4 Boulevard du Palais

- Paris address of La Sainte Chapelle, a Gothic chapel with impressive stained-glass windows, most dating from the thirteenth century. Originally built by Louis IX to house what he believed to be Christ's crown of thorns.

## Four Branches of the Mabinogi

- the main narrative of the Welsh "Mabinogion," the tales of Pwyll, Branwen, Manawydan, and Math.

## Four Brothers

- Nickname of the saxophone section of Woody Herman's big band in the late 1940s (The Second Herd). The band in that period featured the saxes, and was frequently known as the Four Brothers Band. Also the name of their classic 1947 jazz recording, written by saxophonist Jimmy Giuffre, and featuring Zoot Sims, Stan Getz, and Herbie Stewart on tenor saxes, and Serge Chaloff on baritone sax. The song was recorded several times thereafter with variations in the personnel, after Al Cohn replaced Herbie Stewart.

## Four Buttes

- a town in northeastern Montana.

## four-by-four

- a car or truck having four forward speeds.

## four C's of a diamond

- the factors that determine the value of the stone: carats, cut, color, and clarity.

## "four calling birds"

- the gift on the fourth day of Christmas in the carol "The Twelve Days of Christmas."

## Four Cardinal Principles

- declared by Deng Xiaoping in 1979, these were four topics not open for discussion within the People's Republic of China:

1. upholding the socialist path;
2. upholding the people's democratic dictatorship;
3. upholding the leadership of the Communist Party of China
4. upholding Marxist-Leninist-Mao Zedong thought. Implied was that other issues not listed could be argued.

**four cardinal (or heavenly) virtues**
- prudence, temperance, fortitude, and justice.

**four castes**
- social strata in Hinduism: Brahmins or priests, rulers and warriors, merchants and farmers, peasants and laborers.

**four chambers**
- mammalian hearts have four chambers.

**four classes of the top Buddhist pantheon in modern Japan**
1. the Nyorai class: the tathagata or buddhas
2. the Bosatsu class: bodhisattvas
3. the Myoo class: the vidyaraja or Kinds of Science in Tantrism
4. the Ten class: the deva or gods.

*4 Clowns*
- a 1970 Robert Youngson anthology of some of the best silent comedy ever filmed. Includes Buster Keaton's *Seven Chances*, Charley Chase's *Limousine Love*, and selections from Laurel and Hardy.

**four-color process (CMYK)**
- used in printing; four colors (cyan, magenta, yellow, and black), on four separate plates, are combined in different intensities to produce a vast array of other colors.

**four-color theorem**
- any map can be constructed using only four colors so that no two contiguous countries with common borders are represented by the same color.

**four compounds comprising the major arcana of alchemy**
- vitriol, natron, liquor hepatic, and pulvis solaris.

## Four-Corner Method

- a system for encoding Chinese characters using four numerics for each character. The method was invented by Wang Yun-Wuu, editor in chief of the Commercial Press of China in the 1920s.

## Four Corners

- (a) a location in southwestern U.S. where four states, Utah, Colorado, Arizona, and New Mexico, meet at a common point.
- (b) a town in northeastern Wyoming, about 45 miles southeast of Sundance.

## four corners offense

- an offensive strategy for stalling in a college basketball game, whereby four players would position themselves in the corners of the half-court and pass the ball so that none held it for more than the maximum five seconds. The fifth man would take the center and watch for an opportunity to drive the lane while defensive players were spread out, covering the four outside. Fans became so enraged at this technique that the NCAA introduced a shot clock in 1985, disallowing such a freeze on the ball, and a three-point line a year later.

## four corners of the earth

- taken to mean "everywhere."

## four corners rule

- a principle in law that any legal document be interpreted by the language used in its entirety (all "four corners"), ensuring that the intentions of the parties are discerned by the plain meaning of the words chosen.

## Four Crowned Martyrs

- stone carvers of the third-century Yugoslavia who were martyred when they refused to carve an idol of Aesculapius for Diocletian.

## four cycles of Irish sagas:

1. the Mythological Cycle (preserved in the Lebor Gabala) tells of the activities of the pagan Celtic gods.

2.  the Ulster Cycle set in the reign of King Conchobar in pre-Christian Ireland, focuses on stories of the Irish hero Cuchulain.
3.  the Fenian Cycle recounts the adventures of Fionn mac Cumhaill and his band of warriors, the Fianna.
4.  the Historical Cycle (or Cycle of Kings) relates the exploits of characters in the early centuries of the Christian period.

### *Four Daughters*

- a heartwarming 1947 Warner Brothers film about four girls from a small town finding romance. Directed by Michael Curtiz, with three of the four daughters played by the Lane sisters, Priscilla, Lola, and Rosemary.

### Four Diamond Kings

- four deities that protect Buddhist temples and work to defeat the enemies of Dharma: in the east, Dhritarashtra; in the south, Virudhaka; in the west, Virupaksha; in the north, Vaishravana.

### four dimensions

- width, length, height, and time.

### Four Discourses

- those originally written by Indian masters and adopted as the canonical texts of the Buddhist four-discourse school:
  1.  *Discourse on the Middle* by Nagarjuna
  2.  *Twelve-Gate Discourse* by Nagarjuna
  3.  *One-Hundred Verse Discourse* by Aryadeva
  4.  *Great Wisdom Discourse* (Commentary on the Prajna-paramita Sutra) by Nagarjuna.

### four distinctive emblems of Shinto

- the bird's nest, composed of two upright unpainted tree trunks with a third resting on their top and horizontal beam below; the "gohei," a slim wand of unpainted wood with two long pieces of paper notched alternately on opposite sides of it; the mirror; and the rope of rice straw.

### four divisions of a classic tragedy

- the protasis, the epistasis, the catastasis, and the catastrophe.

## four-dollar gold piece

- the "stella," minted by the U.S. from 1879 to 1880.

## four-down territory

- in football, the area between the opposing team's 30- and 40-yard lines, where a field-goal attempt seems risky, and a punt may not cover enough ground to hold back the other team's offense.

## Four Dragon Kings

- in Chinese mythology, the rulers of the seas and the four cardinal compass points: Ao Shun (north), Ao Ch'in (south), Ao Kuang (east), Ao Jun (west).

## Four Elementary Truths of Buddhism

1. both birth and death bring grief, and life is vain.
2. the cause of grief, and hence, the vanity of life, is indulgence of desire.
3. with the ending of desire will come surcease from grief.
4. the best way to end desire is by application of wisdom and intelligence to life.

## four eyes

- disparaging term for a person who wears glasses.

## 4-F

- U.S. Selective Service classification of a candidate unfit for military service.

## four faces of Fu Xi

- in Chinese myth, the divinity Fu Xi has four faces, each overseeing one of the cosmic directions: north, south, east, and west.

## *Four Feathers*

- 1939 classic adventure film about Harry Faversham, a staid Britisher who must vindicate himself from the charge of cowardice, starring John Clements, Ralph Richardson, and C. Aubrey Smith. Rip-roaring battle scenes, and Smith's memorable reenactment of Crimean War skirmishes using nuts and fruits to portray the British "thin red line." Filmed several times, as a silent film in 1915, 1921, and 1929, and with sound again in 1955 (as *Storm Over the Nile*), 1977, and 2002, but the 1939 version is by far the best.

**4 February**

- Kosciuszko Day, celebrated by Polish Americans on the birthday of Tadeusz Kosciuszko (1746), who fought with the colonists in the Revolutionary War.

**four flush**

- poker hand having four of the five cards of the same suit.

**four-flusher**

- a person who can't substantiate his pretensions; a braggart or pretender; a person who doesn't pay his debts.

**the fourfold great struggle**

- in Buddhism, to prevent trouble arising, to obliterate erroneous ideas that have arisen, to produce goodness which did not previously exist, and to increase goodness which is already existent.

**Four Forest Cantons**

- Unterwalden, Schwyz, Uri, and Lucerne: the first four Swiss communities to gain freedom from the Habsburgs after the 1315 battle of Morgarten.

**four forms of fearlessness**

- those attributed to a Buddha in preaching the Dharma:
    1. in asserting that he has attained the perfect enlightenment;
    2. in asserting that he has destroyed all defilements;
    3. in pointing out evil passions of sentient beings; and
    4. in expounding the method of emancipation.

**Four Freedoms**

- espoused by President Franklin D. Roosevelt in his address to Congress on January 6, 1941: freedom of speech, freedom of religion, freedom from want, freedom from fear.

**four fundamental constituents of DNA and RNA**

- adenine, cytosine, guanine, and thymine.

**four fundamental forces of nature**

- gravity, the strong nuclear force, the weak nuclear force, and the electromagnetic force.

**four giant gas planets in the solar system**

• Jupiter, Saturn, Uranus, Neptune.

**four gods of the Lakota Indians**

• Skan, Inyan, Maka, and Wi.

**Four Gospels of the New Testament**

• Matthew, Mark, Luke, and John.

**four great mountain monasteries of China**

• each is associated with a bodhisattva of Mahayana Buddhism; they are P'ut'o Mountain (off the coast of central China), Five Platform Mountain (in northern China), Omei Mountain (in Szechuan Province), and Nine Flower Mountain (in Anhui Province).

**Four Guardians of the Universe**

• also, Four Great Heavenly Kings. In Japanese mythology, four great deities living on Mount Sumeru who protect the world, each responsible for one of the cardinal directions: Jijoku Tenno (east), Komoku Tenno (west), Bishamon Tenno (north), and Zojo Tenno (south).

**4-H Club**

• a youth organization sponsored by the Department of Agriculture offering instruction in home economics and agriculture. (Its name reflects its goals: to improve head, heart, hands, health.)

**Four Heavenly Emperors**

• four Tao deities who are in charge of all things in Heaven and on Earth:
    *1.* The Great Jade, Emperor (master of 10,000 spirits): in charge of the Dao of Heaven.
    *2.* The Middle Heaven Great Emperor of the North Pole Star of Purple Subtlety: administers the sun, moon, and stars, as well as the climates of the seasons.
    *3.* The Great Heavenly Emperor of the Highest Palace of Polaris (at the pivot of Heaven): in charge of the three

powers of Heaven, Earth, and Man, and of wars in the human world.

*4.* Imperial God of Earth: in charge of the land, birds, rivers, and mountains.

## Four Heavens of the Bon religion of ancient Tibet

- Sipa Yesang (home of the procreator god Sangpo Bumtri); Sipa Gunsang (where the gods discuss the fate of the Earth); Barlha Wösel (where the gods purify themselves); and the highest heaven, Gontsun Phyva (where the gods are educated).

## Four Holy Truths of Buddhism

- a set of principles concerning human existence, through which it is said Buddha achieved enlightenment: the holy truth concerning ill, the holy truth concerning the cause of ill, the holy truth concerning the cessation of ill, and the holy truth concerning the path which leads to that cessation.

## four horsemen

- in ten-pin bowling, the pins left standing after the first ball has hit: either 1, 2, 3, and 7, or 1, 3, 6, and 10.

## Four Horsemen of Notre Dame

- football backfield under coach Knute Rockne from 1922 to the Rose Bowl Game in 1925: Harry Stuhldreher, quarterback, Don Miller and Jim Crowley, halfbacks, Elmer Layden, fullback. The name first appeared in an October 18, 1924, column by sports writer Grantland Rice after Notre Dame's 13-7 victory over Army.

## Four Horsemen of the Apocalypse

- four horsemen, one each riding a white, red, black, and pale horse, representing personifications of the evils of war: respectively, conquest, slaughter, famine, and death. (From the Revelation of Saint John the Divine, Revelation 6:2-8).

## *Four Horsemen of the Apocalypse*

- widely read antiwar novel of 1916 by the Spanish author Vicente Blasco Ibáñez, about an Argentinean family that fights on opposing sides in World War I. Made into an effective 1921 silent film directed by Rex Ingram, starring Rudolph Valentino dancing his famous tango. Remade as a talkie in 1962, an ill-advised project of director Vincente Minnelli, with Glenn Ford, Charles Boyer, Lee J. Cobb, Paul Henreid, and Paul Lukas.

**four humors**

- black bile, yellow bile, phlegm, blood.

**"Four Hymns to Love and Beauty"**

- a group of poems written by Edmund Spenser (1596).

**Four Immortal Chaplains**

- when the U.S. Army troopship *Dorchester* was torpedoed off the coast of Greenland on February 3, 1943, four clergymen aboard perished after giving up their own life jackets to waiting servicemen for whom none were available. They were: Father John Washington (Catholic), Reverend Clark V. Poling (Dutch Reformed), Rabbi Alexander Goode (Jewish), and Reverend George L. Fox (Methodist). Of more than 900 men on the ship, only 230 survived.

**four-in-hand**

- (a) a vehicle drawn by four horses; also called coach and four.
- (b) a necktie tied in a slipknot with the two ends hanging loose.

**4-iron**

- golf club equivalent to the earlier mashie iron.

**4 July**

- Independence Day.

**four kings**

- in a deck of cards, the four kings supposedly represent Charlemagne (Western European), David (Jewish), Alexander (Macedonian), and Caesar (Roman) monarchies.

**Four Ladies Bank**

- an area in the Indian Ocean adjacent to Prydz Bay, off Princess Elizabeth Land in Australian Antarctic Territory.

## Four Lakes

- a chain of lakes in southern Wisconsin: Mendota, Monona, Waubesa, and Kegonsa.

## four-leaf clover

- a clover with four leaves, rather than the usual three; considered to be a good-luck symbol.

## four-legged burglar alarm

- slang for a watchdog.

## four-letter man

- obsolete slang for a male homosexual, the four letters indicating H-O-M-O.

## four-letter word

- any of several short words generally regarded as being vulgar or obscene.

## four levels of meaning

- (as per Dante): the literal meaning, the moral meaning, the allegorical meaning, and the anagogical meaning.

## The Four Lings

- four living Buddhas of Tibet, called Hutogtus, each empowered to serve as Prince Regent until the Dalai Lama became eighteen years old and able to assume the reins of leadership. The four were Daingyailing, Gundeling, Cemoinling, and Xedeling.

## four main currents of Tibetan Buddhism

- the four schools: Nyingma (the "ancient school"), dating from the eighth century; Kagyu ("the transmitted word"), from the eleventh century; Sakya ("gray earth"), also eleventh century; and Gelug (the "virtuous way"), founded in the fourteenth century.

## 4 March

- Charter Day in Pennsylvania, commemorating the granting of the charter to William Penn by Charles II in 1681.

## The Four Marys

- (a) the constant companions of Mary, Queen of Scots (friends from her childhood): Mary Seaton, Mary Beaton, Mary Livingston, and Mary Fleming.
- (b) a Scottish folk song about Mary Hamilton, a young woman who is soon to be executed for abandoning to the sea her newborn baby, fathered by the Queen's husband. The ballad appeared in Sir Walter Scott's 1802 *Minstrelsy of the Scottish-Border*.

## Four Masters of Anhui

- a group of Chinese artists painting in Anhui Province in the seventeenth century (Qing dynasty), who, separated from the mainstream currents of Chinese art, developed their own styles. They are usually identified as the little-known Sun Yi and Wang Zhirui, and the well-known Zha Shibiao and Hongren.

## Four Masters of the Yuan Dynasty

- Chinese painters who were active during the Yuan period (1260–1368) but gained stature during the succeeding Ming Dynasty. Adherents of the school of literati painting, which focused on individual expression instead of visual appeal, their names were Wu Zhen, Huang Gongwang, Ni Zan, and Wang Meng.

## 4 May

- Rhode Island Independence Day, celebrated as the day in 1776 when the state declared independence, two months before the Continental Congress did so for the nation as a whole.

## four meanings of poetry

- (per I. A. Richards): the sense (what is actually said), the feeling (the writer's emotional attitude toward it), the tone (the writer's attitude toward his reader), and the intention (the writer's purpose; the effect he hopes to produce).

## four means of attaining knowledge

- (according to the Nyaya school of Indian philosophy): perception, inference, comparison, testimony.

### *The Four Million*

- a 1906 collection of O. Henry stories, containing arguably the author's most famous, "The Gift of the Magi." The title, reflecting the actual population of New York City at the time, was chosen as a corrective to the widely quoted remark by Ward McAllister that there were only 400 people in New York society.

### four minor Arcana face cards in a Tarot deck

- page, knight, queen, king.

### four-minute mile

- once thought to be an unassailable barrier, until runner Dr. Roger Bannister set a new record of 3:59.4 on May 6, 1954, in Oxford, England. The best time has gradually shortened since then; the current record is 3:43.13, set by Hicham El Guerrouj of Morocco, in Rome, Italy, on July 7, 1999.

### "Four Minutes Thirty-Three Seconds"

- 1952 musical composition by John Cage, in which the musicians sit through three silent movements of thirty seconds; two minutes, twenty-three seconds; and one minute, forty seconds with their instruments poised but never playing even one note.

### Four Modernizations

- the goals of Deng Xiaoping's reform program, first introduced by Zho Enlai in 1976: agriculture, industry; science and technology, and the military. These constituted a reversal of Mao Zedong's policy of economic self-reliance.

### Four Noble Truths of Buddhism

- the central core of Buddhist religious doctrine, as taught by Gautama Buddha:
    1. suffering is omnipresent;
    2. its cause is in the indulgence of desire;
    3. elimination of desire will relieve the suffering;
    4. the best way to eliminate desire is through the Noble Eightfold Path (q.v.).

## Four Noes and One Without

- a pledge by President Chen Shui-bian of the Republic of China made at his inauguration address on May 20, 2000, regarding the political status of Taiwan. Provided that the People's Republic of China does not attack Taiwan militarily, Chen's administration will not do the following (the "Four Noes"):
  1. declare Taiwanese independence;
  2. change the national title from "The Republic of China" to "The Republic of Taiwan";
  3. include the doctrine of state-to-state relations in the Constitution of the Republic of China;
  4. promote a referendum on unification or independence.

- the "One Without" was that Chen pledged not to abolish the National Unification Council.

## 4 November

- birthday of Will Rogers in 1879, celebrated as a state holiday in Oklahoma.

## four-o

- slang for "excellent," from U.S. Navy use indicating highest efficiency rating of officers.

## Four Oaks

- a northern suburb of West Bromwich, England.

## four-o'clock

- an ornamental plant native to tropical America; its name derives from its flowers opening in late afternoon, and closing by the following morning.

## four offerings

- prescribed items that can be offered to the Buddha: food and drink, clothes, bed, medicinal drink.

## four on the floor

- designates an automobile with a four-speed transmission operated by a manual shift protruding from the floor next to the driver's seat.

**four orders of mendicant friars**

- Dominicans, Franciscans, Augustinians, Carmelites.

*The Four P's*

- an entertainment featuring a palmer, a pardoner, and a pothecary, with a pedlar acting judge, to determine who can tell the biggest lie. By John Heywood; performed about 1520, published 1544.

**Four Paths**

- a town in the south-central part of the island of Jamaica.

**four-penny nail**

- a carpenter's common nail, 1-½ inches long (coded 4d).

**four-point average**

- the highest level of grade point average; an academic rating system in which A = 4, B = 3, etc. The GPA is the weighted mean average of all grade points in all courses.

**Four-Point Plan**

- Harry Truman's economic plan promoted in his inaugural address on January 20, 1949:
  1. continue support of the UN;
  2. continue the Marshall Plan for world economic recovery;
  3. extend collective defense agreements against communist aggression;
  4. develop a program for aid to undeveloped nations.

**4-point type**

- typographer's measure, known as "brilliant."

**four-poster**

- a bed having four tall corner posts, originally intended to hold a canopy.

**Four-Power Pacific Treaty**

- an agreement of December 13, 1921, among Great Britain, France, Japan, and the United States, by which each country guaranteed to respect the others' rights in island territorial possessions in the Pacific Ocean.

## Four-Power Pact

- agreement signed on July 15, 1933, between Great Britain, France, Italy, and Germany; impelled by Mussolini, who wanted to replace the influence of small countries in the League of Nations by a coalition of major powers.

## four prohibitions of Confucius

- do not be swayed by personal opinion; recognize no inescapable necessity; do not be stubborn; do not be self-centered.

## "Four Quartets"

- a four-part poem by T. S. Eliot, published between 1936 and 1942, considered by many critics as his major poetic work.

## four questions

- in Judaism, the traditional four questions on the meaning of the Seder service (subsumed under the general question, "Why is this night different from all other nights?"), asked by the youngest boy at Passover Seder and answered by reading from the Haggadah:
  *1.* On all other nights of the year we eat leavened bread and matzoh, but why on this night do we eat only matzoh?
  *2.* On all other nights of the year we eat many kinds of herbs, but why on this night do we eat mainly bitter herbs?
  *3.* On all other nights of the year we do not dip even once, but why on this night do we dip twice?
  *4.* On all other nights of the year we eat either sitting upright or reclining, but why on this night do we eat only reclining?

## four rivers in the Garden of Eden

- Pison, Gihon, Hiddekel, and Euphrates (Genesis 2:11-14).

## 4, St. James Place

- London residence of Frederic Chopin.

## *Four Saints in Three Acts*

- an opera in four acts [sic] by Virgil Thomson, with libretto by Gertrude Stein, first produced in 1934. The four saints are St. Theresa, St. Ignatius, and the fictional St. Settlement and St. Chavez; the action is set in sixteenth-century Spain.

### The Four Seasons

- Antonio Vivaldi's most famous composition, consisting of four concertos accompanying four sonnets written by the composer.

### four seasons of the year

- summer, autumn, winter, spring.

### fhe four senses

- the four modes of scriptural interpretation: literal, allegorical, moral, anagogical.

### four sheets to the wind

- slang for "drunk."

### four silver stars

- insignia of a full general of the U.S. Army.

### four sons of Entoria

- Janus, Hymnus, Faustus, and Felix; fathered by Saturn, they provided the four faces on the altar decoration in the temple of Saturn at the foot of the Capitol in ancient Rome.

### four stages of life

- in Indian philosophy: student (period of learning), house-holder (family life, period of emotional development), impending retirement (period of intellect and thought-fulness), meditation (earlier three are melded and the inner man gets attention).

### four-star

- (adj.) of superlative quality.

### Four Star Playhouse

- dramatic TV anthology featuring Dick Powell, with rotating stars, on CBS from September 1952 through September 1956.

### four-striper

- a U.S. Navy captain.

## four suits of a playing card deck

- correspond to social divisions: spades (the military and nobles), hearts (the church), diamonds (commerce), and clubs (the common men).

## fourth

- British slang for a privy, a latrine.

## Fourth Amendment to the Constitution

- provides protection from unreasonable search and seizure.

## Fourth Arrondissement

- section of Paris containing Notre Dame Cathedral.

## Fourth Avenue

- in New York City, the extension of Park Avenue below 32nd Street.

## Fourth Cataract

- a section of white water in the Nile River, north of Merowe in northern Sudan.

## fourth-class mail

- class of mail in the U.S. postal system including merchandise and some printed matter, weighing 8 ounces and unsealed for inspection purposes.

## Fourth Commandment

- Thou shalt keep holy the Sabbath day.

## Fourth Crusade

- initiated by Pope Innocent III in 1202, with the aim of conquering Egypt and then proceeding to the Holy Land, the Crusade leaders were pressured by the Venetians to turn on Constantinople, which they eventually sacked in 1204 for its rich spoils.

## fourth dimension

- time regarded as a scalable dimension, required by relativity theory, along with three spatial dimensions, to specify the location of any event.

## fourth estate

- the press; class of society other than the clergy, the nobility, and the common people.

## Fourth French War of Religion

- provoked by the murder of Gaspard de Coligny, French Protestant leader, on St. Bartholomew's Day (August 24, 1572), and the subsequent three-day massacre of some 2,000 Huguenots in Paris and the provinces. Ended by the Edict of Boulogne, July 8, 1573. The St. Bartholomew's Day Massacre, as it came to be called, radicalized many survivors, created a strong distrust of the king, and led to Huguenots organizing and creating their own armies for defense against the Catholics.

## Fourth International

- a loose federation of small socialist groups attempted by Leon Trotsky in Mexico in 1937, antagonistic to Moscow.

## Fourth Lateran Council

- the most important of the Lateran Councils, held in 1215 under Pope Innocent III. Among its significant decrees, it proclaimed the dogma of transubstantiation, required Catholics to observe annual confession and receive the Eucharist at Easter, set rules for episcopal elections, condemned Albigensian and Catharist heresies, and forbade trial by battle. It also called for a new Crusade.

## Fourth of July

- (a) America's birthday, 1776; Declaration of Independence, written almost totally by Thomas Jefferson, is approved by the Continental Congress, meeting in Philadelphia.
- (b) rhyming slang for a tie.

## Fourth of March Speech

- 1850 address to Congress by John Calhoun attacking Henry Clay's Compromise of 1850. Calhoun asked that the North return fugitive slaves and suggested that a Constitutional Amendment be added, restoring the balance between the North and the South.

## fourth position

- in ballet, the position in which the feet are slightly apart and parallel, at right angles to the direction of the body, toes pointing out in opposite directions, heels in line front to back, with one foot forward of the other.

## fourth president

- James Madison, March 4, 1809–March 3, 1817.

## 4-3

- in football, the most widely used defensive formation: four linemen backed by three linebackers.

## Fourth Republic

- the republic established in France in 1946, following the German occupation 1940–44 and the provisional government of 1944–46; replaced by the Fifth Republic in 1958.

## fourth state of the Union

- Georgia ratified the Constitution on January 2, 1788.

## fourth wall

- in theater, the imaginary fourth side of a box, separating the audience from the actors on stage.

## Fourth Way School

- the esoteric teachings of George Ivanovich Gurdjieff, going beyond the traditional three ways of the fakir, the monk, and the yogi. This new religion uses arduous physical exercise to reach a higher level of consciousness.

## fourth wedding anniversary gift

- linen or silk is customary.

## fourth wheel

- the wheel in a watch that drives the escape pinion.

## 4 to

- quarto: the page size of a book from printer's sheets that are folded into four leaves, about 9-½ x 12 inches.

**4-to-an-em**

- in printing, an obsolete measure of spacing (equal to 1/4 em) used to justify a typographical line (i.e., spread out the words to fill the line to the margins); known also as a "thin space."

**4 to 3**

- the odds against being dealt a pair in five-card poker.

**Four Transcendent Lords**

- the main peaceful Tibetan Bon gods: Satrig Ersang (great mother goddess), Shenlha Okar (god of wisdom or of light), Sangpo Bumtri (the procreator), and Tönpa Shenrab (the teacher).

**4 Wheaton 316**

- citation for the Supreme Court decision in *McCulloch v. Maryland* (1819), in which the Court, under Chief Justice Marshall, ruled that Congress had the authority to establish a national bank, thereby establishing the supremacy of the federal government over the states.

**four-wheel drive**

- an automotive drive system in which engine power is transmitted to all four wheels for greater traction.

**four wide ones**

- in baseball, base on balls.

**Four Wind Gods**

- in Greek mythology, the sons of Astraeus and Eos: Boreas, god of the North Wind; Eurus, god of the East Wind; Notus, god of the South Wind; and Zephyrus, god of the West Wind.

**4-wood**

- golf club, sometimes called a baffie.

**Four-Year Plan**

- Adolf Hitler's agenda, proposed October 19, 1936, to have the German army and the German economy ready for war within four years. To this end he launched a program to build highways for the army, which by 1938 had built 3,000 kilometers of roads costing an estimated six million marks.

## Gang of Four

- a group of Communist functionaries who tried to seize power after the death of Mao Zedong in 1976: his widow, Jiang Qing; and three Shanghai politicians, Zhang Chunqiao, Wang Hongwen, and Yao Wenyuan. The coup failed, and in 1980 they were tried and found guilty of treason.

## Group 4 elements (of the Periodic Table)

- titanium (22), zirconium (40), hafnium (72), rutherford-ium (104).

## Islands of the Four Mountains

- a section of the Aleutian Islands chain in Alaska.

## Lagrange's four-square theorem

- . states that every positive integer can be written as the sum of at most four square numbers. Three are not always sufficient.

## Mendelssohn's Fourth Symphony

- "Italian" symphony (in A major).

## most shutouts pitched in world series games

- total of four (baseball record): by Christy Mathewson for New York Giants against Philadelphia Athletics, 1905 (three times), and 1913, same teams.

## the name of God

- in most languages, God's name is spelled in four letters, e.g., in Hebrew, *JHVH* (JeHoVaH); In Latin, *Deus;* in Greek, *Θεός* (Theos); in French, *Dieu;* in German, *Gott;* in Spanish, *Dios;* in Dutch, *Godt;* in Swedish, *Goth;* in Egyptian, *Deva;* etc.

## number 4 billiard ball

- the ball that is conventionally solid purple.

## number 4 wood screw

- a screw with a screw shank diameter of 7/64 inches.

**on all fours**

• on one's hands and knees.

**Operation T-4**

• the Nazis' October 1939 program for euthanizing Germans who were disabled physically or mentally, and thus deemed "unworthy of life" by not conforming with the Third Reich's concept of "racial integrity."

**the phases of the moon**

• full, waning, new, waxing.

**plus fours**

• long baggy knickers for men, popular in the 1910s through the 1930s, worn as sports clothing, especially for golf. So called because four inches were added to regular knickers to provide the bagginess below the knees.

**quadrilateral**

• a polygon of four sides.

**Quadrilateral**

• four famous fortresses in Italy between Lombardy and Venice: Mantua, Peschiera, Legnano, and Verona.

**quadrivium**

• the four studies of the higher division of the seven liberal arts in the Middle Ages: geometry, arithmetic, astronomy, and music.

**Quadruple Alliance**

• Any one of several European alliances:
   (a) of 1718: Austria, Britain, France, and the United Provinces (Netherlands) allied to prevent Spain from annexing Sardinia and Sicily.
   (b) of March 1814: Austria, Britain, Prussia, and Russia joined forces to defeat the French emperor Napoleon; this alliance was renewed in 1815 and 1818.
   (c) of 1834: Britain, France, Portugal, and Spain joined together to defend the constitutional monarchies of Spain and Portugal against the Carlists.

**quartet**

- a musical group composed of four instruments or four voices.

**Schubert's Fourth Symphony**

- "Tragic" symphony (in C minor).

*The Sign of the Four*

- the second of Arthur Conan Doyle's (he was not yet "Sir") Sherlock Holmes novels, first appearing in 1890. Subsequent editions and serializations frequently shortened the title to four words, *The Sign of Four*. The story has been filmed at least nine times, the earliest in 1913 with Harry Benham as Holmes.

**stella**

- U.S. four dollars gold piece, minted by the U.S. from 1879 to 1880.

## 4-1/4

- 4-¼ inches is the diameter of the hole on a golf green.

## 4-1/2

- point size, typographer's measure, known as "diamond."

## 005

- Decimal ASCII code for [enquiry].
- Dewey Decimal System designation for books on computer programming.
- medical classification code for bacterial food poisoning.

## 05

- number of the French Département of Hautes-Alpes.

## 5

- Article of the New York State Labor Law that deals with hours of labor.

- Article of the U.S. Uniform Commercial Code that deals with letters of credit.
- atomic number of the chemical element boron, symbol B.
- Beaufort number for wind force of 19 to 24 miles per hour, called "fresh wind–fresh breeze."
- dental chart designation for the right first premolar (first bicuspid).
- ephahs in a lethech (ancient Hebrew measure of dry capacity).
- the first nontrivial circular number; i.e., a number whose powers terminate in the same digit(s) as the root number (e.g., 25, 125, etc.). The next circular number is 6.
- five cents, original fare on New York City's Staten Island Ferry, from the Battery on Manhattan Island to St. George on Staten Island.
- five feet, width of a regulation table-tennis surface.
- five grams, weight of a U.S. nickel.
- fluid ounces in a gill (British measure).
- gauge (number) of a wood screw with shank diameter of 1/8 of an inch.
- the highest grade of prostate cancer cells on the Gleason scale.
- in astrology, the number 5 is ruled by the planet Mercury.
- in baseball, 5 represents the third baseman's position.
- in Hebrew, the numerical value of the letter *he.*
- in mineralogy, Mohs Hardness Scale level characterized by apatite; can be scratched by a penknife.
- ISBN group identifier for books published in what had been the USSR, now covering Azerbaijan, Tajikistan, Turkmenistan, and Uzbekistan, and shared by Armenia (with 99930 and 99941), Belarus (with 985), Estonia (with 9949 and 9985), Georgia (with 99928 and 99940), Kazakhstan (with 9965), Kyrgyzstan (with 9967), Latvia (with 9984), Lithuania (with 9986 and 9955), Moldova Republic (with 9975), and Ukraine (with 966).
- lunch-counter code for a large glass of milk.
- milliliters in a teaspoon.
- most fumbles in Super Bowl games, career (football record): by Roger Staubach for the Dallas Cowboys, SB-VI, 1972; SB-X, 1976; SB-XII, 1978; SB-XIII, 1979.

- most hits in a World Series game (baseball record): by Paul Molitor, of the Milwaukee Brewers against St. Louis Cardinals, October 12, 1982. Nonetheless, Milwaukee lost.
- most home runs in a single World Series (baseball record): twice, by Reggie Jackson for the New York Yankees against the Los Angeles Dodgers, 1977; and by Chase Utley for the Philadelphia Phillies against the New York Yankees, 2009.
- most individual goals scored in a World Cup soccer match: by Oleg Salenko in Russia's win over Cameroon in 1994.
- most Tonys won by an actress: by Julie Harris, for *I Am a Camera* (1952), *The Lark* (1956), *Forty Carats* (1969), *The Last of Mrs. Lincoln* (1973), and *The Belle of Amherst* (1977). Nominated ten times, in 2002 Miss Harris also received a Special Tony for Lifetime Achievement in the Theater.
- most touchdown pass receptions in an NFL game (football record): three times, by Bob Shaw for Chicago Cardinals against Baltimore Colts, October 2, 1950; by Kellen Winslow for San Diego Chargers against Oakland Raiders, November 22, 1981; and by Jerry Rice for San Francisco 49ers against Atlanta Falcons, October 14, 1990.
- most touchdowns scored in Pro Bowl games in an NFL career (football record): twice, by Jimmy Smith for the AFC, 1998–2001, and by Marvin Harrison for the AFC, 2000–06.
- most soccer World Cup Championships won, by Brazil, in 1958, 1962, 1970, 1994, and 2002. (The quadrennial Cup competitions started in 1930.)
- name of a character in the *Peanuts* comic strip in the 1970s and 80s. He has two sisters, named "3" and "4."
- number of counters on each side in the Egyptian game, Senet.
- number of counties in each of the states of Hawaii and Rhode Island.
- number of enemy planes a fighter pilot must down in order to qualify as an ace.
- number of fingers on one hand, or toes on one foot.
- number of interlocking rings that make up the symbol for the Olympics, representing the number of inhabited continents (North and South America taken as one).
- the number of "kyu" (pupil) grades in judo.

- number of lines in a *tanka*, a Japanese form of thirty-one-syllable poetry in which the first and third lines consist of five syllables, the rest, of seven.
- number of permanent members with veto power on the UN Security Council: U.S., UK, France, Russia, and China.
- number of Platonic solids (a convex polyhedron whose faces are all the same regular polygon shape with the same number of faces meeting at all its vertices); namely, tetrahedron (its four faces each a triangle); cube, or hexahedron (its six faces each a square); octahedron (its eight faces each a triangle); dodecahedron (its 12 faces each a pentagon); and icosahedron (its 20 faces each a triangle).
- number of presidential electoral votes apportioned to each of the states of Nebraska, Nevada, New Mexico, Utah, and Wyoming. (May be reapportioned for the 2012 election.)
- number of presidents of France under the Fifth Republic: Charles de Gaulle, 1959; Georges Pompidou, 1969; Valéry Giscard d'Estaing, 1974; François Mitterand, 1981; Jacques Chirac, 1995–present.
- number of rings of the planet Neptune, named Galle, Le Verrier, Lassell, Arago, and Adams.
- number of seats in the House of Representatives allotted to each of the states of Connecticut, Iowa, Oklahoma, and Oregon (as of the 2000 Census).
- number of swimming pools on Cunard's *Queen Mary 2*.
- number of tennis players winning all four Grand Slam championships in the same year, namely, Don Budge, 1938; Maureen Connolly, 1953, Rod Laver, twice, 1962 and 1969; Margaret Smith Court, 1970; and Steffi Graf, 1988.
- number of times per day that devout Muslims pray to Allah.
- number of U.S. presidents named John: Adams, Quincy Adams, Tyler, Coolidge (who was born John Calvin), and Kennedy.
- number of U.S. presidents that attended Harvard University: John Adams, John Quincy Adams, Theodore Roosevelt, Franklin D. Roosevelt, and John F. Kennedy.
- number of U.S. states named after monarchs: Georgia, Louisiana, Virginia, North and South Carolina.
- Paris Opera House box number occupied by the Phantom (in the 1908 novel, *Phantom of the Opera*, by Gaston Leroux).

- players on a basketball team.
- record number of balks by a pitcher in one baseball game (baseball record): by Bob Shaw, Milwaukee Braves, May 4, 1963.
- record number of individual goals scored in a Stanley Cup match: by several players, the most recent being Mario Lemieux for Pittsburgh Penguins in a match against Philadelphia Flyers, April 25, 1989.
- record number of U.S. Open tennis doubles titles won by a men's team: Richard Sears and James Dwight, 1882–84 and 1886–87.
- retired baseball uniform number of:
  ⇨ first baseman Jeff Bagwell, retired by the Houston Astros.
  ⇨ team executive Carl Barger, retired by the Florida Marlins.
  ⇨ catcher Johnny Bench, retired by the Cincinnati Reds.
  ⇨ shortstop Lou Boudreau, retired by the Cleveland Indians.
  ⇨ third baseman George Brett, retired by the Kansas City Royals.
  ⇨ outfielder Joe DiMaggio, retired by the New York Yankees.
  ⇨ first baseman Hank Greenberg, retired by the Detroit Tigers.
  ⇨ catcher Willard Hershberger, retired by the Cincinnati Reds.
  ⇨ third baseman Brooks Robinson, retired by the Baltimore Orioles.

- retired basketball jersey number of:
  ⇨ Dick Van Arsdale, retired by the Phoenix Suns.
  ⇨ Mendy Rudolph, retired as a referee (from officials' numbers, by the NBA).

- retired football jersey number of George McAfee, retired by the Chicago Bears.
- retired hockey shirt number of:
  ⇨ Bill Barilko, retired by the Toronto Maple Leafs.
  ⇨ Dit Clapper, retired by the Boston Bruins.
  ⇨ Bernie Geoffrion, retired by the Montreal Canadiens.
  ⇨ Rod Langway, retired by the Washington Capitals.

⇨ Denis Potvin, retired by the New York Islanders.

- sides on a pentagon; notably, the Department of Defense building in Arlington, Virginia.
- sum of the first two squared numbers (1 + 4).
- Title of the New York State Vehicle and Traffic Law dealing with driver's licenses.
- Title of the United States Code dealing with governmental organization and employees.
- value of Roman numeral V.

### The Beast with Five Fingers

- horror flick of 1946, with Peter Lorre besieged by a disembodied hand.

### Beethoven's *Piano Concerto No. 5*

- the "Emperor" (in E-flat major).

### big five

- (a) nickname for the five major powers among the allies in World War II: U.S., Great Britain, USSR, France, and China.
- (b) the basketball teams of five institutions of higher education in the vicinity of Philadelphia, Pennsylvania: University of Pennsylvania, Temple University, Villanova, La Salle University, and St. Joseph's College.

### a bunch of fives

- obsolete slang for a fist.

### C-5

- the four-engine Galaxy, the largest transport in U.S. military service, with maximum cargo capacity of 270,000 pounds. Built by Lockheed-Georgia, in service since June 1970, the 248-foot-long C-5 carries 51,150 gallons of fuel and has a wingspan of 223 feet.

### Canon 5 of the American Bar Association Code of Professional Responsibility

- a lawyer should exercise independent professional judgment on behalf of a client.

## City of Five Flags

- sobriquet of Pensacola, Florida, which has been ruled by five different political authorities: Spain, France, Britain, the Confederacy, and the United States.

## Civil War of the Five Kings

- legendary war in Britain over who would succeed the sons of Gorboduc, who had feuded for the crown, when they both died childless.

## CV-5

- number of the U.S. Navy aircraft carrier, USS *Yorktown*, sunk by a Japanese submarine on June 7, 1942.

## Dionne quintuplets

- the first set of quintuplets known to survive their infancy, born on May 28, 1943, in Ontario, Canada: Annette, Cecile, Emilie, Marie, and Yvonne. Dr. Allan Roy Dafoe, the doctor who delivered them, became an instant celebrity.

## fifth

- the interval between the tonic and the dominant in a diatonic musical scale.

## Fifth Amendment to the Constitution

- establishes provisions for due process of law, denying double jeopardy, protecting against property seizure without just compensation.

## Fifth Avenue

- a major thoroughfare in New York City known for its expensive, world-renowned shops between 49th and 59th Streets. From there, heading north and bordering the east side of Central Park, is some of the most valuable residential property in the City, boasting two miles of opulent apartment buildings and several important museums.

## Fifth Cataract

- a section of white water in the Nile River, north of Berber in northern Sudan.

**fifth column**

- a group of people who work surreptitiously and subversively within a country to foster the aims of an enemy. The term was first used in 1936 to designate rebel sympathizers inside Madrid when four columns of rebel troops were advancing on that city.

***The Fifth Column***

- 1940 play by Ernest Hemingway dealing with fascist and communist espionage in Madrid during the Spanish Civil War.

**Fifth Commandment**

- Thou shalt honor thy father and thy mother.

**Fifth Crusade**

- preached by Pope Innocent III, and impelled by the Fourth Lateran Council, a crusading force from Hungary, Austria, and Bavaria in 1217 set its sights on Egypt, the center of Muslim power. They captured Damietta, in Egypt, in 1219, and then advanced on a foolhardy expedition against Cairo, where they were defeated in 1221. This was the last crusade initiated by the Papacy.

**fifth estate**

- any class of society other than the nobility, the clergy, the common people, and the press.

**Fifth French War of Religion**

- war between the Catholics and the Huguenots (French Protestants), 1574–76, ended by the Peace of Chastenoy, May 6, 1576, and the Edict of Beaulieu, granting freedom of worship throughout France except in Paris. In response, the Catholics formed the Holy League to regain political power.

**Fifth House**

- in astrology, deals with pleasure, romance, and creativity.

**Fifth Lateran Council**

- called by Pope Julius II in 1512 and continued by Pope Leo X, it was attended mainly by Italian bishops. It lasted until 1517, and served to establish peaceful relationships among Christian rulers, and to mobilize more effective military efforts against the Turks. It also forbade the printing of books without ecclesiastical authority.

## fifth monarchy

- the fifth and final monarchy following the Assyrian, Persian, Greek, and Roman monarchies, supposedly prophesied in Daniel 2.

## Fifth Monarchy Men

- a militant religious group during the Pilgrim Revolution in seventeenth-century England, expecting the Second Coming of Christ (as the leader of the fifth monarchy), and believing that they should help institute His reign by force, while rejecting allegiance to any church or civil authority.

## fifth position

- in ballet, position in which the feet are together, pointing to the sides in opposite directions, the heel of each foot touching the joint of the toe of the other.

## fifth president

- James Monroe, March 4, 1817 to March 3, 1825.

## Fifth Republic

- the republic established in France in 1958, successor to the Fourth Republic.

## fifth state of the Union

- Connecticut ratified the Constitution on January 9, 1788.

## fifth wedding anniversary gift

- wood is customary.

## fifth wheel

- an unnecessary or superfluous person or thing.

## first five divinities of Japanese Mythology

- Amenominakanusi (Lord of the Center of Heaven), Takami-musubi (August High Producer), Kami-musubi (August Divine Producer), Umasiasikabipikodi (Pleasant Prince Elder of the Reed Shoot), Amenotokotati (Everlasting Heaven Stander).

## *"5"*

- a 1998 album by Lenny Kravitz, heavy on digital technology.

## The Five (or the Mighty Five)

- designation for a group of young, talented, self-taught composers in late-nineteenth-century St. Petersburg who cooperatively worked toward creating an authentic national school of Russian music: Cesar A. Cui, Alexander P. Borodin, Mily A. Balakirev, Modest P. Mussorgsky, and Nicolay A. Rimsky-Korsakov. Known collectively as *Moguchaya Kuchka,* "the mighty handful," they influenced change in music styles both in and outside of Russia.

## *Five Against the House*

- a 1955 caper film that has five college friends put together a perfect plan to rob a Reno casino. Beautifully directed by Phil Karlson, from a tight screenplay by Stirling Silliphant, the cast features Guy Madison, Kim Novak, and Brian Keith.

## five aggregates

- in Buddhism, the five constituent elements of all existence: matter, perception, conception, volition, and consciousness.

## five-and-ten-cent stores

- variety stores, popular in the first half of the twentieth century, that sold inexpensive household sundries and other general merchandise. Probably the best known was Woolworth's, which opened its first store in 1878. Also called five-and-dimes, they fell out of favor toward the end of the 1900s.

## five bells

- time check on-board ship, signifies 2:30 or 6:30 or 10:30, AM or PM.

## five blessings

- in Chinese culture: long life, wealth, health, love of virtue, and a natural death.

## The Five Bloods

- the principal septs or families of Ireland: The O'Neills of Ulster, the O'Briens of Thomond, the O'Connors of Connaught, the O'Lachlans of Meath, and the M'Murroughs of Leinster.

## 5 by 5

- (adj.) obese; "Mr. Five by Five," a song from the 1942 Universal film *Behind the Eight Ball*, contains the line: "He's five feet tall and five feet wide."

## five-card stud

- a poker game in which the first card is dealt face down, and the following four cards face up. Betting follows each face-up card dealt.

## Five Celestial Buddhas

- the Tathagatas of Tibet: Vairochana (Respendent one), Akshobhya (Imperterable), Amitabha (Boundless Light), Ratnasambhava (Jewel-Born), and Amoghasiddhi (Perfect Accomplishment).

## Five Civilized Tribes

- the Cherokee, Chickasaw, Choctaw, Creek, and Seminole tribes that were forcibly resettled from the Southeast to eastern Oklahoma (the Indian Territory) in the 1830s. (Also known as "Five Civilized Nations.")

## five classes of content in the Vedic text of Hindu philosophy

- injunctions (*vidhi*), hymns (*mantra*), names (*namadheya*), prohibitions (*nisedha*), and explanatory passages (*arthavada*).

## The Five Classics of Chinese literature

- *Book of Historical Records* (Shu Ching), *Book of Poetry* (Shih Ching), *Book of Rites* (Li Chi), *Book of Changes* (I-Ching),

*Spring and Autumn Annals* (Ch'un Ch'iu). These, together with the so-called Four Books (q.v.), comprise the modern Confucian canon and served as the basis for Chinese education until recent times.

## five constant virtues of Confucianism

- humanity, righteousness, propriety, knowledge, and sincerity.

## Five Cross Road

- a town in County Londonderry, Northern Ireland.

## Five Days of Milan

- rebellion by the Italians against Austrian rule in 1848. General Radetzky's Austrian forces, greatly outnumbered by the Milanese, finally retreated from Milan to fortresses between Lombardy and Venice, while at the same time Venice declared itself an independent republic. Thus began the Italian War of Independence, 1848–49.

## five districts in the Burgundy wine-producing area

- Chablis, Côte de Nuits, Côte de Beaune, Côte Chalonnaise, and Mâconnais.

## five dollars

- denomination of U.S. paper money that bears the portrait of Abraham Lincoln.

## Five Dynasties

- Later Liang (907–923), Later T'ang (923–936), Later Jin (935–946), Later Han (947–950), and Later Zhou (951–960): The families that dominated the Yellow River Valley in North China from the end of the T'ang Dynasty until the beginning of the empire's unification under the Sung Dynasty in 960.

## *Five Easy Pieces*

- 1970 film that has Jack Nicholson in a complex character study of a man searching for himself. Good supporting cast includes Karen Black, Billy Bush, Fannie Flagg, and Sally Ann Struthers. Memorable is Nicholson's order of a chicken salad sandwich in a diner.

## Five Elements

- (a) in Chinese mythology, those from which all the world is created: fire, earth, wood, metal, and water.
- (b) in Buddhism, the five constituent elements of one's existence: earth, water, fire, wind, and space.
- (c) in Hindu cosmology, the five constituents of the universe (*bhuta*): *kshiti* (the earth), *apa* (the water), *teja* (the energy), *maruy* (the wind), and *yoma* (the sky).
- (d) in the Tibetan myth of creation, hardness, fluidity, heat, motion, and space. These constitutents came together and combined into a huge egg, from which sprung all earthly values and misfortunes.

## five evidences of gluttony

- (as per Thomas Aquinas): eating too soon, eating too fast, eating too much, eating too eagerly, eating too greedily.

## five-factor model of personality

- five dimensions of personality identified through empirical research: agreeableness, conscientiousness, extraversion, openness, and neuroticism. Sometimes called the "Big Five" personality traits.

## 5 February

- Mexican Constitution Day, commemorating its signing in 1917.

## Five Field Kono

- a board game played in Korea.

## five-finger discount

- a slang term for shoplifting.

## five fingers

- (a) another name for the card game Spoil Five, popular among the Irish.
- (b) fisherman's name for a starfish.
- (c) underworld slang for a five-year prison term.

## Five Fingers Peninsula

- a spit of land off the southwest tip of South Island, New Zealand.

**Five Foot Shelf**
- popular name for the series of books issued as the Harvard Classics.

**Five Forks**
- a site near Petersburg, Virginia, at which Confederate general R. E. Lee made his final assault of the Civil War, on April 1, 1865, only to be repulsed by the Union forces under General Sherman. Lee surrendered at Appomattox a week later.

**five forms of litany authorized for public worship**
- In Catholicism:
  1. Litany of the Saints
  2. Litany of the Blessed Virgin
  3. Litany of the Holy Name of Jesus
  4. Litany of the Sacred Heart
  5. Litany of St. Joseph.

**Five Freedoms of the Air**
- came out of a fifty-four-nation meeting in Chicago in November/December, 1944, dealing with international air transport. Proposed that five principles be accepted as the basis for reciprocal air rights:
  1. freedom of peaceful transit;
  2. freedom of non-traffic stop (to refuel and/or repair);
  3. freedom to take traffic from the homeland to another country;
  4. freedom to bring traffic from any country to the homeland;
  5. freedom to pick up and discharge passengers at intermediate points.

**"five geese in a flock"**
- line from the nursery rhyme, "Intery Mintery":

  > Intery, mintery, cutery corn,
  > Apple seed and apple thorn;
  > Wire, brier, limber-lock,
  > Five geese in a flock,
  > Sit and sing by a spring,
  > O-u-t, and in again.

**Five Glorious Mysteries of the Catholic Faith**
- the Resurrection, the Ascension, Descent of the Holy Ghost on the Apostles, the Assumption, and the Coronation.

**"five golden rings"**
- the gift on the fifth day of Christmas in the carol "The Twelve Days of Christmas."

**Five Good Emperors**
- an era of Rome, from AD 96 to 180, during which the Empire was ruled by a succession of moderate emperors: Nerva (96–98), Trajan (98–117), Hadrian (117–138), Antonius Pius (138–161), and Marcus Aurelius (161–180). This era corresponds with the period known as Pax Romana, a time during which Rome expanded considerably and consolidated its empire.

**Five Graves to Cairo**
- a World War II espionage film with Franchot Tone as a British soldier outsmarting Field Marshall Erwin Rommel (played by Erich von Stroheim) in the North African desert. Billy Wilder directed this taut 1943 movie featuring Anne Baxter, Akim Tamiroff, Peter Van Eyck, and Fortunio Bonanova.

**Five Great Kings**
- in Tibetan Buddhism, five recognized heroes deified as protectors against all enemies:
    1. Pe-har, chief of the Five Great Kings, "King of the Karma," who rides a white lion;
    2. Brgya-byin, "King of the Mind," who rides an elephant;
    3. Mon-bu-pu-tra, "King of the Body," who rides a white lioness;
    4. Shing-bya-can, "King of Virtue," who rides a black horse;
    5. Dgra-lha skyes-gcig-bu, "King of Speech," who rides a black mule.

**Five Great Myoo**
- the five great kings of science in the Japanese Buddhist pantheon: Fudo (the immovable one), Gundari, Gozanze, Daiitoku, and Kongoyasha (or Ususama).

**five great obstacles to the higher Buddhist life**

- sloth, pride, malice, lust and doubt.

**five great vows of Jainism (ahavratas):**

- Ahimsa (non-violence), Aparigraha (non-possession), Asteya (non-stealing), Satya (truth), Brahmacarya (celibacy).

**the five kleshas**

- in yoga, the five hindrances to enlightenment: ignorance, egocentricity, desire, aversion, possessiveness.

**Five Householders**

- as told in *Chandogya Upanishad*, five Hindu seekers of the Universal Self: Pracinashala, Aupamanayava, Satyayajna Paulusi, Indradyumna Bhallaveya, Jana Sharkaraksya, and Budila Ashvatarashvi.

**Five Hu**

- five non-Chinese tribes, originally from outside China, that migrated into areas of northern China from the late third to the early fifth centuries: Xiongnu, Xianbei, Di, Qiang, and Jie. Also known as the Wu Hu ("five non-Chinese races").

**5-iron**

- golf club equivalent to the earlier mashie.

**Five Islands**

- town on Cobequid Bay, Nova Scotia, Canada.

**Five Japanese Kings**

- kings in the Yamato dynasty that were in communication with China in the fifth century: San, Chin, Sai, Jou, and Bu.

**Five Joyful Mysteries of the Catholic Faith**

- the Annunciation, the Visitation, the Nativity, the Presentation, and the Finding of Jesus in the Temple.

**five kinds of suffering**

- in Buddhism:
    1. the pain accompanying one's birth;
    2. the pain of getting old;
    3. the pain of illness;

*4.* the pain of death;

*5.* the pain of separation from those one loves.

### five kingdoms of living organisms

- according to the Whittaker taxonomic system: Monera (bacteria), Protocista (algae and protozoans), Fungi (mushrooms and molds), Animalia (animals), and Plantae (plants).

### Five Lanes

- a town in Cornwall, England.

### Five Laws of Library Science

*1.* books are for use;

*2.* every reader has his or her book;

*3.* every book has a reader;

*4.* save the time of the reader;

*5.* the library is a growing organism.

(Proposed by S. R. Ranganathan, who librarians around the world accept as the father of library science.)

### *Five Little Peppers and How They Grew*

- popular children's classic of 1881, first in the series of Five Little Peppers children's books by Margaret Sidney (pen name of Harriet Mulford Stone Lothrop).

### five liturgical colors

- used by the Catholic Church in her vestments and altar drapes: white, for purity; red, for blood; green, for hope; purple, for penance; black, for sorrow.

### five main wrathful gods of Bon

- (Tibet's first religion): Welse Ngampa (fierce god of body), Lhago Togpa (fierce god of speech), Trowo Tsochag (fierce god of the mind), Purba (fierce god of action), and Welchen Gekho (fierce god of excellence). All dwell on Mount Kailash.

### five major forces of evil in Zoroastrianism

- Fury, Lie, Dearth, the Evil Eye, and Nasu (the female demon of dead matter).

### five major wine districts in the Loire Valley of France

- in the Eastern Loire are Sancerre and Pouilly-Fumé; in the Central Loire, Touraine and Anjou-Saumur; and in the Western Loire is Muscadet.

### 5 May

- Cinco de Mayo, celebrating the victory of Mexican troops over French forces in Puebla, Mexico, on May 5, 1862.

### five meditation Buddhas

- Vairocana, whose color is white and who rides a dragon; Ratnasambhava is yellow, rides a horse, and rules the south; Amitabha is red, is escorted by a peacock, and rules the west; Amoghasiddhi is green, is carried by an eagle, and rules the north; Aksobhya is blue, rides an elephant, and rules the east.

### Five Members

- John Pym, John Hampden, Sir Arthur Haselrig, Denzil Holles, and William Strode: Members of Parliament whom Charles I attempted to arrest on the floor of the House of Commons, in January 1642, for opposing his seizure of command of the armed forces. The five, however, were forewarned and escaped.

### Five Mile Act

- an English penal law passed under Charles II in 1665, intended to force conformity to the Church of England. It forbade any clergyman or schoolmaster from coming within five miles of a city or incorporated town unless he swore that he would "not at any time endeavor any alteration in Government either in Church or State."

### Fivemile Creek

- a waterway in Central Wyoming.

### Fivemiletown

- a town in County Tyrone, Northern Ireland.

**five-minute clause**

- an addition to a deed of separation stipulating that the deed is null and void if the husband and wife remain together for five minutes after the separation is effectuated.

**Five Mythical Emperors of China's prehistory**

- Huang Ti, Chuan Hsiun, K'u, Yao, and Shun.

**Five Nations**

- confederation of Iroquoian Indian tribes of the northeastern United States founded in the late sixteenth century: Mohawk, Seneca, Oneida, Cayuga, and Onandaga. Also known as the Iroquois League.

**5 November**

- Guy Fawkes Day in the UK, commemorating the frustrated plan (known as the Gunpowder Plot) to blow up the houses of Parliament and King James I on the opening day of Parliament on this date in 1605.

**five number summary**

- used in descriptive statistics to define a data set consisting of the minimum value, the lower quartile, the median, the upper quartile, and the maximum value.

**The Five Observances (Niyama) of Yoga**

- cleanliness of body and mind, contentment, body-conditioning (*tapas*), study of self (*swadhyaya*), and attentiveness or purity of mind (*prasadana*) (Ernest Wood, *Yoga Dictionary*).

**five o'clock shadow**

- the stubble that appears late in the afternoon on the face of a man who has shaved in the morning.

**five of clubs**

- slang for a fist.

**"five on it"**

- five dollars for a sack of marijuana (rapper's slang).

**five on three**

- a term used in ice hockey when one team has two players off the ice on penalties, giving the opposing team an advantage of five players against three.

## Five Pecks of Rice

- religious rebellion toward the end of the Han Dynasty in China (206 BC to AD 220), led by Chang Lu, a response to the unremitting poverty among the peasantry of central China. The Five Pecks of Rice movement had been initiated by Chang Lu's grandfather, Chang Ling, and took its name from the five pecks of rice that his followers paid annually as tithe. Chang Lu set up an independent theocratic state which fostered the spread of the Taoist religion, but eventually surrendered to the Han general Ts'ao Ts'ao in 215.

## *The Five Pennies*

- a 1959 music flick with Danny Kaye as Red Nichols, who led a jazz group called The Five Pennies in the 1920s. In supporting roles are Barbara Bel Geddes, Bob Crosby, Louis Armstrong, and several well-known musicians recreating 1920s jam sessions.

## five-penny nail

- a carpenter's common nail, 1 ¾ inches long (coded 5d).

## Five Percent Nation

- a group founded by members of the Nation of Islam. They believe that only 5 percent of the world's population has the knowledge and ability to lead the rest of the world: the 10 percent that have the knowledge, but not the ability, and the 85 percent that do not understand the world because they have neither the knowledge nor the ability.

## Five Pillars of Islam

- the duties of the devout: profession of faith, prayer five times a day, alms from savings, fasting in the month of Ramadan, and pilgrimage to Mecca at least once in a lifetime.

## 5-point card

- in the game of Canasta, any 7, 6, 5, 4, or black 3.

## 5-point type

- typographer's measure, known as "pearl."

## Five Points

- a slum area of lower Manhattan in New York City that existed throughout much of the 1830s and 1840s; it was filled with dilapidated and uninhabitable tenements, disease, flourishing crime of all types, drunkenness, terrorized residents, and corrupt politicians. The district was known as the Sixth Ward, contained by Reade Street on the south, Broadway on the east, Canal Street on the north, and West Street on the west. The familiar name, Five Points, was derived from the convergence of five streets: Mulberry, Anthony (now Worth Street), Cross (now Park), Orange (now Baxter), and Little Water Street (which no longer exists).

## Five Points of Calvinism

- as distinguished from Arminianism, Calvinists believe in five constructs known by the acronym TULIP (also referred to as the Five Articles of the Remonstrants):
  1. Total Depravity (or Original Sin);
  2. Unconditional Election;
  3. Limited Atonement;
  4. Irresistible Grace;
  5. Perseverance of the Saints (or, Once Saved Always Saved).

## Five Points of Fundamentalism

- These points derived from the 1895 Niagara Bible Conference:
  1. divinely inspired infallible scriptures;
  2. Christ's virgin birth and deity;
  3. Christ's substitutionary atonement;
  4. Christ's resurrection;
  5. Christ's imminent second coming.

**Five-Power Constitution**

- Sun Yat-sen's plan of a democratic government for China, first proposed in December 1906, encompassing five branches: executive, legislative, judicial, examination (civil service), and control. The first three were to govern, the latter two to prevent abuses engendered by the division of the first three. A derivative system was finally installed by the Kuomintang in October, 1928, and confirmed by a People's National Convention in May, 1931.

**Five-Power Naval Armaments Treaty**

- an agreement of February, 1922, among the United States, Great Britain, Japan, France, and Italy, providing for a ten-year "holiday" in the building of capital naval ships of over 10,000 tons, and for limiting existing total fleet tonnage of capital ships to a ratio of 5:5:3:1.7:1.7.

**Five Principles**

- (*Pancasila*): the Indonesian state philosophy propounded as early as mid-1945 by the nationalist leader, Sukarno, who argued that Indonesian independence should be based on Five Principles:
    1. Indonesian nationalism;
    2. internationalism, or humanism;
    3. consent, or democracy;
    4. social prosperity;
    5. belief in one God.

The Five Principles, in somewhat different terms and in a difference sequence, were adopted into the constitution of the Republic of Indonesia.

**Five Principles of Peaceful Coexistence**

- a 1954 agreement between China and India covering relations between the two countries in the Tibet region of China. The items include:
    1. mutual sovereignty and integrity respected;
    2. non-aggression;
    3. non-interference;
    4. equality and mutual benefit;
    5. peaceful coexistence.

**fiver**

- slang for a five-dollar bill.

**five races of man**

- those that have lived on Earth, in Greek mythology (from Hesiod's *Works and Days*):
  1. the golden race, created by the Titans;
  2. the silver race, created by the Olympian gods;
  3. the warlike bronze race, fashioned by Zeus;
  4. (without metal designation) the great heroes who fought at Troy and Thebes;
  5. the iron race, the present race of man.

The first race became benevolent spirits (daimones) living on the Earth; the second became underworld spirits; the third passed to the underworld ruled by "chill Hades"; many of the fourth did not die, but moved to the Isles of the Blessed, where they live free from care and sorrow; the fifth is destined to pass away sometime in the future. (R. Cavendish, *An Illustrated Encyclopedia of Mythology*)

**The Five Relationships of social life**

- in Confucianism, those between husband and wife, parent and child, elders and youth, rulers and subjects, friend and friend.

**Five Rivers**

- a town in southern part of South Island, New Zealand.

**Five Rivers of Hades**

- in Greek mythology, Acheron, river of woe; Cocytus, river of lamentation; Lethe, river of forgetfulness; Phlegethon, river of fire; and Styx, river of hate.

**5 rue Dannou**

- Paris address of the celebrated Harry's Bar, known to be a "temple of intemperance."

**fives**

- a British form of handball played on a court with a front wall and two side walls.

**Five Sacred Mountains of China**

- Hengshan (northern peak), Hengshan (southern peak), Huashan (western peak), Songshan (central peak), Taishan (eastern peak).

**the five senses**

- sight, hearing, taste, smell, and touch.

**five silver stars**

- insignia of the highest-ranking military officer, General of the Army.

**Five Sisters**

- a town in Ross and Cromarty County, Scotland.

**Five Sorrowful Mysteries of the Catholic Faith**

- the Agony in the Garden, the Scourging at the Pillar, the Crowning with Thorns, Jesus Carries His Cross, and the Crucifixion.

**five-spice powder**

- an aromatic seasoning for Chinese cuisine, encompassing the five basic flavors of Chinese cooking: sweet, sour, pungent, bitter, and salty. Consists of cinnamon, star anise, ground clove, fennel, and red pepper.

**five-spot**

- a five-dollar bill.

**five stages of grief**

- shock, denial, anger, bargaining, and acceptance. (E. Kübler-Ross)

***Five Star Final***

- director Mervyn LeRoy's 1931 film excoriates the workings of scandal-sheet tabloids similar to those found today at supermarket checkout counters. Edward G. Robinson plays against a whole slew of unsavory muckraking newspaper types.

**fivestones**

- an alternate name for the popular children's game, Jacks.

## 5-to-an-em

- an obsolete measure of spacing (equal to 1/5 em) used to justify a typographical line (i.e., spread out the words to fill the line to the margins); known also as a "thin space."

## 5 to 1

- the odds against tossing a 7 in dice; or the odds in poker against filling a straight open at both ends.

## Five Towns

- (a) an upscale area of Long Island, New York, comprised of Cedarhurst, Hewlett, Inwood, Lawrence, and Woodmere.
- (b) a district in central England known for the manufacture of pottery and china; also called the Potteries. The towns in this district were combined in 1910 to form Stoke-on-Trent.

## five varieties of grape used in Alsace wines

- Pinot Blanc, Riesling, Gewurztraminer, Pinot Gris, and Sylvaner.

## five vital airs

- in yoga, the forces of nature that vitalize the parts of the human body: *prana* (the heart), *apana* (the anus), *samana* (the navel), *udana* (the throat), and *vyana* (the genitals).

## five vowels in the English language

- a, e, i, o, u.

## five Ws and an H

- points to be covered by good newspaper reporting: Who, what, when, where, why, and how.

## five wits

- common sense, imagination, fantasy, estimation, and memory.

## Five World Regions

- in Aztec lore, the four cardinal points associated with the four sons of Ometecuhti who created the world: Xipe Totec, Camaxtli, Quetzalcoatl, and Huitzilopochtli. The fifth and central region is that of the present world, managed by Xiuhtecuhtli.

## five yamas of Yoga

- five "restraints" important in Patanjali's system of classical "Ashtanga Yoga": nonviolence, truthfulness, nonstealing, periods of celibacy, and nonhoarding.

## Five-Year Plan

- standard feature of Soviet economic planning, the first covering the years 1928 to 1932.

## "Five years have passed; five summers, with the length of five long winters!"

- start of "Lines Composed a Few Miles Above Tintern Abbey" (July 13, 1798), by William Wordsworth (1770–1850).

## Gautama's Five Types of Fallacy

1. the erratic or inconclusive reason;
2. the contradictory reason;
3. the neutralized reason;
4. the unknown or unproven reason;
5. the inopportune reason.

## Group 5 elements (of the Periodic Table)

- vanadium (23), niobium (41), tantalum (73), dubnium (105).

## *Hawaii Five-O*

- highly successful CBS-TV police series set in Honolulu, starring Jack Lord and James MacArthur. Enjoying a twelve-year run, from September 1968 to April 1980, it popularized the phrase "Book 'em, Danno." Reborn in September 2010 with Alex O'Loughlin as Detective Garrett and Scott Caan as Danno.

## *Henry V*

- (a) film adaptation of Shakespeare's play, co-produced, co-directed, co-adapted by, and starring Laurence Olivier in a masterful translation to the screen. England's most expensive film to date, shot in 1944, served as a propaganda piece, rallying the British people to the same united patriotic fervor that Henry induced in the fifteenth cCentury. Released in the U.S. in 1946, after the war, Olivier lost the Best Actor Oscar to Fredric March for his special Academy Award, for his outstanding achievement as actor, producer, and director in bringing *Henry V* to the screen.

- (b) 1989 version directed by and starring Kenneth Branagh, different but no less stirring than the Olivier interpretation. Highlighted by his passionate speech to his men before leading them into battle on the fields of Agincourt.

## high five

- a gesture or triumph or shared understanding in which one person slaps his open hand against the raised palm of another.

## I-5

- major north-south interstate highway from San Diego to Los Angeles, Sacramento, Portland, Seattle, and into Vancouver, Canada. Known locally as "the 5."

## Mayan God of Five Unlucky Days

- named Uayeb, who lives in a snail shell.

## Mendelssohn's Fifth Symphony

- "Reformation" symphony (in D minor).

## nickel

- five-cent piece; U.S. coin in circulation since 1866.

## nickel bag

- five dollars' worth of narcotics (underworld slang).

## number 5 billiard ball

- the ball that is conventionally solid orange.

## *Party of Five*

- a one-hour TV series on the Fox Network from September 1994 to May 2000, about the five siblings of the Salinger family, determined to stay together after the death of their parents. The three sisters were played by Neve Campbell, Lacey Chabert, and Jennifer Love Hewitt; the two brothers, by Scott Wolf and Matthew Fox.

## pentathlon

- athletic competition comprising five events: long jump, javelin throw, 200-meter sprint, discus throw, and 1,500-meter race.

## The Pentateuch

- the first five books of the Old Testament: Genesis, Exodus, Leviticus, Numbers, and Deuteronomy.

## plead the fifth

- to refuse to testify by invoking the Fifth Amendment provision against self-incrimination.

## quintet

- a musical group composed of five instruments or five voices.

## Red Nichols and the Five Pennies

- a jazz group of the 1920s.

## *Slaughterhouse-Five*

- Kurt Vonnegut's 1969 offbeat fantasy antiwar novel, subtitled "The Children's Crusade," takes us along on Billy Pilgrim's odyssey through time (in an apparent search for meaning, especially of the allies' fire-bombing of Dresden in February 1945). Made into a film in 1972, directed by George Roy Hill, with Michael Sacks, Ron Leibman, Eugene Roche, Sharon Gans, and Valerie Perrine in her first film.

## Spoil Five

- Also called Five Cards or Five Fingers, this is a card game popular in Ireland. The object is to take three of the five tricks played in a hand.

## take five

- take a short break from activity (approximately five minutes).

## weight of regulation professional baseball

- between 5 and 5 ¼ ounces.

## 5-1/2

- point size, typographer's measure, known as "agate."

## 006

- Decimal ASCII code for [acknowledgment].

## 06

- number of the French Département of Alpes-Maritimes.

## 6

- atomic number of the chemical element carbon, symbol C.
- Beaufort number for wind force of 25 to 31 miles per hour, called "strong wind–strong breeze."
- bottles in a Rehoboam (champagne measure).
- dental chart designation for the right canine tooth (cuspid).
- faces on a cube.
- feet in a fathom (measure of water depth).
- the first perfect number (its divisors being 3, 2, and 1).
- gauge (number) of a wood screw with a shank diameter of 9/64 of an inch.
- hins in a bath (ancient Hebrew measure of liquid capacity).
- in astrology, the number 6 is ruled by the planet Venus.
- in baseball, 6 represents the shortstop's position.
- in football, the number of points scored for a touchdown.
- in Hebrew, the numerical value of the letter *vav*.
- in mineralogy, Mohs Hardness Scale level characterized by feldspar; can be scratched by quartz.
- in rugby union, the position of blindside flanker.
- magnums in a Salmanazar (champagne measure).
- mathematical value of 3 factorial $(3!) = (3 \times 2 \times 1)$.
- most British Open golf tournaments won: by Harry Vardon of the UK in 1896, 1898–99, 1903, 1911, and 1914.
- most consecutive shutouts pitched (baseball record): by Don Drysdale, Los Angeles Dodgers, 1968.
- most Cy Young Awards as league's outstanding pitcher for the year (baseball record): by Roger Clemens, for Boston Red Sox, 1986, 1987, and 1991; for Toronto Blue Jays, 1997 and 1998; for New York Yankees, 2001.

- most individual points scored in an NHL All-Star game (hockey record): by Mario Lemieux, for the Wales Conference, 1988 (3 goals, 3 assists).
- most Masters titles won by any golfer: by Jack Nicklaus in 1963, 1965–66, 1972, 1975, and 1986.
- most runs batted in a World Series game: twice, by Robert Richardson for New York Yankees, October 8, 1960, against the Pittsburgh Pirates; and by Hideki Matsui for New York Yankees against Philadelphia Phillies, November 4, 2009.
- most Super Bowl victories (football record): by Pittsburgh Steelers, SB-IX, 1975; SB-X, 1976; SB-XIII, 1979; SB-XIV, 1980; SB-XL, 2006; and SB-XLIII, 2009.
- most touchdown passes in a Super Bowl game (football record): by Steve Young for San Francisco 49ers against San Diego Chargers in SB-XXIX, 1995.
- most touchdowns scored in an NFL game (football record): three times, by Ernie Nevers (rushing) for Chicago Cardinals against Chicago Bears, November 28, 1929; by Dub Jones for Cleveland Browns against Chicago Bears, November 25, 1951; and by Gale Sayers for Chicago Bears against San Francisco 49ers, December 12, 1965.
- number of Bach's "Brandenburg Concertos."
- number of carbon atoms in a benzene ring.
- number of categories in which Nobel prizes are awarded.
- number of commandments of the (Catholic) Church
  1. to assist at Mass on all Sundays and holy days of obligation;
  2. to fast and to abstain on the days appointed;
  3. to confess at least once a year;
  4. to receive Holy Communion during the Easter time;
  5. to contribute to the support of the Church;
  6. to observe the laws of the Church concerning marriage.

- number of counties comprising Northern Ireland: Antrim, Armagh, Down, Fermanagh, Londonderry, and Tyrone.
- number of cups each player has in the African game of Wari.

- number of decimal places in the inverse multiple designated by the International System prefix micro-, denoted by the Greek letter "μ-".
- number of English kings named George.
- number of eyes on Azi Dahaka, the most powerful demon in Zoroastrianism, along with three heads and three mouths.
- number of Hanoverian monarchs of Great Britain: George I in 1714; George II, 1727; George III, 1760; George IV, 1820; William IV, 1830; Victoria 1837–1901. (Succeeded by Edward VII, the only British monarch of the House of Saxe-Coburg-Gotha.)
- number of heads on Skanda, a Hindu warrior-god, son of Shiva.
- number of known quarks (subatomic particles): Up, Down, Strange, Charm, Bottom, and Top.
- number of legs on an insect.
- number of presidential electoral votes apportioned to each of the states of Arkansas, Kansas, and Mississippi. (May be reapportioned for the 2012 election.)
- number of quarts of blood in the average human body.
- number of seats in the House of Representatives allotted to each of the states of Kentucky and South Carolina (as of the 2000 Census).
- number of sides to a snowflake.
- number of strings on a standard guitar.
- number of symphonies composed by Peter Ilyich Tchaikovsky.
- number of Tudor rulers of England: Henry VII in 1485; Henry VIII, 1509; Edward VI, 1547; Lady Jane Grey, 1553; Mary I, 1553; Elizabeth I, 1558–1603. (The Stuart Dynasty followed.)
- number of U.S. presidents named James, the most common first name for U.S. chief executives: Madison, Monroe, Polk, Buchanan, Garfield, and Carter.
- number of white stripes on the American flag.
- number of 0s in a million.
- number of 0s in the multiple designated by the International System prefix mega-, written "M-".
- players on a hockey team.

- record number of bases on balls taken in a baseball game: twice, by Walt Wilmot, Chicago White Sox, August 22, 1981, and by Jimmie Foxx, Boston Red Sox, June 16, 1938.
- record number of batters hit by a pitcher in one baseball game: several times, the last by John Grimes, St. Louis Browns, July 31, 1897.
- record number of grand-slam home runs in a baseball season: twice, by Don Mattingly, New York Yankees, 1987, and Travis Hafner, Cleveland Indians, 2006.
- record number of men's French Open singles tennis titles: won by Sweden's Bjorn Borg, in 1974–75 and 1978–81.
- record number of overtime periods in a hockey game: twice, Detroit Red Wings (1) against Montreal Maroons (0), March 24, 1936; and Toronto Maple Leafs (1) against Boston Bruins (0), April 3, 1933.
- retired baseball uniform number of:
  - ⇨ first baseman Steve Garvey, retired by the San Diego Padres.
  - ⇨ outfielder Al Kaline, retired by the Detroit Tigers.
  - ⇨ outfielder/first baseman Stan Musial, retired by the St. Louis Cardinals.
  - ⇨ outfielder Tony Oliva, retired by the Minnesota Twins.
  - ⇨ infielder Johnny Pesky, retired by the Boston Red Sox.

- retired basketball jersey number of:
  - ⇨ Walter Davis, retired by the Phoenix Suns.
  - ⇨ Julius Erving, retired by the Philadelphia 76ers. Erving's number 32 with the New Jersey Nets has also been retired.
  - ⇨ Avery Johnson, retired by the San Antonio Spurs.
  - ⇨ Bill Russell, retired by the Boston Celtics.
  - ⇨ The Fans, "the 6th man," retired by the Orlando Magic and the Sacramento Kings.

- retired hockey shirt number of Ace Bailey, retired by the Toronto Maple Leafs.
- sextan (ancient bronze coin of the Roman republic) equals one-sixth of an *as.*
- sides on a hexagon.
- six cubits and a span (approximately nine feet, nine inches): height of Goliath (2 Samuel 17:4).

- six feet: width of the goal cage on a regulation ice hockey rink.
- six seconds: minimum duration for a bull rider to stay on his animal in order to qualify.
- six sextans equal one as (early Roman coinage)
- six years: length of a standard term in office for a U.S. senator.
- sum of the first three integers $(1 + 2 + 3)$.
- teaspoons in a fluid ounce.
- Title of the New York State Vehicle and Traffic Law dealing with accidents and accident reports.
- Title of the United States Code dealing with domestic security.
- value of Roman numeral VI.

**at sixes and sevens**

- in a state of turmoil or confusion; untidy, unpredictable.

**Big Six**

- (a) a gambling game (also called "Wheel of Fortune") played on a large wheel with dollar bills of various denominations inset as spokes. The player bets on the denomination he expects will be under the arrow when the wheel stops spinning.
- (b) a bet in craps that the shooter will throw a 6 before he throws a 7. It is an even-money bet, even though the odds are 6 to 5 in favor of the house.
- (c) six of the world's largest ISPs and e-mail providers who joined forces in 2004 to fight spam: AOL, BT, Comcast, Earthlink, Microsoft, and Yahoo!.

**Birmingham Six**

- six men unjustly accused of setting bombs in two pubs in Birmingham, England, in November 1974, killing twenty-one people. The six—Hugh Callaghan, Patrick Hill, Gerard Hunter, Richard McIlkenny, William Power, and John Walker—were found guilty of murder in August 1975 and were sentenced to life terms. On appeal, new evidence of police mismanagement and coercion led the court to overturn the conviction and the men were freed, later being granted large compensatory cash awards.

**Beethoven's Sixth Symphony**

- "Sinfonie Pastorale" ("Pastoral symphony" in F major).

## Canon 6 of the American Bar Association Code of Professional Responsibility

- s lawyer should represent a client competently.

## CV-6

- number of the U.S. Navy aircraft carrier, USS *Enterprise*, scrapped in September 1958.

## deep six

- to bury at sea; by extension, to throw something away, as if tossing it overboard.

## ESP

- extrasensory perception, called the "sixth sense."

## fastest score in a hockey game

- on April 17, 1972, in a Stanley Cup Playoff, Don Kozak of the Los Angeles Kings, playing against the Boston Bruins, scored a goal six seconds after the start of the game. Los Angeles won, 7–4.

## F6F

- U.S. shipborne fighter plane, the Grumman Hellcat; had the highest kill/loss ratio of any American fighter aircraft in any service in World War II.

## Group 6 elements (of the Periodic Table)

- chromium (24), molybdenum (42), tungsten (74), seaborgium (106).

## Haydn's *Symphony No. 6*

- "Le Matin" symphony (in D major).

## hexameter

- a poetic line having six feet.

## *Inn of the Sixth Happiness*

- Ingrid Bergman is a domestic servant in London who becomes a missionary in China in this 1958 Cinemascope production featuring Robert Donat and Curt Jurgens.

### International Six-Day Enduro (ISDE)

- a motorcycle enduro event that lasts for six days, called "the Olympics of Motorcycling." Begun in 1913, held in Chile in 2007; Greece in 2008; Portugal in 2009, and in Morelia, Mexico in November 2010.

### Jena Six

- six black teenagers at the high school in Jena, Louisiana, accused of beating up a white classmate, on December 4, 2006, after three nooses were hung from "the white tree"— where typically only white students congregated—after a black student chose to sit there. Their names: Robert Bailey Jr., Mychal Bell, Carwin Jones, Bryant Purvis, Theo Shaw, and Jesse Ray Beard. The six were initially charged with attempted murder, which was later scaled back to lesser offenses.

### League of Six Nations

- the Iroquois Confederacy after it was joined by the Tuscaroras in the eighteenth century. The League was friendly to the English and hostile to the French.

### Les Six

- a group of avant-garde composers—guided by Jean Cocteau as their spokesman and Erik Satie as their spiritual leader—who rejected the Romanticism of Wagner and Strauss, as well as the Impressionism of Debussy, and brought a new simplicity to the music of the 1920s. They were Darius Milhaud, Francis Poulenc, Louis Durey, Arthur Honegger, Germaine Tailleferre, and Georges Auric.

### Mahler's Sixth Symphony

- "Tragic" symphony (in A minor).

### Mark Six

- a lottery-type game in which the players guess which balls, numbered 1 to 49, will be drawn from the lottery machine. The game is played in Hong Kong as a source of income for the Hong Kong government.

### *The Moon and Sixpence*

- 1942 biopic of Paul Gauguin, with George Sanders in a sharp portrayal of the self-indulgent, immoral Gauguin. Film adapted from W. Somerset Maugham's novel by Albert Lewin, who also directed.

### Muhammadan Six Tenets

- belief in the unity of God, in His angels, in His Scriptures, in His Prophets, in the Resurrection and the Day of Judgment, and in God's absolute and irrevocable decree and irrevocable decree and predetermination of good and evil.

### *Now We Are Six*

- 1927 book of children's verse by A. A. Milne.

### No. 6

- cocktail made with gin, sweet vermouth, lemon peel, orange peel, and Curaçao.

### number 6 billiard ball

- the ball that is traditionally solid green.

### number 6 wood screw

- a screw with a screw shank diameter of 9/64 inches.

### original six

- the six teams of the National Hockey League in the period from 1942 to 1967, before expansion doubled the roster; namely, Montreal, Toronto, New York, Boston, Chicago, and Detroit.

### "Page Six"

- name of the gossip column featured in the *New York Post* newspaper.

### senators' term of office

- members of the United States Senate are elected for six years.

## Secret Six

- a group of dedicated abolitionists who financed John Brown in his fight against slavery: Thomas Wentworth Higginson, Gerrit Smith, Dr. Samuel Gridley Howe, George Luther Stearns, Reverend Theodore Parker, and Franklin Sanborn. Also known as the Committee of Six.

### *The Secret Six*

- Wallace Beery in one of his best performances as a ruthless Prohibition Era gangster vying with other crime lords for control of the city's alcohol supply. George Hill directed this violent 1931 film that has Clark Gable and Jean Harlow in secondary roles.

### sextet

- A musical group composed of six instruments or six voices.

### sextuplets

- the first known surviving set of six infants, the Rosenkowitz sextuplets, was born on January 11, 1974, in Cape Town, South Africa.

### *Short Sixes*

- subtitled "Stories to be Read While the Candle Burns," thirteen tales by H. C. Bunner, published in 1891.

### Six

- outdated Oxford University slang for a privy.

### six accomplishments (*shatsampatti*)

- according to *Yoga Dictionary*, the attainments to which self-discipline and self-training are to be directed by an aspirant to liberation:
    *1.* Shama: control of mind, resulting in calmness;
    *2.* Dama: control of body;
    *3.* Uparati: cessation from eagerness to have certain things and persons around one, and therefore a willing acceptance of what the world offers as material for living with;

4. Titiksha: patience; the cheerful endurance of trying conditions and the sequence of karma;

5. Shraddha: fidelity and sincerity, and therefore confidence in oneself and others;

6. Samadhana: steadiness or onepointedness, with all one's forces gathered together and turned to the definite purpose in hand.

## the Six Acts

- the English government's repressive response to national unrest in the second decade of the 1800s. These so-called "Gag Acts" of 1819 included the Training Prevention Act, the Seizure of Arms Act, the Seditious Meanings Act, the Blasphemous and Seditious Libels Act, the Misdemeanours Act, and the Newspaper and Stamp Duties Act.

## six-and-eightpence

- in old English currency, a third of a pound, once called a "noble."

## six aspects of religion

- (for religions in general, with the notable exception of Buddhism): according to philosophy professor Huston Smith, the six aspects include: authority, ritual speculation, tradition, God's sovereignty, God's grace, and mystery.

## 6 August

- Christian feast day commemorating the Transfiguration.

## 6 August 1926

- Gertrude Ederle of New York, age nineteen, is the first woman to swim the English Channel; her time was fourteen hours and thirty-one minutes.

## six bells

- time check on-board ship, signifies 3:00 or 7:00 or 11:00 AM or PM.

## six bits

- slang for seventy-five cents.

## six bulls in Pamplona

- known as the Pamplona (Spain) Running of the Bulls Festival, each day, six bulls are run through the streets during the Fiesta of Saint Fermin, July 7–14.

## six-card stud

- a poker game in which the first and last cards are dealt face down; all others are dealt face up. The player selects his best five cards as his poker hand.

## six categories of reality

- according to the Vaisheshika school of Hindu philosophy: substance, quality, activity, generality, particularity, and inherence.

## Six Chakras

- in yoga, the centers of spiritual power in the body, which can be released by proper exercise: Muladhara (at the base of the spine, near the anus), Swadhisthana (at the level of the genitals), Manipuraka (at the level of the navel), Anhata (at the level of the heart), Vishuddhi (at the level of the throat), and Ajna (at the level of the eyebrows).

## *Six Characters in Search of an Author*

- a 1921 play-within-a-play by Luigi Pirandello that vacillates between the related worlds of reality and fantasy as the six characters insist on acting out their harrowing life experiences on the bare stage.

## six characteristics of Buddhist teachings

- focus on human dignity, attitude of non-attachment, tolerance, compassion, tendency to meditation, practicality.

## six cities of refuge

- towns which by Jewish law provided asylum for anyone who had unintentionally slain another person (Numbers 35 and Joshua 20): three west of the Jordan River—Kedesh, Shechem, and Hebron—and three east of the Jordan—Bezer, Ramoth, and Golan (Joshua 20:7–8).

## six classes of human beings

- Tibetan myth tells of a tortoise that laid six eggs of different colors that became six kinds of serpents, which became the six classes of human beings.

## Six Classics

- works in the Confucian tradition; at times in Chinese history, these six were considered to be crucial sources, encompassing the Five Classics (q.v.) plus the *Book of Music.*

## six-day bicycle races

- America's most popular spectator sport from about 1890 to 1935. The biggest draw in the period of multiday races, the six-day bicycle event usually ran from Monday through Saturday on an indoor track and offered substantial prizes.

## Six-Day War

- started June 5, 1967, between Israel and the adjoining states of Egypt, Syria, and Jordan, in which Israel gained control of large sectors of Arab territory, including the Gaza Strip, the Sinai Peninsula to the Suez Canal, East Jerusalem, Syria's Golan Heights, and Jordan's West Bank. The UN arranged a cease-fire on June 10.

## six days of creation

- first day: creation of light, division of light from darkness;
- second day: creation of the sky (the firmament);
- third day: creation of dry land and the seas, plants and trees;
- fourth day: creation of the sun, moon, and stars;
- fifth day: creation of creatures in the water and birds in the air;
- sixth day: creation of beasts of the Earth, including man and woman; following, God blessed the seventh day and rested from his work.

## Six Degrees of Kevin Bacon

- a game played with the names of actors and actresses (based on the following play).

## *Six Degrees of Separation*

- a Broadway play and a subsequent movie; proposes that everyone on the planet is related to everyone else on the planet by no more than six separate connections; e.g., A knows B, who knows C, who knows D, who knows E, who knows F; thus A relates to F through six sequential connections.

## Six Dorchester Laborers

- the six, who had formed a lodge under the Grand National Consolidated Trade Union in England, were sentenced to seven years' imprisonment by the British government. Their aim was a general strike, which, along with the growth of trade unionism, seriously alarmed the government, which decided on drastic action.

## Six Dynasties

- in China, the period of AD 220 to 589, from the end of the Han dynasty to the reunion of the empire through conquest of the South. In this period six dynasties followed in quick succession, each with their capital at the central city of Nanking: the Wu dynasty (222–280), the Eastern Chin (317–420), the Former Sung dynasty (the Liu family) (420–479), the Southern Ch'i (479–502), the Southern Liang (502–557), and the Southern Ch'en (557–589).

## Six Edicts

- a set of decrees made by French minister of finance Turgot in 1776, applying his theories of free trade to improve the economic conditions of the state: reform taxation, abolish the guilds and promote of freedom of occupation, apply scientific agricultural methods, and do away with unpaid labor by vassals for public work. He was opposed by the guilds, landowners, and Queen Marie Antoinette, and was dismissed, his reforms largely unaccomplished.

## six elements

- in Japanese Tantric Buddhism, the universe was believed to be composed of six elements: earth, fire, water, wind, space, and knowledge.

**sixer**

- in cricket, a shot in which the ball clears the boundary without bouncing.

**Sixes**

- a town in southwestern Oregon.

*Sixes and Sevens*

- a posthumous collection of O. Henry stories published in 1911.

**six feet under**

- slang for "buried"; thus, refers to someone who is dead.

*Six Feet Under*

- TV drama on HBO from June 2001 to August 2005, about a dysfunctional Los Angeles family in the funeral business. Created by Alan Ball, this highly lauded series captured three Golden Globe awards and nine Emmys.

**six gates of Troy**

- (from the Prologue to *Troilus and Cressida*): Darden, Tymbria, Helias, Chetas, Troien, and Antenorides.

**six gauge**

- a measure of shotgun bore diameter = 23.34 mm.

**"six geese a-laying"**

- the gift on the sixth day of Christmas in the carol "The Twelve Days of Christmas."

**six-gun**

- a handgun, usually a revolver that has six bullet chambers.

**six holy days of obligation**

- in the Catholic Church, in the United States
  1. Christmas Day (December 25)
  2. the Octave of the Nativity of Jesus (January 1)
  3. Ascension Thursday (40 days after Easter)
  4. the Assumption (August 15)
  5. All Saints' Day (November 1)
  6. the Immaculate Conception (December 8).

In Canada, the Epiphany (January 6) is a holy day of obligation, while the Assumption (August 15) is not.

## 6-iron

• golf club equivalent to the earlier spade mashie.

## 6 January

• Epiphany (feast commemorating the coming of the Magi).

## 6 June, 1944

• D-Day; date of the Allied invasion of Europe (The Longest Day) during World War II.

## six main schools of Hindu philosophy

• the darshanas (in Sanskrit), the primary belief systems of Hinduism: Vaisheshika, Nyaya, Samkhya, Yoga, Purva Mimamsa, and three systems of Uttara Mimamsa or Vedanta.

## Six Masters of the Early Ch'ing Period

• a group of Chinese scholar-artists who produced subtle, conservative paintings in the seventeenth and early eighteenth centuries: the flower painter, Yün Shou-p'ing, and landscape artists Wu Li; and the Four Wangs, Wang Shih-min, Wang Chien, Wang Yüan-ch'i, and Wang Hui.

## 6 Meter Club

• list of all pole vaulters who have cleared 6.00 meters or higher. Highest on record is Sergei Bubka of the Ukraine, who cleared 6.14 meters outdoor in July 1994 and 6.15 meters indoor in February 1993.

## Sixmilebridge

• a town in County Clare, Ireland.

## Six Mile Creek

• a town in County Tyrone, Northern Ireland.

## Six Mile Lake

• body of water in southern Louisiana.

### *Six Million Dollar Man*

- adventure show, starring Lee Majors as one of TV's first cyborgs—part man, part machine—invested with superhuman capabilities. First telecast on ABC-TV, January 18, 1974; last episode aired on March 6, 1978.

### Six Months' War

- Franco-Prussian conflict that lasted from July 28, 1870, to January 28, 1871.

### Six National Histories of Japan

- in the year 720, a detailed history of Japan (the *Nihonshoki*) was compiled in Chinese. It was continued by five other official histories written in Chinese. Together, they comprise the Six National Histories (*Rikkokushi*).

### Six Nations

- the five-nation Iroquoian federation (Mohawk, Seneca, Oneida, Cayuga, and Onandaga) after the addition of the Tuscaroras.

### "Six O'Clock"

- a verse tribute to the end of the day, by Trumbull Stickney (1874–1904).

### six-pack

- convenient packaging for cans or bottles of beer or soda; six units boxed or bound together.

### "sixpence"

- in the nursery rhyme:

  > I've got sixpence, jolly jolly sixpence,
  > I've got sixpence to last me all my life.
  > I've got sixpence to spend, and sixpence to lend,
  > And sixpence to send home to my wife.

### six-penny nail

- a carpenter's common nail, 2 inches long (coded 6d).

### six-point star

- the Star of David, a symbol for Judaism.

## 6-point type

- typographer's measure, known as "nonpareil."

## Six-Power Agreement

- the London Agreement on Western Germany, by the three Western Powers (U.S., UK, and France) and the three Benelux countries (the Netherlands, Belgium, and Luxembourg), signed June 1, 1948. The accord, which set the stage for the eventual unification of Germany, provided for international control of the Ruhr, creation of a federal constitution for Western Germany, German representation in the European Recovery Program, and the formation of a military security authority.

## six practices of perfection

- in Buddhism, the bodhisattva's conduct is characterized by:
    1. charity (*dana*);
    2. virtue (*sila*);
    3. perseverance (*kshanti*);
    4. vigor (*virya*);
    5. meditation (*dhyana*);
    6. wisdom (*prajna*).

## Six Principles of Chinese Painting

- formulated by Hsieh Ho, sixth-century Chinese figure painter:
    1. creativity (or "spiritual resonance");
    2. structural use of the brush;
    3. proper representation of objects;
    4. specific coloration of those objects;
    5. good composition;
    6. transmitting the old masters by copying.

These principles later became the standards of painters' training and critics' assessment.

## six-shooter

- a handgun, usually a revolver that holds six bullets; or, a child's toy gun.

## Six Sigma

- a quality-improvement methodology, copyrighted by Motorola; intended to increase the level of performance, reduce defects, and maximize reliability.

## Six's thermometer

- a device used in meteorology, employing mercury and alcohol, for recording the maximum and minimum air temperature between two readings, ordinarily twenty-four hours apart.

## Sixth Amendment to the Constitution

- establishes right to a speedy trial and other rights under criminal prosecution.

## Sixth Cataract

- a section of white water in the Nile River, north of Khartoum in north-central Sudan.

## Sixth Commandment

- Thou shalt not murder.

## Sixth Crusade

- in 1228, Holy Roman Emperor Frederick II constructed a peaceful settlement with the Muslims, by which Jerusalem, Nazareth, and Bethlehem were surrendered to the Crusaders for a period of years. Frederick crowned himself King of Jerusalem, but did little in that role, not even responding when the Muslims later reoccupied the city.

## Sixth French War of Religion

- in response to pressure from Catholics following the Edict of Beaulieu, which granted certain rights to the Huguenots after the Fifth Religious War, King Henry III forbade the exercise of the Protestant religion. In the Sixth War (1577), the Huguenots were defeated. But at the Peace of Bergerac (September 17, 1577, published as the Edict of Portiers), the Edict of Beaulieu was reaffirmed, although Henry III failed to put the provisions of the Edict into practice.

## Sixth House

- in astrology, deals with health and work.

## sixth president

- John Quincy Adams, March 4, 1825 to March 3, 1829.

## sixth sense

- a power of perception apparently independent of the five senses; keen intuition.

## *The Sixth Sense*

- 1999 film about a boy who communicates with spirits; a complex mixture of drama, horror, and action, starring Haley Joel Osment as the boy and Bruce Willis as the psychologist who attends him.

## sixth state of the Union

- Massachusetts ratified the Constitution on February 6, 1788.

## sixth wedding anniversary gift

- iron is customary.

## "six white horses"

- from the children's song:

    She'll be comin' 'round the mountain when she comes,
    She'll be riding six white horses when she comes . . .

## six wives of Henry VIII

- Catherine of Aragon (married 1509, divorced 1533), Anne Boleyn (married 1533, beheaded 1536), Jane Seymour (married 1536, died in childbirth 1537), Anne of Cleves (married and divorced 1540), Catherine Howard (married 1540, beheaded 1542), Catherine Parr (married 1543, outlived him).

## six-year molar

- the maxillary first molar, the first permanent tooth, appearing at age six.

**Six-Year Plan**

- an agenda of social legislation and economic development adopted in 1933 Mexico under provisional president General Abelardo Rodriguez, but really the program of ex-president Plutarco Calles and his National Revolutionary Party, to move the government to the Right.

**Statute of the Six Articles**

- Parliament's "Act Abolishing Diversity of Opinions" (1539), by which Henry VIII attempted to uphold Catholic orthodoxy but at the same time establish an English church. Death was prescribed for anyone denying any of the Articles:
  *1.* confirmed transubstantiation;
  *2.* supported the sufficiency of one kind of Communion;
  *3.* affirmed clerical celibacy;
  *4.* affirmed vows of chastity;
  *5.* supported private Masses;
  *6.* accepted auricular confession.

The law, called "the bloody whip with six strings" by many Protestants, was repealed by Henry VI in 1547.

**Tchaikovsky's Sixth Symphony**

- "Pathetique" symphony (in B minor).

**the Vedanga**

- the six sciences regarded as auxiliary to the sacred Hindu text, the Veda: phonetics (*saksa*), meter (*chandas*), grammar (*vyakarana*), etymology (*nirukta*), astronomy (*jyotisa*), and religious ceremony (*kalpa*); these six are intended to teach how to understand, recite, and apply Vedic texts.

## *6.4*

- number of Earth days in a day on the dwarf planet Pluto.

## *6.8*

- percentage of the U.S. population below five years of age, according to the 2000 census.

## *007*

- ID number of British secret agent James Bond, from the fertile mind of author Ian Fleming. The "00" prefix classifies the agent as authorized to kill.

## *07*

- number of the French Département of Ardèche.

## *7*

- Article of the U.S. Uniform Commercial Code that deals with warehouse receipts.
- atomic number of the chemical element nitrogen, symbol N.
- Beaufort number for wind force of 32 to 38 miles per hour, called "near gale–moderate gale."
- colors of the spectrum: red, orange, yellow, green, blue, indigo, violet.
- days of the week.
- dental chart designation for the right lateral incisor 72.
- gauge (number) of a wood screw with a shank diameter of 5/32 of an inch.
- in astrology, the number 7 is ruled by the planet Neptune.
- in baseball, 7 represents the left fielder's position.
- in Cantonese, the number 7 is pronounced the same as the word for male genitalia, presenting many opportunities for puns and jokes in that language.
- in Chinese culture, seven reincarnations and seven weeks of mourning follow a death.
- in Hebrew, the numerical value of the letter *zayin*.
- in mineralogy, Mohs Hardness Scale level characterized by quartz; can be scratched by a steel file.
- international telephone calling code for Russia and Kazakhstan.
- in the Bible the number 7 appears frequently in different contexts:
  ⇨ seventh day, the day of rest (Genesis 2:2).

⇨ seven years of plenty in Egypt under Joseph (Genesis 41:47).

⇨ seven days, the feast of unleavened bread for the Israelites in Egypt (Exodus 12:17).

⇨ seven lamps of pure gold adorned the tabernacle (Exodus 37:23).

⇨ seven times Moses sprinkled anointing oil on the altar to sanctify it (Leviticus 8:11).

⇨ seven days and seven nights Aaron was to stay in the tabernacle for his consecration (Leviticus 8:33).

⇨ seven days a woman was designated "unclean" after giving birth (Leviticus 12:2).

⇨ seven days Miriam was shut out of camp for speaking against Moses (Numbers 12:14).

⇨ Balak built seven altars for sacrifice (Numbers 23:1).

⇨ seven nations occupied the land before the Israelites: the Hittites, the Girgashites, the Amorites, the Canaanites, the Perizzites, the Hivites, and the Jebusites (Deuteronomy 7:1).

⇨ every seventh year shall be a year of release (Deuteronomy 15:1).

⇨ seven priests instructed to blow seven trumpets on the seventh day to blow down the walls of Jericho (Joshua 6:4).

⇨ number of Samson's locks that Delilah had shaved, sapping his strength (Judges 16:19).

⇨ Naaman must wash seven times in the Jordan (2 Kings 5:10, 14).

⇨ number of Job's sons (Job 1:2).

⇨ seven pillars of Wisdom (Proverbs 9:1).

⇨ seven loaves of bread with which Jesus fed four thousand men (Matthew 15:34–38).

⇨ number of devils Jesus cast out of Mary Magdalene (Mark 16:9).

⇨ the seven churches in Asia: Ephesus, Smyrna, Pergamos, Thyatira, Sardis, Philadelphia, and Laodicea (Revelation 1:3).

⇨ the book of seven seals, which only the lamb is worthy to open (Revelation 5:1).

⇨ seven angels sounding seven trumpets accompany the opening of the seventh seal (Revelation 8:1-2).

> ⇨ seven heads, with ten horns, on the red dragon (Revelation 12:3)
>
> ⇨ seven golden vials contain the seven last plagues, the wrath of God (Revelation 15:6–7).

- ISBN group identifier for books published in the People's Republic of China.
- minutes in a chukka of a polo match. A full game must be no more than eight chukkas.
- minutes in a water polo period, four periods in a game.
- most career no-hit games pitched (baseball record): by Nolan Ryan, playing with seven different teams in both leagues, 1973–91.
- most field goals scored in Super Bowl games, career (football record): by Adam Vinatieri for New England Patriots against Green Bay Packers, SB-XXXI, 1997; against St. Louis Rams, SB-XXXVI, 2002; against Carolina Panthers, SB-XXXVIII, 2004; against Philadelphia Eagles, SB-XXXIX, 2005; and for Indianapolis Colts against Chicago Bears, SB-XLI, 2007.
- most fumbles in an NFL game (football record): by Len Dawson for Kansas City Chiefs against San Diego Chargers, November 15, 1964.
- most goals scored by one player in an NHL game (hockey record): by Joe Malone for Quebec Bulldogs against Toronto St. Pats, January 31, 1920.
- most sacks in an NFL game (football record): by Derrick Thomas for Kansas City Chiefs against Seattle Seahawks, November 11, 1990.
- most touchdown passes thrown in one NFL game by the same player (football record): four times, by Sid Luckman for Chicago Bears against New York Giants, November 14, 1943; by Adrian Burk for Philadelphia Eagles against Washington Redskins, October 17, 1954; by George Blanda for Houston Oilers against New York Titans, November 19, 1961; by Y. A. Tittle for New York Giants against Washington Redskins, October 28, 1962; and by Joe Kapp for Minnesota Vikings against Baltimore Colts, September 28, 1969.
- natural, in craps.
- a number seen as lucky in many cultures.
- number of Bourbon rulers of France, from Henry IV in 1589; Louis XIII, 1610; Louis XIV (the Sun King), 1643; Louis XV,

1715; Louis XVI, 1774–92; interrupted by the First Republic and the first empire, then Louis XVIII, 1814; Charles X, 1824–30.

- number of castaways on *Gilligan's Island* (on CBS-TV, September 1964 to September 1967): Gilligan, the skipper, the millionaire and his wife (Mr. and Mrs. Howell), the movie star (Ginger), the professor, and Mary Ann.

- number of cities that claimed to have been the birthplace of the Greek poet, Homer: Argos, Athens, Chios, Colophon, Rhodes, Salamis, and Smyrna.

- number of games in a baseball World Series or a hockey Stanley Cup final.

- number of people murdered in the St. Valentine's Day Massacre in Chicago, 1929.

- number of presidential electoral votes apportioned to each of the states of Connecticut, Iowa, Oklahoma, and Oregon. (May be reapportioned for the 2012 election.)

- number of red stripes on the American flag.

- number of *Road* films made by Bob Hope and Bing Crosby: *Road to . . . Singapore* (1940), *Zanzibar* (1941), *Morocco* (1942), *Utopia* (1945), *Rio* (1947), *Bali* (1952—the only one in color) and *Hong Kong* (1962).

- number of seats in the House of Representatives allotted to each of the states of Alabama, Colorado, and Louisiana (as of the 2000 Census).

- number of Stuart monarchs of Great Britain, starting with James I in 1603; then Charles I, 1625, until his beheading in 1649; the throne regained after the Interregnum by Charles II in 1660; James II, 1685; William III and Mary II, 1689; Anne 1702–14. (The Hanoverian Dynasty followed.)

- number of symphonies composed by Sergei Prokofiev.

- number of tiles in a "hand" of Scrabble.

- number of U.S. presidents from Ohio, the second most frequent: Grant, Hayes, Garfield, B. Harrison, McKinley, Taft, and Harding.

- the number worn by Dodge Connelly (George Clooney) on his Duluth Bulldogs football team jersey in the 2008 film, "*Leatherheads*."

- the number worn on Paul Newman's hockey jersey in the 1977 movie, *Slap Shot.*

- pH of water, a neutral substance.

- players on a water polo team.
- record number of consecutive wins of the biennial Tour de France bicycle race: by American Lance Armstrong, his first in 1999, and the most recent completed on July 24, 2005, 4 minutes and 40 seconds ahead of his nearest competitor, Ivan Basso of Italy. His time for the 2,233 miles was 86 hours, 15 minutes, 2 seconds, an average speed of 25.9 miles per hour.
- record number of men's U.S. Open singles tennis titles: three times, won by Robert Sears, 1981–87; William Larned, 1901–02, and 1907–11; and Bill Tilden, 1920–25 and 1929.
- record number of men's Wimbledon singles tennis titles: twice, won by William Renshaw of the UK, 1881–86 and 1889; and by American Pete Sampras, 1993–95 and 1997–2000.
- record number of pinch-hit home runs in a baseball season: by Dave Hansen for the Los Angeles Dodgers in 2000, and Craig Wilson for the Pittsburgh Pirates in 2001.
- record number of runs batted in, in one inning (baseball record): twice, by Ed Cartwright for the St. Louis Browns, September 23, 1890; and by Alex Rodriguez for the New York Yankees, October 4, 2009.
- record number of stolen bases in a baseball game: by Billy Hamilton for the Philadelphia Phillies, August 31, 1894, and by George Gore for the Chicago White Stockings, June 25, 1881.
- record number of women's French Open singles tennis titles: won by American Chris Evert, 1975, 1975, 1979, 1980,1983, 1985, and 1986.
- retired baseball uniform number of:
  - ⇨ second baseman and occasional outfielder Craig Biggio, retired by the Houston Astros.
  - ⇨ outfielder Mickey Mantle, retired by the New York Yankees.

- retired basketball jersey number of:
  - ⇨ Kevin Johnson, retired by the Phoenix Suns.
  - ⇨ Pete Maravich, retired by both the New Orleans Hornets and the Utah Jazz.
  - ⇨ Bingo Smith, retired by the Cleveland Cavaliers.

- retired football jersey number of:
  - ⇨ Dutch Clark, retired by the Detroit Lions.
  - ⇨ John Elway, retired by the Denver Broncos.
  - ⇨ George Halas, retired by the Chicago Bears.
  - ⇨ Mel Hein, retired by the New York Giants.
  - ⇨ Bob Waterfield, retired by the St. Louis Rams.

- retired hockey shirt number of:
  - ⇨ Bill Barber, retired by the Philadelphia Flyers.
  - ⇨ Neal Broten, retired by the Dallas Stars.
  - ⇨ Paul Coffey, retired by the Edmonton Oilers.
  - ⇨ Phil Esposito, retired by the Boston Bruins.
  - ⇨ Rod Gilbert, retired by the New York Rangers.
  - ⇨ Yvon Labre, retired by the Washington Capitals.
  - ⇨ Ted Lindsay, retired by the Detroit Red Wings.
  - ⇨ Rick Martin, retired by the Buffalo Sabres.
  - ⇨ Howie Morenz, retired by the Montreal Canadiens.

- seven days and seven nights, period of the great flood on Earth in Mesopotamian myth.
- seven feet:
  - ⇨ height of the goal cage on a regulation hockey-playing field.
  - ⇨ diameter of circle position in shotputting.

- seven inches, diameter of the largest hailstone ever to fall in the U.S., at Aurora, Kansas, on June 22, 2003. It measured 18.75 inches in circumference.
- seven primary AOC (Appellation contrôlée) subregions within the Médoc wine-producing area: Haut-Médoc, St. Estéphe, Pauillac, St. Julien, Margaux, Moulis, and Listrac.
- seven yards of wool, length of the average Scottish kilt.
- sides on a heptagon.
- sum of the digits on the opposite sides of a six-sided die: 1 & 6 or 2 & 5 or 3 & 4.
- sum of the first five Fibonacci numbers $(0 + 1 + 1 + 2 + 3)$.
- Title of the United States Code dealing with agriculture.
- tones in a whole-tone scale (e.g., C, D, E, F#, G#, A#, C).
- value of Roman numeral VII.

## "Assertion of the Seven Sacraments"

- tract written by Henry VIII in 1821 in reply to Luther's ninety-five theses, for which he received the title of Defender of the Faith from Pope Leo X.

## Battle of Seven Pines

- Civil War battle (May 31–June 1, 1962) in which Confederate forces were repulsed by the arrival of General Sumner's reinforcements. The Confederate general, Joseph Johnston, was killed in the battle, and Robert E. Lee was made the new commander of the Confederate armies of Eastern Virginia and North Carolina.

## canonical hour

- according to Psalm 118:164, "Seven times a day do I praise thee." These hours were called, sequentially, matins and lauds, prime, tierce, sext, nones, vespers, and compline.

## Canon 7 of the American Bar Association Code of Professional Responsibility

- A lawyer should represent a client zealously within the bounds of the law.

## Chapter 7

- a section of the U.S. bankruptcy code providing for the sale of company assets to meet debt obligations to its creditors, following which the company ceases operations.

## Chicago Seven

- political radicals (originally eight) arrested for violent anti–Vietnam War protests at the Democratic National Convention in August 1968 in Chicago. The seven, tried for inciting to riot, were Abbie Hoffman, Jerry Rubin, Tom Hayden, David Dellinger, Rennie Davis, Lee Weiner, and John Froines. Lead defense attorney was William Kunstler. Presiding was Judge Julius Hoffman. An eighth defendant, Black Panther activist Bobby Seale, was severed from the case for his verbal attacks on the judge, and was sentenced to prison for contempt before the trial ended. Five of the seven were found guilty, but the convictions were reversed by the appeals court.

## Dance of the Seven Veils

- a voluptuous performance by Salome to entice King Herod before John the Baptist is beheaded. Feature of the 1905 opera, *Salome*, by Richard Strauss. Also, subject of *Salome*, an 1893 play by Oscar Wilde on which the opera is based.

## feast of the seven fishes

- traditional southern Italian meal on Christmas Eve. The choice of fish varies, but usually includes baccala, calamari, scungilli, shrimp, and some type of big fish (snapper, tuna, sea trout, or salmon).

## *Friendship 7*

- name of the first American manned space capsule to achieve orbit, in which Lieutenant Colonel John H. Glenn orbited the earth three times on February 20, 1962.

## The Göttingen Seven

- seven professors—Albrecht, Dahlmann, Ewald, Gervinus, Weber, and the two Grimm Brothers (Jacob and Wilhelm)—dismissed from Göttingen University in December 1837 for refusing to pledge fidelity to the new king of Hanover, Ernest Augustus, eldest son of George III, who ascended to the throne on the death of William IV of England. The new king set aside the constitution of 1833 as too liberal and replaced it. The seven professors protested to the king, explaining that they considered his act the breaking of an oath, and they were dismissed.

## Group of Seven

- seven Canadian landscape artists with common interests who held group exhibits from 1920 to 1931 and had a significant impact on Canadian art. Included were J. E. H. MacDonald, Franklin Carmichael, Frank Johnston, Arthur Lismer, Lawren S. Harris, Frederick Varley, and A. Y. Jackson.

## Group 7 elements (of the Periodic Table)

- manganese (25), technetium (43), rhenium (75), bohrium (107).

## Guguletu Seven

- seven black youths who, in March 1986, were set up by South African Security Police to plan an act of violence against a local police station in a township near Cape Town, and were ambushed and massacred as a warning of retribution for anti-apartheid activity. The Truth and Reconciliation Commission, established to make amends for the government's treatment of blacks, discovered the truth—that the police had supplied the group's guns and were waiting to kill them.

## Haydn's Seventh Symphony

- "Le Midi" symphony (in C major).

## heptad

- a group of seven.

## *The House of the Seven Gables*

- Nathaniel Hawthorne's 1851 novel traces the effect of Matthew Maule's curse on several generations of the Pyncheon family, the ingenious plot examining how events of the past reverberate in the future. Adapted as a film in 1940, with George Sanders, Vincent Price, and Margaret Lindsay.

## *The Magnificent Seven*

- 1960 film directed by John Sturgis, an Americanized version of the 1954 Akira Kurosawa film, *The Seven Samurai*. The seven were played by Yul Brynner (Chris), Steve McQueen (Vin), Horst Buchholz (Chico), Charles Bronson (O'Reilly), Robert Vaughn (Lee), James Coburn (Britt), and Brad Dexter (Harry). Film has a notable score by Elmer Bernstein.

## Northern Seven Years' War

- also known as the Nordic Seven Years' War: The war between Sweden and an alliance of Denmark-Norway, Lubeck, and

Polish-Lithuanian Commonwealth, from 1563 to 1570. The result of the war was indecisive, neither side gaining any distinct advantage or additional land.

## number 7 billiard ball

- the ball that is conventionally solid maroon.

## "Rover"

- the seventh player on an early hockey team, at the time when there were seven players on a side.

## "seven wives"

- from the children's riddle-poem, "St. Ives":

  As I was going to St. Ives
  I met a man with seven wives;
  Each wife had seven sacks,
  Each sack had seven cats.
  Each cat had seven kits.
  Kits, cats, sacks, and wives,
  How many were going to St. Ives?

  (The answer, of course, is one, "I.")

## Seagram's 7

- familiar name for Seagram's 7 Crown, a blended whiskey produced by the Seagram Company of Montreal, Canada.

## Septennial Act

- 1715 act of the UK Parliament fixing a maximum parliamentary term of seven years, which replaced the three-year term under the Triennial Acts of the late seventeenth century.

## septet

- a musical group composed of seven instruments or seven voices.

## septuplets

- the first known surviving set of seven infants: the McCaughey septuplets, born in Blank Children's Hospital, Des Moines,

Iowa, on November 19, 1997. Parents Bobbi (age 29) and Kenny (age 27) named their children Kenneth, Alexis, Natalie, Kelsey, Brandon, Nathan, and Joel. The mother had taken the fertility drug Metrodin.

## Se7en

- 1995 police drama about two cops tracking a serial killer whose MO involves brutal killings. Cast includes Brad Pitt, Morgan Freeman, Gwyneth Paltrow, and Kevin Spacey.

## Seven

- the first installment (1964) of the award-winning Michael Apted documentary on British TV following the lives of a group of British children in seven-year intervals.

## 7-A

- Article of the New York State Education Law covering standardized testing.

## Seven Against Thebes

- seven heroes in Greek legend: Adrastus, Amphiaraus, Capaneus, Hippomedon, Parthenopaeus, Polynices, Tydeus. The legend provided the basis for plays by Aeschylus (*Seven Against Thebes*) and Euripides (*The Phoenician Women*).

## seven ages of man

- as described by Jaques in Act II, Scene VII, of Shakespeare's play, *As You Like It:* The infant, the schoolboy, lover, soldier, the justice, the pantaloon (retiree), and finally, second childhood, leading to "mere oblivion; sans teeth, sans eyes, sans taste, sans everything."

## Seven Angry Men

- 1955 film that evenhandedly tells the story of abolitionist John Brown in his attempt to free the slaves. Raymond Massey gives a strong portrayal of a complex, obsessed man.

## Seven Beauties

- Lisa Wertmuller's disturbingly explicit 1976 film about a small-time crook (Giancarlo Giannini) in Naples during

World War II who is subjected to a string of indignities in order to ensure his and his seven sisters' survival.

## seven bells

- time check on-board ship, signifies 3:30 or 7:30 or 11:30, AM or PM.

## Seven Blocks of Granite

- the offensive line of Fordham University's football team in 1936–37: right end John Druze, right guard Vince Lombardi, right tackle Al Babartsky, center Alex Wojciechowicz, left tackle Ed Franco, left guard Nat Pierce, and left end Leo Paquin. Head coach was Jim Crowley; line coach was Frank Leahy, who later coached at Notre Dame.

## seven bodies in alchemy

- Sun is gold; the moon, silver; Mars, iron; Mercury, quicksilver; Saturn, lead; Jupiter, tin; and Venus, copper.

## Seven Bounteous Immortals

- the Amahraspands of Zoroastrian myth: Good Mind, Righteousness, Devotion, Kingdom, Obedience, Wholeness, and Immortality; auxiliaries of Ohrmazd, Creator of the World, leader of the heavenly forces of good.

## *Seven Brides for Seven Brothers*

- Stanley Donen's 1954 film has songs by Johnny Mercer and Gene DePaul and the voices of Jane Powell and Howard Keel, but the real stars of the film are the incredible dance sequences staged by Michael Kidd. The story of seven ranch hands who kidnap a like number of country maids is adapted from "The Sobbin' Women," a short story by Stephen Vincent Binet (punning "The Sabine Women").

## seven-card high-low stud poker

- played the same as seven-card stud, but in this variation, each player declares, after the last betting round but before the hole cards are exposed, whether he is playing for high, low, or both. At the showdown, each player selects five of his seven cards to form either a high or low hand, or he may use two groups of five if he has chosen both high and low.

## seven-card stud

- a poker game in which the first two cards are dealt face down; cards three, four, five, and six are dealt face up; and the seventh, face down. In the showdown, each player selects his best five cards as his competing hand.

## seven categories of knowledge

- as defined by the Vaisheshika system of Hindu philosophy: substance, quality, motion, generality, particularity, inherence, and nonexistence.

## Seven Champions of Christianity

- St. George of England, St. Andrew of Scotland, St. Patrick of Ireland, St. David of Wales, St. Denis of France, St. James of Spain, and St. Anthony of Italy—the national patron saints of the several countries.

## Seven Charities

- one of Buster Keaton's best films, this 1925 silent film has him inheriting seven million dollars—if he marries before seven o'clock that night. When he loses his girlfriend due to a clumsy proposal, he advertises for a wife and is beset by thousands of bride hopefuls.

## seven chemicals comprising the minor arcana of alchemy

- sulfuric acid, iron oxide, sodium carbonate, sodium nitrate, liquor hepatis, red pulvis solaris, and black pulvis solaris.

## the seven Christian seasons

- Advent, Christmas, the Season After Epiphany, Lent, Holy Week, Easter, and the Season after Pentecost.

## Seven Churches of Rome

- (Catholic): St. Peter, St. Paul, St. Sebastian, St. John Lateran, Santa Croce, San Lorenzo fuori le Mura, Santa Maria Maggiore.

## Seven Cities of Cíbola

- a legend from the mid-sixteenth century about cities of gold and silver in New Spain (Colonial Mexico). The viceroy's emissary, friar Fray Marcos de Niza, returned from his 1539 expedition with exaggerated reports of seeing pueblos of gold among the Zuñi Indians, which attracted Francisco Coronado and others to find this El Dorado, all futilely.

## seven continents

- North America, South America, Europe, Asia, Africa, Australia, and Antarctica.

## seven contrary virtues

- meant to counterbalance the seven deadly sins: humility, kindness, abstinence, chastity, patience, liberality, diligence. From *Psychomachia* ("Battle for the Soul"), an epic poem by Prudentius (c. 410).

## seven corporal works of mercy

- in Catholicism:
  1. feed the hungry.
  2. give drink to the thirsty.
  3. clothe the naked.
  4. house the homeless.
  5. visit the sick.
  6. comfort the imprisoned.
  7. bury the dead.

## Seven Days' Battles

- with Union forces under General McClellan nearing Richmond, capital of the Confederacy, Confederate General Lee made several attacks (June 25–July 1, 1962) on the federal army to prevent a siege of the city. Lee's final attempt to destroy the Union Army of the Potomac was at Malvern Hill, where he lost almost 5,600 men but saved Richmond.

### Seven Days in May

- this 1964 political thriller has Burt Lancaster, as head of the Joint Chiefs of Staff, leading a coup to unseat the president (Fredric March) and seize the government because they object to his signing a nuclear disarmament treaty with the Soviets. Kirk Douglas, aide to Lancaster, discovers the plot and is instrumental in suppressing it. Taut direction by John Frankenheimer of the screenplay by Rod Serling, adapted from a novel by Fletcher Knebel and Charles W. Bailey.

### Seven Days to Noon

- a British atomic scientist (Barry Jones) has become disillusioned with the direction of nuclear policy and threatens to detonate a device that will destroy London unless the government stops production of atomic weapons by the following Sunday. A 1950 film directed with unremitting suspense by John Boulting; Oscar to Paul Dehn and James Bernard for their original story.

### seven deadly sins

- pride, avarice, wrath, envy, gluttony, sloth, lust.

### 7 December 1941

- date of the Japanese attack on Pearl Harbor, drawing the U.S. into World War II. "A day that shall live in infamy," according to then-president, Franklin D. Roosevelt.

### Seven Devils Mountains

- a range in western Idaho, bordering Hell's Canyon.

### seven dwarfs

- they of the 1937 Disney film, *Snow White and the Seven Dwarfs:* Doc, Bashful, Sneezy, Grumpy, Sleepy, Happy, and Dopey.

### seven Ecumenical Councils of the early church

- First Council of Nicaea (325), First Council of Constantinople (381), Council of Ephesus (431), Council of Chalcedon (451), Second Council of Constantinople (553), Third

Council of Constantinople (680–681), Second Council of Nicaea (787).

## 7-Eleven
- a chain of convenience stores throughout the U.S. and Canada.

## Seven Emu
- a town in the Northern Territory of Australia.

## Seveners
- also known as Waqifiyah: a branch of the Isma'ilite sect of Shi'ite Muslims who believe that Isma'il, the eldest son of Ja'far ibn Muhammad, the sixth imam, was the seventh and last imam (as contrasted with the Twelvers, who recognize Ja'far's younger son, Musa, as the seventh imam).

## *Seven Figures and a Head*
- sculpture by Alberto Giacometti, completed 1950.

## Seven Fountains
- a town in the Union of South Africa, about 90 miles northeast of Port Elizabeth.

## seven gates of hell
- portals that Ishtar, Assyrian goddess of love and war, had to enter to find her beloved spouse, Tammuz.

## seven gifts of the Holy Ghost
- in Roman Catholicism: wisdom, understanding, counsel, knowledge, fortitude, piety, and the fear of the Lord. These are transmitted to the person at the time of baptism.

## The Seven Graces (of the Virgin Mary)
- in Roman Catholicism:
    1. I will grant peace to their families.
    2. they will be enlightened on the divine mysteries.
    3. I will console them in their pains and I will accompany them in their work.

*4.* I will give them as much as they ask for as long as it does not oppose the adorable will of my divine Son or the sanctity of their souls.

*5.* I will defend them in their spiritual battles with the infernal enemy and I will protect them at every instant of their lives.

*6.* I will visibly help them at the moment of their death; they will see the face of their Mother.

*7.* I have obtained [this Grace] from my divine Son, that those who propagatethis devotion to my tears and dolors will be taken directly from this earthly life to eternal happiness, since all their sins will be forgiven and my Son and I will be their eternal consolation and joy.

## seven grades of initiate to Mithraism

- the lower three, Servitors, made up of Raven, Soldier, and Bridegroom; the upper four, or Participants, were Lion, Persian, Courier of the Sun, and Father. Initiation at each level involved tests of endurance.

## Seven Hathors

- the deities of fate; a group of women who predicted the destiny of each infant born in ancient Egypt.

## seven-headed dragon

- that which is slain by Baal in the mythology of Ugarit. Also, a character in mythologies of India, Persia, Cambodia, and the Celts.

## Seven Head

- point of land east of Clonakilty Bay on the south coast of County Cork, Ireland.

## seven hills of Rome

- Aventine, Caelian, Capitoline, Esquiline, Palatine, Quirinal, and Viminal.

## The Seven Hills of Westwald

- a scenic grouping on the Rhine River six miles above Bonn, known as the Siebengebirge: Drachenfels (1,067 feet), Wolkenburg (1,076 feet), Petersberg (1,096 feet), Olberg (1,522 feet), Lowenburg (1,056 feet), Lohrberg (1,444 feet), Nonnenstromberg (1,107 feet).

## The Seven Hogs Island

- also known as Magharee Island; tract of land outside Tralee Bay, in County Kerry, Ireland.

## seven holy angels

- Michael, Gabriel, Raphael, Uriel, Chamuel, Jophiel, and Zadkiel.

## seven holy days of the Jewish calendar

- the holy festivals of the Jewish year: Rosh Hashanah (Jewish New Year), Yom Kippur (Day of Atonement), Sukkot (Feast of the Tabernacles), Hanukkah (Feast of Lights), Purim (Feast of Deliverance), Pesach (Feast of the Unleavened Bread), and Shavuot (Feast of Weeks).

## Seven Holy Founders

- seven young men from prominent Florentine families who, after withdrawing to a life of renunciation and spiritual austerity, in 1240 shared a vision of the Virgin Mary and founded an order known as the Friar Servants of St. Mary, or Servites. They adopted the names Brothers Bonfilius, Alexis, Amadeus, Hugh, Sostenes, Magnettus, and Bonagiunta, and all but Brother Alexis became fully ordained priests. In 1304 their order received formal approbation of Benedict XI. Also called Seven Servite Founders.

## seven holy sacraments of the Catholic Church

- baptism, confirmation, Holy Eucharist, confession, extreme unction, holy orders, and matrimony. These were ratified by the Council of Trent in 1545.

## 7-iron

- golf club equivalent to the earlier mashie-niblick.

## Seven Islands

- (Sept Iles) a town on Seven Islands Bay at the mouth of the St. Lawrence River, seat of Côte-Nord (North Shore) region, eastern Quebec Province.

## 7 January

- Eastern Orthodox Christmas Day.

## Seven Japanese Gods of Happiness (or of Good Fortune)

- Hotei, Jurojin, Fukurokiya, Daikoku, Ebisu, Bishamon, and Benten (the sole female).

## Seven Joys of Mary

- in Catholicism:
  1. The Annunciation;
  2. The Visitation;
  3. The Nativity;
  4. The Epiphany;
  5. The Finding in the Temple;
  6. The Resurrection;
  7. The Ascension.

## 7 July

- Star festival (Hoshi Matsuri) of Shinto.

## *Seven Keys to Baldpate*

- a 1913 novel by Earl Derr Biggers recounting the odd experiences of a writer who isolates himself in a vacant hotel to complete a novel he is working on. Adapted into a stage play by George M. Cohan, also produced in 1913, encompassing a series of surprising twists; Cohan also appeared in the melodrama. Five film adaptations followed, the first in 1917, with Cohan; the second silent film in 1925, starring Douglas MacLean; Richard Dix starred in the first talkie in 1930, followed by a 1935 version with Gene Raymond, and the last in 1947 with Philip Terry.

## seven kinds of wisdom

- in Buddhism, zeal, intelligence, meditation, investigation, joy, repose, and serenity.

## seven kings of Rome

- tradition says that Rome was initially ruled by a succession of seven kings: Romulus (founder of the city), Numa Pompilius, Tulus Hostilius, Ancus Marcius, Lucius Tarquinius Priscus, Servius Tullius, and Lucius Tarquinius Superbus.

## Seven Lakes

- a town in northwest New Mexico, south of the Chaco Canyon National Monument.

## *Seven Lamps of Architecture*

- an influential book-length essay by John Ruskin, 1849, detailing the author's beliefs on the role and aesthetics of architecture, illuminated by the principles (lamps) of sacrifice, truth, power, beauty, life, memory, and obedience.

## Seven Last Words

- the last utterance of Christ on the Cross: "My God, why hast thou forsaken me?" (Mark 15:34).

## seven-league boots

- a means of speedy conveyance in many fairy tales, permitting the wearer to advance seven leagues with each step.

## seven liberal arts

- the basis of education in the Middle Ages, comprising the trivium (grammar, rhetoric, and logic) and the more advanced quadrivium (geometry, arithmetic, astronomy, and music).

## *The Seven Little Foys*

- comedian Bob Hope stars in this 1955 screen biography of vaudevillian Eddie Foy and his seven children. High point occurs near the end when James Cagney, as George M. Cohan, joins Hope for some jokes and a dance routine at a Friars Club dinner.

## *Seven Lively Arts*

- book by Gilbert Seldes (1924), who argued that pop culture was as valid as the more-traditional elitist art forms and propounded the virtues of such "arts" as motion pictures, jazz music, comic strips, vaudeville, popular songs, etc.

## Seven Masters

- The Seven Patriarchs of Jodoshinshu (Buddhist): Nagarjuna, Vasubandhu, T'an-luan, Tao-ch'o, Shan-tao, Genshin, and Honen.

## Seven Mile Creek

- a waterway in the western part of Australia.

## Seven Mules

- linemen for the Four Horsemen on the 1924 Notre Dame football team: ends Charles Collins and Ed Hunsinger, tackles Joe Bach and Edgar "Rip" Miller, guards Noble Kizer and John Weibel, and center Adam Walsh.

## seven names of God

- used by the ancient Hebrews: El, Elohim, Adonai, YHWH (interpreted by modern scholars to be either Jehovah or Yahweh), Shadai, Ehyeh-Asher-Ehyeh, and Zebaot.

## Seven Oaks

- name of towns in town in Kent County, southeast of London, England, and in Natal Province, Union of South Africa.

## *Sevenoaks*

- 1875 novel by Josiah G. Holland recounting the dealings of one Robert Belcher, an unscrupulous millionaire believed to have been patterned on Jim Fisk. The novel was dramatized by J. L. Gilder in 1876.

## Seven Oaks Massacre

- June 1816 destruction of the Hudson's Bay Company's Red River settlement, near present-day Winnipeg, Manitoba, by followers of the competing North West Company, killing twenty settlers and soldiers. Some five years later, the two rival fur-trading companies merged.

## The Seven Odes

- considered to be the finest of pre-Islamic Arabic odes, all by different poets of the sixth and early seventh centuries. Comprise a book entitled Al-Mu'allaqat.

## 7 on 7 (or 7 & 7)

- a mixed drink composed of equal parts Seagram's 7 Crown whiskey and 7UP carbonated soda, with or without a lemon wedge.

## seven original astronauts

- first group in the NASA Space Program: M. Scott Carpenter, L. Gordon Cooper Jr., John H. Glenn Jr., Virgil I. (Gus) Grissom, Walter M. Schirra Jr., Alan B. Shepard Jr., and Donald K. (Deek) Slayton.

## Seven Pagodas

- Mahabalipuram, historic town in southeastern India, on the Bay of Bengal, 37 miles south of Madras.

## Seven Penitential Psalms

- Biblical Psalms 6, 31, 37, 50, 101, 129, and 142, which express sorrow for sin and desire for pardon.

## seven-penny nail

- a carpenter's common nail, 2 -1/4 inches long (coded 7d).

## *The Seven-Per-Cent Solution*

- impelled by Dr. Watson (Robert Duvall), Sherlock Holmes (Nicol Williamson) consults Sigmund Freud (Alan Arkin) to find a cure for Holmes's cocaine addiction in this 1977 film directed by Herbert Ross. Laurence Olivier appears as Holmes's nemesis, Professor Moriarty. The title refers to the relative concentration of cocaine in water that Holmes injects.

## Seven Persons

- a town in the southeast corner of Alberta, Canada, about 25 miles southwest of Medicine Hat.

### Seven Pillars of Wisdom

- 1926 book by T. E. Lawrence (he of Arabia) recounting his exploits with the Arab forces fighting the Turks in World War I. Basis of the film *Lawrence of Arabia*.

## Seven Pleiades

- also known as the Seven Daughters of the Sea; in Greek mythology, daughters of the Titan Atlas and the ocean nymph, Pleione: Alcyone, Asterope, Celaeno, Electra, Maia, Merope, and Taygete, who became the constellation bearing their name.

### 7 Plus Seven

- the second installment (1970) of Michael Apted's *7 Up* documentary series on British TV, following the lives of a group of British children in seven-year intervals.

## 7-point type

- typographer's measure, known as "minion."

## seven primary mental abilities

- proposed by psychologist L. L. Thurstone from his factor analytic studies of temperament: activity, impulsiveness, emotional stability, sociability, athletic interest, ascendance, and reflectiveness.

## seven regions of the underworld

- in Hindu cosmology, atala, vitala, sutala, talatala, mahatala, rasatla, and patala. In Jain myth, each is inhabited by a different type of demon.

## seven regions of the upper world

- in Hindu cosmology, *bhurloka* (Earth), *bhuvarloka* (sky), *svarloka* (heaven), *maharloka* (middle region), *janarloka* (place of rebirths), *taparloka* (mansion of the blessed) and *satyaloka* (abode of truth).

## seven rites associated with the Lakota Indian pipe

- the first rite, the Keeping and Releasing of the Soul, is used to "keep" the soul of a dead person for a number of years until it is properly released, ensuring a proper return to the spirit world. The second ritual is the Sweat Lodge, a purification rite used before any other major ritual. The third, Crying for a Vision, lays down the ritual pattern of the Lakota vision quest, when an individual goes out alone to seek a sacred vision. The fourth ritual is the communal recreation ceremony known as the Sun Dance. The fifth is the making of Relatives, a ritual joining two friends into a sacred bond. The sixth is the girl's puberty ceremony. A final ritual is called Throwing the Ball, a game representing Wakan Tank and the attaining of wisdom. (Roy Willis, ed., *World Mythology*.)

## Seven Robbers

- martyrs on the Island of Corfu in the second century: Saturninus, Insischolus, Faustianus, Januarius, Marsalius, Euphrasius, and Mammius.

## sevens

- card game also known as fan-tan (not to be confused with Chinese Fan-Tan), in which the players, in turn, starting with sevens, discard the next highest or lowest card in suit. The first to empty his hand is the winner.

## The Seven Sacred Books

- Christian Bible, Eddas of the Scandinavians, Five Kings of the Chinese, Tri Pitikes of the Buddhists, Quran of Muhammadans, Three Vedas of the Hindus, and Zenda of the Persians.

## Seven Sages of Ancient Greece

- Plato, in his *Protagoras*, mentions seven extraordinary wise men from the seventh and sixth centuries BC who "were lovers and emulators and disciples of the culture of the Spartans": Thales of Miletus, Pittacus of Mitylene, Bias of Priene, Solon of Athens, Chilon of Sparta, Cleobulus the Lindian, and Myson the Chenian. (Some authorities replace Myson with Periander of Corinth.)

## Seven Sages of Prehistory

- in Babylonian myth, powerful beings who were summoned in magical ceremonies.

## Seven Sages of the Bamboo Grove (or Seven Worthies of the Bamboo Grove)

- a group of seven Chinese scholars and artists living in the mid-third century who regularly met in a bamboo grove to enjoy conversation, poetry, and wine. The most prominent was the poet and musician Hsi K'ang; his companions were Yuan Chi and his nephew Juan Hsien, Liu Ling, Hsiang Hsiu Wang Jung, and Shan Tao.

## *Seven Samurai*

- 1956 classic film by Akira Kurosawa, in which a group of ex-samurai (unemployed in the social changes of the 1600s) agree to defend a small town as a matter of honor. With an outstanding performance by Toshiro Mifume, it is one of the best films ever made. Adapted for Western viewers as *The Magnificent Seven*, directed by John Sturgis.

## 7, Saville Row, Burlington Gardens

- residence of Phileas Fogg in London, the man who went *Around the World in Eighty Days* in the 1873 Jules Verne novel.

## seven seas

- proverbial number of seas on the planet: North Pacific, South Pacific, North Atlantic, South Atlantic, Indian, Arctic, Antarctic.

## seven senses

- the ancients believed man's soul is composed of seven properties, influenced by nature. Earth provides a sense of feeling, water gives speech, air gives taste, mist gives sight, flowers give hearing, the south wind provides smell, fire animates. Thus, the seven senses are feeling, speech, taste, sight, hearing, smell, and animation.

## Seven Sisters

- (a) women's Ivy League colleges: Barnard (New York City), Mount Holyoke (South Hadley, Massachusetts), Bryn Mawr (Bryn Mawr, Pennsylvania), Radcliffe (Cambridge, Massachusetts), Smith (Northampton, Massachusetts), Vassar (Poughkeepsie, New York), and Wellesley (Wellesley, Massachusetts).

- (b) (in Russian, *vysotki*); the so-called "Stalin's skyscrapers," a group of seven tall buildings built in Moscow in the 1950s (during the last ten years of Stalin's rule) in an elaborate style combining Baroque and Gothic elements:
  1. the main building of Moscow University
  2. Hotel Leningrad
  3. Hotel Ukraine
  4. the Ministry of Foreign Affairs
  5. the Ministry of Heavy Industry
  6. an apartment building on Kotelnicheskaya waterfront
  7. a building on Kudrinskaya Square.

- (c) (*Kungarankalpa*) the ancestral heroines of native Aboriginals who fled across Australia and leapt into the sky to become the constellation Kurialya, the Pleiades.

- (d) a town in southern Texas, about 70 miles northeast of Laredo.

**Seven Sisters Falls**

- (in Norwegian, *Syv Søstrefosser*): waterfall in western Norway.

**Seven Sleepers of Ephesus**

- seven Christians in the third century who fled to a cave on Mount Celion during the Decian persecution, and there they slept for 200 years. Upon awakening, they convinced the emperor of life after death and then fell into a second sleep that would last until the Resurrection. Their names are usually given as Maximianus, Malchus, Martinianus, Dionysius, Johannes, Serapion, and Constantius.

**Seven Sorrows of Mary**

- In Catholicism:
  1. the prophecy of Simeon
  2. the flight into Egypt
  3. the loss of the child Jesus in the temple
  4. the meeting of Jesus and Mary on the way of the cross
  5. the crucifixion
  6. the taking down of the body of Jesus from the cross
  7. the burial of Jesus

## Seven Spirits of God

*1.*    the Spirit of Wisdom
*2.*    the Spirit of Understanding
*3.*    the Spirit of Counsel
*4.*    the Spirit of Power
*5.*    the Spirit of Knowledge
*6.*    The Spirit of Righteousness
*7.*    The Spirit of Divine Awfulness

    (from Brewer, *Dictionary of Phrase and Fable*)

## seven spiritual works of mercy

- In Catholicism:
  *1.*    instruct the ignorant;
  *2.*    counsel the doubtful;
  *3.*    admonish the sinner;
  *4.*    bear wrongs patiently;
  *5.*    forgive offenses willingly;
  *6.*    comfort the afflicted;
  *7.*    pray for the living and the dead.

## seven states in the United Arab Emirates

- Abu Dhabi, Dubai, Sharjah, Ajman, Umm al-Qaiwain, Ras al-Khaimah, and Fujairah; federated since December 2, 1971.

## seven steps of Buddha

- at birth, the Buddha is said to have taken seven steps, and at each, a lotus flower bloomed.

## "seven swans a-swimming"

- the gift on the seventh day of Christmas in the carol "The Twelve Days of Christmas."

## Seventh Amendment to the Constitution

- establishes the right of trial by jury.

## Seventh Arrondissement

- section of Paris containing the Eiffel Tower.

## 7th Avenue

- "Fashion Avenue"; historically thought of as the center of the fashion industry in New York City.

## *Seventh Book of Moses*

- an anonymous guide to powwowing used in Pennsylvania Dutch witchcraft practice.

## Seventh Commandment

- Thou shalt not commit adultery.

## *Seventh Cross*

- 1944 anti-Nazi film directed by Fred Zinnemann, starring Spencer Tracy, with husband-and-wife team Hume Cronyn and Jessica Tandy, in her first film.

## Seventh Crusade

- in 1244 Palestine was returned to the Christians by a treaty with Damascus, but later that year Egyptian Muslims and their Turkish allies retook the city and routed Christian forces at Gaza, which provoked Louis IX of France to organize a new campaign against Egypt. Lasting from 1248 to 1254, the Seventh Crusade proved to be a failure, Louis himself having been captured at Cairo in 1250.

## seventh day

- a day of rest, following the six days of labor during which God created the world.

## Seventh-Day Adventists

- a Protestant denomination that has its Sabbath on Saturday. They believe the end of the world is at hand, but that obeying the Ten Commandments is not necessary for Salvation, which is achieved by grace alone. Keeping the commandments is an act of love, not duty. They eschew a ministry.

## Seventh French War of Religion

- King Henry III's failure to carry out the terms of the Peace of Bergerac, which ended the Sixth French War of Religion, led to the Seventh War (1580). Nothing much happened until Henry of Navarre and Catherine de Medici ended the War by the Treaty of Fleix (November 26, 1580), confirming the rights granted the Huguenots in former treaties.

## seventh heaven

- a state of great happiness or satisfaction, bordering on bliss.

## *Seventh Heaven*

- in Islam, the highest Heaven one goes to after death as a reward for having lived an exemplary life on Earth. Ruled by Abraham, it is where God dwells with the most exalted angels.

## *Seventh Heaven*

- 1927 silent film starring Janet Gaynor and Charles Farrell in one of the best-known screen romances of all time. Gaynor won the very first Best Actress Oscar for her portrayal of Diane, a young woman of the Parisian slums, and Frank Borzage won Best Director. Remade with sound in 1937, leading roles by Simone Simon and James Stewart. Story is based on a highly successful 1922 Broadway play by Austin Strong, which starred actress Helen Mencken.

## Seventh House

- in astrology, deals with relationships, partnerships, and legal matters.

## seventh-inning stretch

- the time between the two halves of the seventh inning of a baseball game when the fans take a break, stand, and stretch their muscles to avoid cramping from sitting in the same position for too long a period.

## seven-thirties

- a name used for U.S. Treasury notes issued during the Civil War, bearing interest at the rate of $7\frac{3}{10}$ (thirty hundredths) percent annually.

## Seventh of March Speech

- "On the Constitution and the Union," Daniel Webster's defense of Henry Clay's Compromise of 1850, especially the

Fugitive Slave Bill. Webster proposed that abolitionists give up some of their demands in order to preserve the Union in practicality.

## seventh president

- Andrew Jackson, March 4, 1829 to March 3, 1837.

## *Seventh Seal*

- Ingmar Bergman's 1958 allegory, still a cult classic, starring Max von Sydow, Gunnar Bjornstrand, and Nils Poppe, with Bengt Ekerot playing Death.

## seventh son of a seventh son

- in the tradition of folklore, the seventh son of a seventh son is believed to have overwhelming magical powers and the wondrous ability to predict the future and to cast awesome spells. In some folklores, the seventh son of a seventh son will be a werewolf.

## seventh state of the Union

- Maryland ratified the Constitution on April 28, 1788.

## *The Seventh Veil*

- gripping psychological drama about a concert pianist (Ann Todd) who enters therapy with psychiatrist Herbert Lom after attempting suicide. Her troubled past is confronted and she is put in the care of James Mason, who commits himself to nurturing her talent. Compton Bennett directed this 1946 film, one of the first to explore psychotherapy.

## seventh wedding anniversary gift

- wool or copper is customary.

## seven tinctures of heraldry

- the colors used in heraldic emblems: Or (= gold/yellow), Argent (= silver/white), Azure (= blue), Gules (= red), Purpure (= purple), Sable (= black), Vert (= green).

## seven topics of a Hindu tantra (scripture)

- creation, destruction of the universe, worship of the gods, spiritual exercise (yoga), rituals and ceremonies, the six "magical" powers, and meditation.

## Seven Treasures

- among Buddhists, the seven materials representing material wealth: gold, silver, lapis lazuli, crystal, agate, pearl, and carnelian.

## Seven Troughs

- a town in northwestern Nevada.

## 7UP

- a carbonated drink.

## Seven Valleys

- a town in southeastern Pennsylvania.

## Seven Virtues

- three theological: faith, hope, charity; and four heavenly: prudence, temperance, fortitude, and justice.

## Seven Virtues of Bushido

- rectitude, courage, benevolence, respect, honesty, honor, and loyalty.

## seven voyages of Sinbad the Sailor

- in the *Arabian Nights* tale: To the Indian Ocean, to the Valley of Diamonds, to an encounter with the Cyclops, to the Cavern of the Dead, to the island of the Old Man of the Sea, to Serendip, and to the land of elephant tusks.

## Seven Weeks' War

- or Austro-Prussian War (June 8–July 26, 1866): Dispute over administration of Schleswig-Holstein, provoked by Bismarck's intent to expel Austria from the German Federation. Prussia quickly defeated Austria and the German states, leading to the Treaty of Prague on August 23, 1866. Made possible the founding of the German Empire in 1871.

## Seven Wise Masters

- an ancient tale about false charges brought against Prince Lucien by his spurned step-mother, in which seven wise men tell stories to the king illustrating the consequences of ill-conceived punishment. King Dolopathus, who had originally favored the queen's version of events, reconsiders and sentences her to death.

## Seven Wonders of the Ancient World

- Great Pyramid of Khufu (Cheops) or all the pyramids, with or without the Sphinx, Hanging Gardens of Babylon, Pharos (Lighthouse) of Alexandria, Colossus of Rhodes, Olympian Statue of Zeus by Phidian, Mausoleum of Halicarnassus, and the Temple of Artemis (Diana) at Ephesus.

## Seven Wonders of the Middle Ages

- Coliseum of Rome, Catacombs of Alexandria, Great Wall of China, Stonehenge in England, Leaning Tower of Pisa, Porcelain Tower of Nanking, Mosque of Hagia Sophia at Constantinople.

## Seven Wonders of the Modern World

- Mount Everest, Victoria Falls, the Grand Canyon, the Great Barrier Reef, the Northern Lights, Paricutin, and the Harbor at Rio de Janeiro, Brazil.

## seven-year itch

- a married man's (or woman's) urge to have an extramarital affair after seven years of marriage. Popularized by the 1955 film, *The Seven Year Itch*, starring Marilyn Monroe and Tom Ewell, directed by Billy Wilder.

## seven years' lease

- a lease for a period of seven years or its multiple, from the ancient notion of "climacteric years," in which life was considered to be in special peril.

## seven years of bad luck

- period proverbially following the breaking of a mirror.

## Seven Years' War

- the war (1756–63) in which Frederick the Great of Prussia (aided by England) beat France, Austria, Russia, Sweden, and Saxony. At the close, Prussia took possession of Silesia. In the American phase, which we know as the French and Indian War, Great Britain took possession of Canada by the Treaty of Paris in 1763.

## seven youths, seven maidens

- in classical myth, the Athenians were committed to pay tribute to Minos, King of Crete, which tribute consisted of seven youths and seven maidens to be sent each year to be devoured by the Minotaur, a half-bull, half-human monster. Theseus, with a sword supplied by Ariadne, daughter of Minos, managed to slay the Minotaur and escape with the other youths and maidens. (The story is related in Bullfinch's *Mythology*.)

## Shichi-Go-San

- literally, "seven-five-three," a traditional rite-of-passage celebration for Japanese children, for three- and seven-year-old girls and three- and -five-year-old boys, observed on November 15, or the nearest weekend.

## Shiva

- in the Jewish religion, the seven-day mourning period that begins after the funeral of any one of seven relatives: father, mother, brother, sister, son, daughter, or spouse.

## *Snow White and the Seven Dwarfs*

- Disney's 1937 animated masterpiece, a retelling of the Grimm brothers' classic fairy tale, with several memorable songs including "Heigh Ho," "Whistle While You Work," and "Someday My Prince Will Come." Snow White's voice was dubbed by Adriana Caselotti, the Wicked Queen by Lucille LaVerne, and Prince Charming by Harry Stockwell. For his truly historic accomplishment, Disney was awarded one large Oscar, and seven smaller ones, all presented by Shirley Temple at the Academy Award ceremonies.

## Shostakovich's Seventh Symphony

- "Leningrad" symphony (in C major).

## The Siberian Seven

- seven Russian Pentecostal Christians of two families who were denied emigration from the Soviet Union and in 1979 took refuge in the basement of the American Embassy in Moscow. In January 1982, they went on a hunger strike to gain recognition, and were finally allowed to leave the USSR in 1983.

## Trial of the Seven Bishops

- In 1688 seven clergymen who were charged with seditious libel for refusing to read to their congregations the Declaration of Indulgence issued by King James II, which they saw as an illegal order. Their names: William Sancroft (of Canterbury), Thomas Ken (of Bath & Wells), William Lloyd (of St. Asaph), Jonathon Trelawney (of Bristol), John Lake (of Chichester), Francis Turner (of Ely), and Thomas White (of Peterborough). At trial before the Court of the King's Bench, June 29–30, they were acquitted.

## Vaughan Williams's *Symphony No. 7*

- "Sinfonia Antarctica," based on his music for the film, *Scott of the Atlantic.*

## Watergate Seven

- seven men indicted for conspiracy, burglary, and violation of federal wiretapping laws, the result of a break-in of Democratic campaign offices in the Watergate in Washington, D.C., in June 1972. Five were arrested at the Watergate on June 17: Frank Sturgis, Bernard Baker, Eugenio Martinez, Birgilio Gonzalez, and James McCord; the others, G. Gordon Liddy, a member of Nixon's campaign finance committee, and E. Howard Hunt, a White House consultant, tied the break-in to the Nixon re-election campaign.

# 7-1/2

## Seven and a Half

- italian version of blackjack; some say the forerunner of the modern game of "21." With a deck from which eights, nines, and tens have been eliminated, cards have numerical face value, pictures valued at one-half, ace at one. The king of diamonds is wild, the player getting it assigning a value to it. The object is to get close to seven and a half without going over.

## *008*

- Decimal ASCII code for [backspace].

## *08*

- number of the French Département of Ardennes.

## *8*

- Article of the U.S. Uniform Commercial Code that deals with investment securities.
- atomic number of the chemical element oxygen, symbol O.
- Beaufort number for wind force of 39 to 46 miles per hour, called "gale–fresh gale."
- bottles in a Methuselah (champagne measure).
- chi in a gui, Chinese unit for measuring the width of roads.
- considered a lucky number in Chinese culture because the word (in pinyin, "ba") sounds like the Chinese word for "prosper" (in pinyin, "fa").
- the cube of 2.
- dental chart designation for the right central incisor.
- drams in an apothecary's ounce.
- eight faces on an octahedron.
- eight feet:
    ⇨ height of the crossbar on a soccer goal.
    ⇨ width of the goal on a polo field.

- eight-foot: designation for the size of a home pool table measuring 8 feet by 4 feet.
- eight inches: height of a standard step riser.
- eight meters: width of a beach doubles volleyball court.
- eight seconds: the minimum time a bronc rider must stay in his saddle in order to qualify.
- eighth day after birth was when an Israelite child was to be circumcised (Leviticus 12:3).

- eight weeks: the interval between successive donations of whole blood, as recommended by the Red Cross.
- eight yards: distance between the goal posts on a soccer field.
- fewest team losses in an NHL season (hockey record): by Montreal Canadiens in the 1976–77 season, against sixty wins and twelve ties.
- fewest team wins in an NHL season (hockey record): by Washington Capitals, 1974–75.
- fluid drams in a fluid ounce.
- fluid ounces in a cup.
- furlongs in a statute mile.
- gauge (number) of a wood screw with a shank diameter of 11/64 of an inch.
- Henry VI, King of England, acceded to the throne at eight months old.
- in astrology, the number 8 is ruled by the planet Saturn.
- in baseball, 8 represents the center fielder's position.
- in Hebrew, the numerical value of the letter *cheth*.
- in mineralogy, Mohs Hardness Scale level characterized by topaz; scratched by corundum.
- magnums in a Balthazar (champagne measure).
- most consecutive birdies in one 9-hole golf round on the PGA Tour: six times, the most recent by Jerry Kelly in the third round of 2003 Las Vegas International.
- most field goals made in an NFL game (football record): by Rob Bironas for the Tennessee Titans against Houston Oilers, October 21, 2007.
- most individual touchdowns scored in Super Bowl games, career (football record): by Jerry Rice for the San Francisco 49ers, SB-XXIII, 1989; SB-XXIV, 1990; SB-XXIX, 1995; and for the Oakland Raiders, SB-XXXVII, 2003.
- most interceptions thrown in an NFL game (football record): by Jim Hardy, Chicago Cardinals against the Philadelphia Eagles, September 24, 1950.
- most interceptions thrown in Super Bowl games, career (football record): by John Elway, for the Denver Broncos, SB-XXI, 1987; SB-XXII, 1988; SB-XXIV, 1990; SB-XXXII, 1998; SB-XXXIII, 1999.
- most kickoffs returned for touchdowns, career (football record): by Joshua Cribbs for the Cleveland Browns, 2005–09.

- most personal fouls committed in an NBA playoff game (basketball record): by Jack Toomay for the Baltimore Bullets against the New York Knicks, March 26, 1949.

- most personal fouls committed in an NBA regular season game (basketball record): by Don Otten for the Tri-Cities Blackhawks against the Sheboygan Redskins, November 24, 1949.

- most points scored in an NHL playoff game (hockey record): twice, by Patrik Sundstrom for the New Jersey Devils against the Washington Capitals, April 22, 1988; and by Mario Lemieux for the Pittsburgh Penguins against the Philadelphia Flyers, April 5, 1989.

- most singles wins in the Championnat de France International de Tennis, the forefunner to the French Open, originally for professionals only: by Max Decugis, in 1903–04, 1907–09, and 1912–14.

- most Super Bowl games played: by the Dallas Cowboys, SB-V (lost), VI (won), X (lost), XII (won), XIII (lost), XXVII (won), XXVIII (won), and XXX (won).

- most team touchdowns scored in a Super Bowl game (football record): by the San Francisco 49ers against the Denver Broncos (55–10), in SB-XXIV, 1990.

- most walks pitched in an inning (baseball record): by Dolly Gray, for the Washington Senators, in second inning against the Chicago White Sox, August 28, 1909.

- number of AOC (Appellation contrôlèe) wine zones in Provence: Côte de Provence, Bandol, Bellet, Coteaux d'Aix-de-Provence, Coteaux des Baux-de-Provence, Cassis, Coteaux, Varois, and Palette.

- number of both rows and columns on a chess board.

- number of counties in the state of Connecticut.

- number of days in the Jewish holiday of Passover (Pesach), celebrating the liberation of the children of Israel from slavery in Egypt over 3,000 years ago, led by Moses.

- number of English kings named Edward.

- number of films starring the team of Nelson Eddy and Jeanette MacDonald, filmdom's original operetta duo: *Naughty Marietta* (1935), *Rose Marie* (1936), *Maytime* (1937), *The Girl of the Golden West* (1938), *Sweethearts* (1938), *Bitter Sweet* (1940), *New Moon* (1940), *I Married an Angel* (1942).

- number of "fits" (verses) in the 1876 Lewis Carroll poem "The Hunting of the Snark." (It is subtitled "An Agony, in Eight Fits.")
- number of kings in Macbeth's vision in the cave (Act IV, Scene I).
- number of legs on a spider.
- number of main Crusades to the East, starting, respectively, in 1096, 1147, 1189, 1202, 1217, 1228, 1248, 1270. Some authorities separate the eighth from a later ninth, 1271.
- number of pawns on each side in chess.
- number of planets in our solar system, in sequence of their distance from the sun: Mercury, Venus, Earth, Mars, Jupiter, Saturn, Uranus, Neptune. Pluto had been included as a planet until demoted to the status of "dwarf planet" by the International Astronomical Union on August 24, 2006.
- number of Plantagenet rulers of England, from Henry II (Curtmantle), 1154; then Richard I (the Lionheart), 1189; John, 1199; Henry III, 1216; Edward I (Longshanks), 1272; Edward II, 1307; Edward III, 1327; Richard II, 1377–99. (The Lancastrian Dynasty followed.)
- number of presidential electoral votes apportioned to each of the states of Kentucky and South Carolina. (May be reapportioned for the 2012 election.)
- number of seats in the House of Representatives allotted to each of the states of Arizona, Maryland, Minnesota, and Wisconsin (as of the 2000 Census).
- number of separate parries and thrusts in the sport of fencing.
- number of tentacles on an octopus.
- number of "tiny reindeer" pulling Santa's sleigh in Clement C. Moore's 1822 poem, "A Visit from Saint Nicholas." They were named Dasher, Dancer, Prancer, Vixen, Comet, Cupid, Donder (or Donner), and Blitzen.
- number of U.S. presidents born in Virginia, the state producing the most chief executives: Washington, Jefferson, Madison, Monroe, W. H. Harrison, Tyler, Taylor, and Wilson.
- number of U.S. presidents who died in office, of which four were assassinated: Abraham Lincoln (April 5, 1965), James Garfield (September 19, 1881), William McKinley (September 14, 1901), and John F. Kennedy (November 22,

1963). The other four were William Henry Harrison (April 4, 1841), Zachary Taylor (July 9, 1850), Warren G. Harding, August 2, 1923), and Franklin Delano Roosevelt (April 12, 1945).

- number of warm-up pitches allowed a relief pitcher coming into a baseball game.
- pints in a gallon.
- quarts in a peck (dry measure).
- record number of gold medals won in a single Olympic year: by American swimmer Michael Phelps at the Beijing games in 2008: 100-m and 200-m butterfly, 200-m and 400-m individual medley, 200-m freestyle, 4 x 100-m and 4 x 200-m freestyle relay, and 4 x 100 medley relay. Phelps, or his team, established new world records in all but the 100-m butterfly.
- record number of Grammys won by a single performer in a single year: twice—once at the 42nd Grammy Awards, February 23, 2000, by fifty-two-year-old singer/guitarist Carlos Santana for his 1999 album, *Supernatural*, plus Best Record, Best Rock Album, Best Instrumental Performance, and four others; and in 1983 at the 26th Grammy Awards, by Michael Jackson for his album, *Thriller*, Record of the Year, *Beat It*, and six more.
- record number of individual medals won in a single Olympic year: twice, in 1980 by Soviet gymnast Aleksandr Dityatin (three gold, four silver, one bronze); and in 2004 by American swimmer Michael Phelps (six gold and two bronze).
- record number of times home plate was stolen in a baseball season: by Ty Cobb, Detroit Tigers, 1912.
- record number of Wimbledon tennis doubles titles won by a men's team: Laurie and Reggie Doherty, 1897–1901 and 1903–05. But T. A. Woodbridge, playing with different partners, won nine doubles championships: with M. R. Woodforde, 1993–97, 2000, with J. L. Bjorkman, 2002–04.
- record number of women's U.S. Open singles tennis titles: won by Norway's Molla Bjurstedt Mallory, 1915–18, 1920–22, and 1926. She also won two women's doubles and three mixed doubles titles.
- retired baseball uniform number of:
  ⇨ both catchers Yogi Berra and Bill Dickey, retired by the New York Yankees.

> ⇨ catcher Gary Carter, retired by the Montreal Expos.
> ⇨ second baseman Joe Morgan, retired by the Cincinnati Reds.
> ⇨ shortstop Cal Ripken Jr., retired by the Baltimore Orioles.
> ⇨ first baseman Willie Stargell, retired by the Pittsburgh Pirates.
> ⇨ outfielder Carl Yastrzemski, retired by the Boston Red Sox.

- retired football jersey number of:
  - ⇨ Larry Wilson, retired by the Arizona Cardinals.
  - ⇨ Steve Young, retired by the San Francisco 49ers.
- retired hockey shirt number of:
  - ⇨ Frank Finnegan, retired by the Ottawa Senators.
  - ⇨ Bill Goldsworthy, retired by the Minnesota North Stars.
  - ⇨ Cam Neely, retired by the Boston Bruins.
  - ⇨ Barclay Plager, retired by the St. Louis Blues.
  - ⇨ Marc Tardif, retired by the Quebec Nordiques.

- (*ya*) sacred number in Japanese mythology, meaning "many."
- sides on an octagon.
- sleipnir, horse of the Norse god Odin, had eight legs.
- Title of the California Penal Code, Part I, which deals with violent crimes.
- Title of the United States Code dealing with aliens and nationality.
- value of Roman numeral VIII.

## After Eight

- a Nestle candy bar made of dark chocolate filled with mint creme.

## Ancient Eight

- the Ivy League universities: Brown, Columbia, Cornell, Dartmouth College, Harvard, Pennsylvania, Princeton, Yale. Formally federated since 1945.

## behind the eight ball

- (adj.) in an untenable position; in a serious predicament from which there is no obvious relief.

**Beethoven's** *Sonata No. 8*

- the "Pathetique" (in C minor).

**Big 8**

- a bet in craps that the shooter will throw an 8 before he throws a 7. It is an even-money bet, even though the odds are 6 to 5 in favor of the house.

**Big Eight Conference**

- Former NCAA-affiliated college athletic alliance comprising Colorado, Iowa State, Kansas, Kansas State, Missouri, Nebraska, Oklahoma and Oklahoma State. Dissolved in 1996, when the eight members combined with four former members of the defunct Southwest Conference (Baylor, Texas, Texas A&M, and Texas Tech) to form the Big Twelve Conference.

*Butterfield 8*

- 1960 film adaptation of John O'Hara's 1935 novel about a high-end call girl (Elizabeth Taylor) who believes she's found her Mr. Right (Laurence Harvey), directed by Daniel Mann. Taylor's performance won the Best Actress Oscar.

**Canon 8 of the American Bar Association Code of Professional Responsibility**

- a lawyer should assist in improving the legal system.

**City of the Eight**

- Hermopolis, in ancient Egyptian myth, where out of chaos arose four gods, each accompanied by his feminine counterpart: Nun and Naunet, god and goddess of the primeval waters; Heh and Hehet, god and goddess charged with raising the sun; Kek and Keket, god and goddess of darkness; and Amun and Amunet, god and goddess of mystery, also known as Niu and Niut, who controlled the life-giving air. The eight are known collectively as the Ogdoad.

**crazy eights**

- a card game for two or more players in which the object of the game is to dispose of all your cards by matching, in turn, the number or the suit of the previous card played.

**CV-8**

- number of the U.S. Navy aircraft carrier, USS *Hornet*. Immobilized by torpedoes and bombs, sunk off the Santa Cruz Islands, October, 1942. CV-12, built to avenge CV-8, commissioned November 29, 1943, was also named the *Hornet*.

**DC-8**

- the Douglas Skybus, a replacement for the popular but small DC-3 in commercial short- to medium-range air routes. One of the earliest jet-powered commercial passengers aircraft, the DC-8 was designed as a twin-engine, low-wing monoplane that could carry twice the number of passengers at half the seat-mile cost of the DC-3.

***Dinner at Eight***

- 1933 first-rate comedy produced and directed by George Cukor, with a stellar cast including Marie Dressler, John Barrymore, Wallace Beery, Jean Harlow, Lionel Barrymore, Billie Burke, Lee Tracy, Edmund Lowe, and Madge Evans. Interweaves several separate stories about the guests at a formal dinner party. Based on a 1932 stage play by George S. Kaufman and Edna Ferber.

**The Eight**

- a group of American painters who eschewed the academicism in art at the beginning of the twentieth century in favor of more realistic subjects: Arthur B. Davies, William Glackens, Robert Henri, Ernest Lawson, George Luks, Maurice Prendergast, Everett Shinn, and John Sloan. The widespread disapproval of their first exhibit in 1908 earned them the disparaging name of "The Ashcan School."

## 8-Ball

- popular name for Old English 800, a well-liked beer in the 'hood.

## eight ball

- a game of pocket billiards in which one of the opposing players must pocket the solid-color balls in succession, the other the striped balls in succession, before being allowed to sink the 8-ball to win. If the 8-ball is pocketed out of sequence, the shooter loses.

## eightball

- a socially inept, maladjusted, or functionally incapable person.

## The Eight Beatitudes

- the declarations made in the Sermon on the Mount, as per the Gospel of St. Matthew (Matthew 5:3–10):
  1. Blessed are the poor in spirit: for theirs is the kingdom of heaven;
  2. Blessed are they that mourn: for they shall be comforted;
  3. Blessed are the meek: for they shall inherit the earth;
  4. Blessed are they which do hunger and thirst after righteousness: for they shall be filled;
  5. Blessed are the merciful: for they shall obtain mercy;
  6. Blessed are the pure in heart: for they shall see God;
  7. Blessed are the peacemakers: for they shall be called the Children of God;
  8. Blessed are they which are persecuted for righteousness' sake: for theirs is the kingdom of heaven.

## eight bells

- time check on-board ship, signifies 4:00 or 8:00 or 12:00 AM or PM.

## Eight Buddhas of Medicine

- in Tibetan Buddhism, Bhaishajyaguru is the chief of the eight Buddhas of Medicine.

## Eight-Circuit Model of Consciousness

- a paradigm created by Timothy Leary to explain the workings of the human mind. Eight "circuits" (also called "gears" or "mini-brains"), each of which represents a higher stage of evolution than the previous one, are named:
  1. Bio-Survival Circuit
  2. Emotional
  3. Dexterity-Symbolism
  4. Social Sexual
  5. Neurosomatic
  6. Neuroelectric
  7. Neurogenetic
  8. Neuro-Atomic.

  Leary associated several circuits with specific drugs, respectively:
  1. opiates
  2. alcohol
  3. caffeine and cocaine
  4. MDMA and Ecstasy
  5. marijuana
  6. peyote and psilocybin
  7. LSD
  8. ketamine.

## Eight Cold Hells

- (in Vedic cosmology of ancient India): Arbuda (blistering), Nirarbuda (broken blisters), Atata (chattering of teeth), Hahava (sound of shivering), Huhuva (sound of shivering), Utpala (blue lotus), Padma (red lotus), Mahapadma (deep red lotus).

## 8 December

- Christian feast day commemorating the Immaculate Conception.

## Eight Degree Channel

- a passage between the Laccadine and the Maldive Islands, in the Indian Ocean west of Ceylon.

## eight divine wives of Zeus

- (in Greek mythology): Metis, Themis, Eurynome, Demeter, Mnemosyne, Leto, Maia, Hera.

## Eight Eccentric Painters of Yangzhou

- Chinese artists who painted in the area of Yangzhou, in Jiangsu Province, during the reign of Qianlong (1735–96) in the Qing Dynasty. Moving away from the standard brushstroke techniques of the time, they were influential in creating new painting methods. They are remembered as: Zheng Banqiao, Jin Nong, Huang Shen, Gao Xiang, Li Shan, Luo Ping LiFangying, and Wang Shishen.

## Eightfold Way

- system for classifying subatomic particles known as hadrons into groups based on their symmetrical properties, the number in each group being 1, 8 (the most frequent), 10, or 27. The system was devised in 1961 by American physicist Murray Gell-Mann and Israeli physicist Yuval Ne'eman.

## eight great gods of ancient Egypt

- Neph, Amun, Pthah, Khem, Sati, Neith, Maul, and Bubastis. (A. S. Murray, *The Manual of Mythology*)

## Eighth Amendment to the Constitution

- protects against excessive bail or fines and cruel and unusual punishments.

## eighth bend

- a pipe bend or junction piece used to connect two pipes at 22-½ degrees, i.e., one-eighth of a complete reversal of direction.

## Eighth Commandment

- Thou shalt not steal.

## Eighth Crusade

- Organized by Louis IX in 1270, the Eighth Crusade was aborted when Louis died in Tunisia.

## Eighth French War of Religion

- in 1585 King Henry III named Henry of Navarre (known to be a follower of the Protestant religion) as heir presumptive to the French throne, which was not acceptable to the Catholic League. This precipitated, in 1585, the Eighth French War of Religion (known as the War of the Three Henrys: Henry III, a royalist; Henry of Navarre, a Protestant; and Henry of Guise, of the Catholic League). Henry III had Henry of Guise killed, and in response Henry III was stabbed to death in July 1589. Henry of Navarre became King Henry IV, but only after he returned to the Catholic Church. This led to the Wars of the League (1589–98), which ended with the Edict of Nantes, April 15, 1598, granting Huguenots a large share of political and religious freedom—at least until Louis XIV revoked it in 1685.

## Eighth House

- in astrology, deals with sex, death, and money.

## *The Eighth of January*

- 1829 play by R. P. Smith, about an Englishman whose loyalties are divided between his native and his adopted countries, relating to Jackson's victory at New Orleans in January 1815.

## Eight Hot Hells

- in Vedic cosmology of ancient India, Samjiva (reviving), Kalasutra (black rope), Samghata (crushing), Raurava (howling), Maharaurava (great howling), Tapana (hot), Pratapana (very hot), Avichi (uninterrupted).

## Eight Hour Law

- a federal statute of 1868 limiting the work day to eight hours for laborers and workmen employed by the federal government.

## eighth president

- Martin Van Buren, March 4, 1837 to March 3, 1841.

## Eighth Route Army

- the larger of two Chinese communist legions that fought the Japanese from 1937 to 1945. Although ostensibly positioned under the Chinese Nationalist government, as part of the anti-Japanese alliance between the Chinese Nationalists under Chiang Kai-shek and the Chinese Communists, the army was under the control of Chu Teh, a close associate of Mao Zedong. After the end of World War II, the Eighth Route Army was folded into the new People's Liberation Army.

## eighth state of the Union

- South Carolina ratified the Constitution on May 23, 1788.

## eighth wedding anniversary gift

- bronze is customary.

## Eight Immortals

- the Pa Hsien of Taoist mythology; unacquainted in real life, the eight immortals constitute a heterogeneous group of holy Taoists: Cao Guo Jiu (patron saint of actors), Han Xiang Zi (patron deity of musicians), Lan Cai He (patron deity of florists), He Xian Gu (shown with a lotus blossom), Li Tie Guai (symbol of the sick), Lu Dong Bin (honored as a scholar, but patron deity of barbers), Zhang Guo Lao (represents old men), and Zhongli Quan (symbol of the military man). It is said they achieved immortality by following the Way. They are also known as the "roaming immortals."

## 8-iron

- golf club equivalent to the earlier pitching mashie.

## *Eight is Enough*

- family TV series starring Dick Van Patten and Betty Buckley. The female lead was originally Diana Hyland, but she died after completing only four shows. Series first run played on ABC-TV from March 1977 to May 1981.

## 8 January

- Jackson Day in Louisiana; commemorates the 1815 Battle of New Orleans when Andrew Jackson commanded the forces that defeated the British in the last battle of the War of 1812.

### eight kinds of superior human beings

- those believed to be the protectors of Buddhism: devas, dragons, yaksas, gandharvas, asuras, garudas, kimnaras, and mahoragas.

### eight levels of consciousness

- according to the Buddhist consciousness-only doctrine, we each have the following eight consciousnesses: The first through the fifth correspond to the five sense perceptions; sixth is mental consciousness, the ability to discriminate objects; seventh is ego-consciousness; and eighth is Alaya-consciousness, the fundamental consciousness of one's existence.

### "eight maids a-milking"

- the gift on the eighth day of Christmas in the carol "The Twelve Days of Christmas."

### Eight Masters of Nanking

- a group of Chinese artists in Nanking in the late seventeenth century, painting mostly landscapes. Working in a variety of artistic styles, they were: Kung Hsien, Fan Ch'i, Tsou Che, Wu Hung, Kao Ts'en, Hu Tsao, Yeh Hsien, and Hsieh Sun. Few of their paintings survive.

### 8 May

- (a) V-E (Victory in Europe) Day, marking the end of World War II in Europe, 1945.
- (b) Truman Day, in Missouri.

### eight men in the box

- football term for a defensive formation in which eight players—usually the defensive linemen and linebackers, plus one defensive back playing unusually close to the ball—are lined up tight in "the box," the area bound by where the offensive tackles line up on the opposite side of the ball.

### Eight Men Out

- finely crafted 1988 film depiction of the infamous 1919 World Series, in which members of the Chicago White Sox baseball team took money to throw the games. John Sayles wrote and directed this highly effective period piece from a book by Eliot Asinof.

## *8MM*

- 1999 film directed by Joel Schumacher, with Nicholas Cage as a private investigator hired to discover if a "snuff film" is really authentic. Supported by Joaquin Phoenix and James Gandolfini.

## "Eight O'Clock"

- a 1924 poem by Sara Teasdale (1884–1933), about a woman awaiting the arrival of her lover.

## eight parts of speech

- nouns, verbs, adjectives, adverbs, pronouns, conjunctions, prepositions, and interjections.

## eight-penny nail

- a carpenter's common nail, 2-½ inches long (coded 8d).

## 8-point type

- typographer's measure, known as "brevier."

## Eight Principles of Yong

- explains how to create the eight strokes needed to write Chinese characters, all of which happen to occur in the one character of "yong." By practicing the principles used in writing this character, the novice can refine the elegance of his calligraphy.

## Eight Rebellious Acts

- crimes set forth in the Taiho Statute of eighth-century Japan:
    1. attempts on the emperor's life;
    2. plots to destroy imperial graves or places;
    3. treason;
    4. murder of an older relative, such as a grandparent, parent, brother, or sister;
    5. murder of other senior relatives or one's wife;

6.  disrespectful acts against the emperor or imperial shrines;
7.  unfilial acts against a grandparent;
8.  killing one's master, teacher, or superior.

## eights

- (a) the sport of intercollegiate rowing, so called because each boat is powered by eight oars in the hands of an equal number of crewmen.
- (b) love and kisses (CB slang).

## 8 Sabbats

- the major seasonal holidays celebrated by various neopagan religions; four Greater Sabbats (the Celtic cross-quarter days–on or about February 2, May 1, August 2, and October 31) and the four Lesser Sabbats (the English quarter-days–the solstices and equinoxes).

## Eights Coast

- a stretch of Ellsworth Land in Lesser Antarctica between Cape Waite and Phrogner Point.

## Eight O'Clock Coffee

- a store brand of whole-bean, light-roast coffee introduced by the A&P supermarket chain in 1859.

## 8 September

- Christian feast day commemorating the birthday of the Virgin Mary.

## 8 to 1

- the odds against either tossing a 5 or tossing a 9 in dice.

## eight to the bar

- a descriptive phrase relating to the rhythm of boogie-woogie music, which is structured with eight beats to a musical measure. Popularized by the Andrews Sisters' 1940 recording, "Beat Me, Daddy, Eight to the Bar."

## 8vo

- eightvo or octavo; the page size of a book from printer's sheets that are folded into eight leaves, each about 6 x 9 inches.

## Elizabeth Taylor's eight marriages

- hotel heir Nicky Hilton (1950–51, divorced), actor Michael Wilding (1952–57, divorced), producer Mike Todd (1957–58, widowed), singer Eddie Fisher (1959–64, divorced), actor Richard Burton (1964–74, divorced), Richard Burton a second time (1975–76, divorced), Virginia politician John Warner (1976–82, divorced), and construction worker Larry Fortensky (1991–96, divorced).

## figure eight

- a shape resembling the Arabic number 8, consisting of two loops formed by a continuous line crossing itself, as in an ice-skating pattern.

## Group 8 elements (of the Periodic Table)

- iron (26), ruthernium (44), osmium (76), hassium (108).

## Group of Eight (G8)

- coalition of eight of the world's leading industrialized nations holding an annual economic and political summit of the heads of state. Included are the United States, United Kingdom, Canada, France, Germany, Italy, Japan, and Russia.

## Haydn's Eighth Symphony

- "Le Soir" symphony (in G major).

## heptathlon

- women's equivalent to the men's decathlon; eight events in the competition. Day 1: 100-m, hurdles, long jump, shot put, and 200-m dash; Day 2: long jump, javelin, and 800-m running event.

## Magic Eight Ball

- an American toy, a giant ball that contains a shape with twenty sides, suspended in liquid, that answers yes-or-no questions. A player asks a question, shakes the ball, and waits for the answer to float into the window.

## Mahler's Eighth Symphony

- "Symphony of a Thousand" (in E-flat major).

## Mickey Rooney's eight wives

- Ava Gardner (1942–43, divorced), Betty Jane Rase (1944–49, divorced), Martha Vickers (1949–51, divorced), Elaine Mahnken (1952–58, divorced), Barbara Ann Thomasen (1958–66, she was murdered), Margaret Lane (1967, divorced after 100 days), Carolyn Hockett (1969–74, divorced), Jan Chamberlain (1978–still married as we go to press).

## Noble Eightfold Path

- in Buddhism, the way of life leading to salvation:
  1. correct understanding, free from superstition and delusion;
  2. correct aims and motives, worthy of intelligent, earnest men;
  3. correct use of speech, truthful and open;
  4. correct behavior, peaceful and honest;
  5. correct mode of livelihood, bringing injury to no living thing;
  6. correct effort toward self-knowledge and self-control;
  7. correct thought, the active mind;
  8. correct contemplation, earnest meditation on the deep mysteries of life.

## number 8 billiard ball

- the ball that is conventionally solid black.

## number 8 wood screw

- a screw with a screw shank diameter of 11/64 inches.

## octet

- a musical group composed of eight instruments or eight voices.

## octuplets

- the Chukwu octuplets, the first set of octuplets born alive, to mother Nkem Chukwu and father Iyke Louis Udobi, American citizens of Nigerian ancestry. The first baby was delivered at St. Luke's Episcopal Hospital in Houston, Texas, on December 8, 1998, the remaining seven on December 20. The smallest, Odera, weighing 10.3 ounces at birth, died on December 27. The others were girls Ebuka, Chidi, Echerem, Chima, and Gorom, and boys Ikem and Jioke.

## Ogdoad

- (See *City of the Eight*)

## one fat lady

- slang for the number 8 among bingo players, before PC.

## one over the eight

- (adj.) slang for "inebriated," since it was once believed that eight glasses of beer were the limit for most men.

## "On the Eight Evil Thoughts"

- ascetic Evagrius Ponticus's list of cardinal sins (which later formed the basis of the seven deadly sins): gluttony, fornication, avarice, dejection, anger, weariness, vainglory, and pride.

## PDD-8

- Presidential Decision Directive 8 deals with Declassification of POW/MIA Records, issued by President Bill Clinton on June 10, 1993.

## piece of eight

- the pisaster, a former silver coin of Spain and Spanish America equal to eight reals.

## *The Private Life of Henry VIII*

- Charles Laughton in an outstanding Oscar-winning performance as the dissolute monarch in this 1933 epic directed by Alexander Korda. With Merle Oberon as Anne Boleyn, Wendy Barrie as Jane Seymour, Elsa Lancaster as

Anne of Cleves, Binnie Barnes as Catherine Howard, and
Everley Gregg as Catherine Parr. Laughton's was the first
Oscar ever awarded to a film made in England.

## Protocol of the Eight Articles

- agreement of June 21, 1814, between William, Prince of
Orange, and the Allied powers after the defeat of Napoleon,
creating the Kingdom of the Netherlands, which united
Belgium and Holland. The Kingdom was intended to serve
as a bulwark against France.

## quaver

- eighth note, in musical notation.

## Rebellion of the Eight Kings

- (or, War of the Eight Princes): a civil war for power among
kings (or princes) of the Chinese Jin Dynasty from AD 291 to
306. The eight: Sima Liang, Prince of Runan; Sima Wei, Prince
of Chu; Sima Lun, Prince of Zhao; Sima Jiong, Prince of Qi;
Sima Ying, Prince of Wu; Sima Yi, Prince of Changsha; Sima
Jong, Prince of Hejian; and Sima Yue, Prince of Donghai.

## Schubert's Eighth Symphony

- the "Unfinished" symphony (in B minor).

## Section Eight

- a discharge from the U.S. military service for reason of military
ineptitude or mental incompetence or derangement.

## V8

- a noncarbonated drink containing the juice of eight
vegetables.

## War of the Eight Saints

- in 1375 Florence incited a revolt in the Papal States against
Pope Gregory XI in Avignon. Gregory excommunicated the
Florentines in 1376, but their war council, known as the
Eight Saints, continued its defiance. The following year the
pope sent an army to regain control of the region while he
himself left for Italy to protect his lands. The War ended
with a compromise peace in 1378, and the Papal See left

France after seventy years to return to Rome. The Great Schism followed, during which there were two or three rival popes.

## water of eight excellent qualities

- in Buddhism, water of the ponds in the Land of Utmost Bliss possesses eight qualities: It is pure, cool, sweet, smooth, moistening, comforting, thirst-quenching, and nourishing.

## The Wheel of Life

- a disc with eight spokes representing *samsara*, the continuous cycle of birth, life, death. The symbol is accepted by all three Dharmic religions: Hinduism, Buddhism, and Jainism.

## Yamata-no-orochi

- the eight-tailed, eight-headed dragon of Japanese myth, killed by Susano, god of storms.

## 8 ½

- Federico Fellini's self-indulgent but brilliant 1963 movie about making a movie, starring Marcello Mastroianni, Claudia Cardinale, and Anouk Aimee. The title comes from a tally of Fellini's films: Previous to this, he had directed six full-length pictures and three short pictures (the equivalent of one and a half), which made a total of seven and a half; adding this one, he reached the total of the title.

## 009

- Decimal ASCII code for [horizontal tab].

## 09

- number of the French Departement of Ariège.

## 9

- according to Benet (*The Reader's Encyclopedia*), Lars Porsena, king of Clusium, Etruria, swore by the nine gods: Juno,

Minerva, and Tinia (the three main), Vulcan, Mars, Saturn, Hercules, Summanus, and Vedius.

- atomic number of the chemical element fluorine, symbol F.
- Beaufort number for wind force of 47 to 54 miles per hour, called "strong gale–severe gale."
- candles in a Hanukkah menorah (used on the Jewish Festival of Lights).
- CB radio channel reserved for emergency use.
- considered an unlucky number ("ku") in Japanese culture, but a good number in Chinese because it sounds the same as the Chinese word for "long-lasting" (in pinyin, "jiu").
- dental chart designation for the left central incisor.
- gauge (number) of a wood screw with a shank diameter of 11/64 of an inch.
- in astrology, the number 9 is ruled by the planet Uranus.
- in baseball, 9 represents the right fielder's position.
- inches in a quarter (cloth measure).
- inches in a span (English unit of length).
- in Hebrew, the numerical value of the letter *teth*.
- in mineralogy, Mohs Hardness Scale level characterized by corundum; can be scratched by diamond.
- most goals scored by one team in one period of an NHL game (hockey record): by the Buffalo Sabres, second period, March 19, 1981.
- most penalties served in one period of an NHL game (hockey record): by Randy Holt for the Los Angeles Kings against the Philadelphia Flyers, March 11, 1979.
- nine days:
  - ⇨ in Greek mythology, the length of time Deucalion and his wife Pyrrha were adrift at sea during the great flood, until their ark landed on Mount Parnassus.
  - ⇨ in superstition, the abracadabra is worn nine days, then tossed in a river.

- nine days of "pease porridge in the pot," in the children's nursery rhyme:

  > Pease porridge hot, pease porridge cold,
  > Pease porridge in the pot, nine days old.
  > Some like it hot, some like it cold,
  > Some like it in the pot, nine days old.

- nine feet:
  - ⇨ average depth of the Grand Canal in Venice.
  - ⇨ length of a pool table used in professional matches.
  - ⇨ length of a regulation table tennis surface.

- nine meters (= 29-½ feet): Width of a regulation volleyball court.
- nine seconds: Shortest overtime in an NHL playoff game (hockey record). On May 18, 1986, Brian Skrudland for the Montreal Canadiens scored against the Calgary Flames in the first OT.
- nine things of which Great Man must be mindful (according to Confucius): to see when he looks, to hear when he listens, to have a facial expression of gentleness, to have an attitude of humility, to be loyal in speech, to be respectful in service, to inquire when in doubt, to think of the difficulties when angry, to think of justice when he sees an advantage.
- nine years: average life of a hundred dollars or a fifty dollars bill.
- ninth hour, when Jesus cried out, "My God, my God, why hast thou forsaken me?" (Matthew 27:46).
- number of Beethoven symphonies; an early work, the "Jena" symphony, is attributed to him, but is not counted among the nine major works.
- number of both rows and columns on the playing board for the game of Shogi (Japanese chess).
- number of Carolingian rulers of France: Charles II (the Bald) in 840; Louis II (the Stammerer), 877; Louis III, 879–82, jointly with Carloman 879–84; Charles (the Fat), 884–88; [Odo of Paris (Robertian, non-Carolingian), 888]; Charles III (the Simple), 898; Robert I, 922; Raoul, 923; Louis IV (d'Outremer), 936; Lothar, 954; and Louis V (the Indolent), 986–87. (The Capetian Dynasty followed.)
- number of charms granted by the witch Groa to her son Svipdag, according to Norse mythology.
- number of decimal places in the inverse multiple designated by the International System prefix nano-, written "n-".
- number of heads on the hydra that Hercules had to destroy.
- number of hobs (pegs) in the game of quoits.
- number of innings in a regulation baseball game.

- number of presidential electoral votes apportioned to each of the states of Alabama, Colorado, and Louisiana. (May be reapportioned for the 2012 election.)
- number of seats in the House of Representatives allotted to each of the states of Indiana, Missouri, Tennessee, and Washington (as of the 2000 Census).
- number of Supreme Court judges; increased from seven by Act of Congress, March 3, 1837.
- number of underworlds in the Mayan religion, ruled by nine lords of the night.
- number of virgin priestesses (Gallicenae) of the ancient Gallic oracle.
- number of 0s in a billion (American) or a milliard (British).
- number of 0s in the multiple designated by the International System prefix giga-, written "G-".
- players on a baseball team or softball team.
- possession is reputed to be nine points of the law (the other points are not mentioned).
- proverbial number of a cat's lives.
- record number of prime-time Emmys won: by Carl Reiner, as both writer and actor, from 1957 for *Caesar's Hour*, to 1995 for *Mad About You*; the most (five) for *The Dick Van Dyke Show* in the 1960s.
- record number of hockey Olympic gold medals: won by Canada, in 1920, 1924, 1928, 1932, 1948, 1952, and 2002 by men's teams; 2002 and 2006 by women's teams.
- record number of Wimbledon singles tennis titles, men's or women's: won by Martina Navratilova, 1978–79, 1982–87, and 1990.
- retired baseball uniform number of:
  - ⇨ outfielder Reggie Jackson, retired by the Oakland Athletics. Jackson's number 44 has also been retired by the New York Yankees.
  - ⇨ outfielder Roger Maris, retired by the New York Yankees.
  - ⇨ second baseman Bill Mazeroski, retired by the Pittsburgh Pirates.
  - ⇨ outfielder Minnie Minoso, retired by the Chicago White Sox.
  - ⇨ outfielder Enos Slaughter, retired by the St. Louis Cardinals.
  - ⇨ outfielder Ted Williams, retired by the Boston Red Sox.

- retired basketball jersey number of:
  - ⇨ Dan Majerle, retired by the Phoenix Suns.
  - ⇨ Bob Pettit, retired by the Atlanta Hawks.

- retired hockey shirt number of:
  - ⇨ Glenn Anderson, retired by the Edmonton Oilers.
  - ⇨ Andy Bathgate, retired by the New York Rangers.
  - ⇨ John Bucyk, retired by the Boston Bruins.
  - ⇨ Clark Gillies, retired by the New York Islanders.
  - ⇨ Adam Graves, retired by the New York Rangers.
  - ⇨ Gordie Howe, retired by both the Detroit Red Wings and the Hartford Whalers.
  - ⇨ Bobby Hull, retired by both the Chicago Blackhawks and the Winnipeg Jets.
  - ⇨ Lanny McDonald, retired by the Calgary Flames.
  - ⇨ Maurice Richard, retired by the Montreal Canadiens.

- a sacred number to the ancients, related to the number of months required to give birth to a baby.
- sides on a nonagon (or, to a mathematician, enneagon).
- slang for 9mm gun.
- square feet in a square yard.
- the square of 3.
- sum of the first two cubes $(1 + 8)$.
- Title of the California Penal Code, Part I, that deals with offenses against public morals and decency.
- Title of the United States Code dealing with arbitration.
- value of Roman numeral IX.

### Article 9

- the "no war" clause in the Japanese constitution of 1946 that renounces the threat or use of force in the furtherance of its foreign policy.

### back nine

- refers to holes 10 through 18 on an 18-hole golf course.

### Beethoven's Ninth Symphony

- "Choral" symphony (in D minor).

## Canon 9 of the American Bar Association Code of Professional Responsibility

- a lawyer should avoid even the appearance of professional impropriety.

## cat-o'nine-tails

- a whip made of nine strips of leather each knotted at the end and fastened to a handle, named so because the scars it produces look like scratches of a cat.

## Catonsville Nine

- nine Catholic antiwar activists associated with the Interfaith Peace Mission who entered the Selective Service office in Catonsville, Maryland, on May 17, 1968, and proceeded to burn draft files, demanding that America reassess its policy in Vietnam. The nine—Father Philip Berrigan (himself a veteran), his brother Daniel Berrigan (a Jesuit priest), Tom Lewis, David Darst, Mary Moylan, John Hogan, George Mische, and Tom and Marjorie Melville—were sentenced to federal prison for terms of one to three years.

## cloud nine

- a place of perfect happiness; usually in the phrase "on cloud nine."

## Curse of Scotland

- familiar name for the nine of diamonds.

## DC-9

- Douglas Aircraft passenger plane designed for short runways and short- to medium-range routes. First flown in 1965, the twin-engine craft, smaller than the DC-8, has a cruising speed exceeding 500 mph and altitudes over 30,000 feet. Production ended in 1982, but a large number remain in service. Economy class has five seats per row, unlike the more-common six-seat configuration in other single-aisle jetliners.

## divisibility by 9

- to determine if a number of any length is divisible by 9: Add the several digits sequentially until only one digit remains. If that last digit is a 9, the original number is divisible by 9.

## Dvorak's Ninth Symphony

- "From the New World" symphony (in E minor).

## Earlier Nine Years' War

- marks the 1051–1062 (sic!) destruction of the powerful Abe family, a military clan of northern Japan, by Minamoto Yoriyoshi and his son Yoshiiye, on imperial command. The Abe family had collected taxes and confiscated property in Mutsu province without authority. By his victory, Minamoto gained prestige for himself and his clan, and in 1083 Yoshiiye was appointed governor of Mutsu.

## the Feynman point

- in the decimal expansion of pi, the digit 9 appears six times consecutively in the 762nd through 767th places.

## front nine

- refers to holes 1 through 9 on an 18-hole golf course.

## Group 9 elements (of the Periodic Table)

- cobalt (27), rhodium (45), iridium (77), meitnerium.

## heraldic crowns

- the nine crowns of heraldry (as per Brewer's *Dictionary of Phrase and Fable*):
  1. the oriental;
  2. the triumphal or imperial;
  3. the diadem;
  4. the obsidional;
  5. the civic;
  6. the vallary;
  7. the mural;
  8. the naval;
  9. the celestial.

## Little Rock Nine

- name given to the nine black high school students who made national news while trying to integrate Central High School in Little Rock, Arkansas, in 1957. The Supreme Court had ruled in *Brown v. Board of Education* (1954) that segregation in public schools was unconstitutional, but Arkansas governor Orval Faubus defied that ruling and activated the National Guard to prevent the nine from entering the school building. After several more unsuccessful attempts, President Dwight Eisenhower federalized the Guard and ordered them to escort the students into the school, which they did on September 25, 1957, and several days thereafter. Their names were Minnijean Brown, Elizabeth Eckford, Ernest Green, Thelma Mothershed, Melba Patillo, Gloria Ray, Terrence Roberts, Jefferson Thomas, and Carlotta Walls.

## "make up nine . . ."

- from Shakespeare's *Macbeth*, Act I, Scene III: "Thrice to thine, thrice to mine, and thrice again to make up nine . . ."

## marks of cadency

- in heraldry, nine symbols used to designate the several sons of a family in order of age: Label (the first born), crescent (second), mullet (third), martlet (fourth), annulet (fifth), fleur-de-lis (sixth), rose (seventh), cross moline (eighth), double quatrefoil (ninth).

## *Nine*

- a five-Tony-winning musical by Arthur Kopit and Maury Yeston, adapted from the Fellini autobiographical film, *8 ½* Debuted May 9, 1982, directed by Tommy Tune, with Raul Julia, Karen Akers, Liliane Montevecchi, and Anita Morris. Closed in February 1984, but was revived in April 2003, with Antonio Banderas, Jane Krakowski, Mary Stuart Masterson, and Chita Rivera. The revival ran for ten months. Transmogrified to film in 2009, to mixed reviews; directed by Rob Marshall, with Daniel Day-Lewis, Marion Cotillard, Judi Dench, Penelope Cruz, Nicole Kidman, and Sophia Loren.

## 9 April

- Appomattox Day; commemorates the ending of the Civil War, when Lee surrendered to Grant at Appomattox in 1865.

## 9 August 1974

- the day on which President Richard M. Nixon resigned as President of the United States. The resignation became effective the moment that Secretary of State Henry Kissinger initialed Nixon's letter at 11:35 AM that morning.

## nine ball

- a variant game of pool, in which the players alternately sink the numbered balls in sequence. Whoever sinks the nine ball wins, unless it is pocketed before the other eight.

## *Nine Chapters of the Mathematical Art*

- an anonymous Chinese mathematics text dating from about the time of Christ, the earliest to have survived from China. Offered practical solutions to problem solving rather than deducing propositions, as did Greek mathematics.

## nine circles

- the regions of Hell in Dante's *Inferno*: Limbo, wherein reside the virtuous (pagans and unbaptized infants), then circles of the lustful, the gluttonous, the avaricious, the wrathful (and sullen), the heretical, the violent, the fraudulent, and the treacherous.

## Nine Classics

- at times in Chinese history these nine were considered to be the canonical books of Confucianism: *Book of Changes, Book of History, Book of Odes, Zuo's Commentary on Spring and Autumn Annals, Gongyáng's Commentary on "Spring and Autumn Annals," Guliáng's Commentary on "Spring and Autumn Annals," Book of Rites, Book of Ritual,* and *Rites of Chou.*

## the nine-day queen

- refers to the reign of sixteen-year-old Lady Jane Grey, unwillingly proclaimed Queen of England in 1553 as successor to Edward VI, resulting in a struggle for the throne in which Mary I won the crown. Jane Grey was imprisoned and later beheaded.

## nine days' wonder

- something or someone that causes fascination or excitement for a short time and is then quickly forgotten.

## Nine Degree Channel

- a passage through the southern end of the Laccadine Islands.

## nine elements of virtue

- to a Buddhist: forebearance, great mercy, great compassion, wisdom, mindfulness, resolute mind, absence of greed, absence of anger, absence of stupidity.

## 9/11

- appellation for the September 11, 2001 concerted suicide attacks by al-Qaeda on the United States. Terrorists hijacked four commercial aircrafts, crashed two into the Twin Towers of New York City's World Trade Center and one into the Pentagon in Arlington, Virginia. Passengers on the fourth, immortalized as Flight 93 (q.v.), heard reports of the other incidents and took control of their plane, which was intended to hit at the White House, and crashed it near Shanksville, Pennsylvania. Almost 3,000 people, including the nineteen hijackers, died in the attacks.

## Nine Elms

- an industrial district of southwest London, between Battersea and Vauxhall; site of the New Covent Garden Market and the Battersea Power Station.

## nine External Realities

- according to the Vaisheshika system of Indian philosophy, there are nine external realities (*dravyas*) with special properties that make possible our learning the known phenomena of Nature: earth (*prthivi*), water (*apas*), fire (*tejas*), air (*vayu*), ether (*akasa*), time (*kala*), space (*dik*), soul (*atman*), and mind (*manas*).

## 9 February 1964

- the Beatles appear on *The Ed Sullivan Show*.

## nine gates of hell

- According to Milton's *Paradise Lost*, "thrice three-fold; three folds are brass, three iron, three of adamantine rock."

## nine gods of Heliopolis

- the Ennead: Egyptian dieties involved in the myth of creation: Re-Atum, who spat out Shu (the god of air) and Tefnut (the god of moisture), who together begat Geb (the earth god) and Nut (the sky goddess), who in turn begat two pair of twins; one, Osiris and Isis, the other, Set and Nephthys.

## nine gods of the Etruscans

- Juno, Minerva, Tinia (the three main gods), and Vulcan, Mars, Saturn, Hercules, Vedius, and Summauus.

## nine gods of the Sabines

- Hercules, Romulus, Aesculapius, Bacchus, Aeneas, Vesta, Santa, Fortuna, and Fides.

## nine groups of indigenous gods and demons

- in the Tibetan Buddhist pantheon: gNod-sbyin (believed to cause epidemics), bDud (evil-tempered demons), Srin-po (man-eating giants), Klu (serpent dieties of the underworld), btsan (large group of gods inhabiting nature), Lha (pre-Buddhist celestial beings, generally benevolent), dMu (evil demons), Dre (malign beings who often cause mortal diseases), Gan'dre (other harmful beings). (R. Cavendish, ed., *An Illustrated Encyclopedia of Mythology*.)

## Nine Heavens Above Asgard

- in Nordic mythology, according to the "Gylfaginning," nine realms exist above Asgard: Vindlblain (Darkwind), Hregg-Mimir (Storm-Mimir), Andlang, Vidblain (Dark Distance), Vidfedmir (World-Encompasser), Hriod (Concealer),

Hlymir (Twinlight), Gimir (Fiery), Vet-Mimir (Winter-Mimir), and Skatymir.

## 9-iron

- golf club equivalent to the earlier niblick or baffing spoon.

## nine knots

- in folklore, nine knots made on black wool served as a charm for a sprained ankle.

## "nine ladies dancing"

- the gift on the ninth day of Christmas in the carol "The Twelve Days of Christmas."

## Nine Lords of the Dark

- gods of the night in the Toltec tradition (along with thirteen gods of the day). In the religion of the Mayans, the nine lords of the underworld were known as the *Bolon ti ku*

## nine maidens

- in Celtic myth, the nine maidens whose breath warmed the cauldron of Cerridwen, which produced a brew that imparted inspiration. The poem, "The Spoils of Annwn," by Taliesin, describes how the breath of the nine maidens kindled the fire beneath the cauldron.

## *Nine Men*

- effective 1943 British war film, written and directed by Henry Watt, telling the story of nine men who fight off enemy forces until help arrives. Inspired the Hollywood film *Sahara*.

## Nine Men's Morris

- a board game with elements of both checkers and go, played with nine pieces on each side. The game, which is referred to in Shakespeare's *A Midsummer Night's Dream*, dates back to ancient Egypt.

## Ninemile

- a town in Alaska, about 100 miles northeast of Ketchikan.

## Ninemilehouse

- a town in County Tipperary, Ireland, about 25 miles southwest of Kilkenny.

## Nine Mile Lake

- (a) a body of water northeast of the town of Parry Sound in Ontario, Canada.
- (b) a body of water north of Wilcannia in northwest New South Wales, Australia.

## Ninemile Peak

- a mountain in Central Nevada.

## ninemsn

- Australia's main interactive media company, providing news, information, and communication services; a joint venture between Microsoft Company and Australia's Publishing and Broadcasting Limited (PBL). Delivers content from the group of Nine Network television programs and Australian Consolidated Press (ACP) magazines.

## Nine Muses

- Greek goddesses, daughters of Zeus and Mnemosyne, patrons of arts and sciences: Erato, of love poetry; Calliope, of epic poetry; Clio, of history; Euterpe, of music and lyric poetry; Melpomene, of tragedy; Polyhymnia, of sacred poetry and oratory; Terpsichore, of dance and choral song; Thalia, of comedy; and Urania, of astronomy.

## Nine Mythical Emperors of China

- Fu Xi (ca. 2900 BC), Shen Nong (ca. 2800 BC), Xian Yuan (ca. 2700 BC), Jin Tian (ca. 2600 BC), Gao Yang (ca. 2500 BC), Gao Xin (ca. 2400 BC), Yao (ca. 2300 BC), Shun (ca. 2300 BC), and Yu (ca. 2000 BC)—all mythical dates. The first three are known as the Three Sovereigns or Three Emperors, and the last three as the Sage Kings.

## nine orders of angels

- in the first circle of the heavenly hierarchy are the Seraphim, Cherubim, and Thrones; in the second circle are Dominions,

Virtues, and Powers; in the third circle, Principalities, Archangels, and Angels.

## nine-pence pieces

- obsolete British coins, in circulation until 1696.

## nine-penny nail

- a carpenter's common nail, 2- ¾ inches long (coded 9d).

## Ninepin Islands

- island group east of Hong Kong.

## ninepins

- a game of bowls using nine wooden pins, set in a triangle, as the target.

## Nine Point Mesa

- a mountain in southwest Texas.

## 9-point type

- typographer's measure, known as "bourgeois."

## Nine Power Treaty

- a compact among Belgium, China, France, Great Britain, Italy, Japan, the Netherlands, Portugal, and the United States, by which the signatory nations agreed to guarantee the territorial integrity and independence of China while endorsing the "Open Door Policy." The treaty came out of the International Conference on Naval Limitation called by President Harding and held in Washington, D.C., opening on Armistice Day, 1921, and running through February 6, 1922.

## nine principal passions

- anger, decency, courage, admiration, love, hope, esteem, generosity, and joy.

## nine principal passions of Sufis

- mirth, grief, anger, passion, sympathy, attachment, fear, bewilderment, and indifference.

## Nine Principles of the Karma Theory in Jainism

- those required to progress spiritually to liberation: Jiva (soul), Ajiva (non-living substances), Asrava (cause of the influx of karma), Bandh (bondage of karma), Punya (virtue), Papa (sin), Samvara (arrest of the influx of karma), Nirjara (exhaustion of the accumulated karma), and Noksha (total liberation from soul by elimination of all karma).

## nine privy councillors to William III

- four Whigs—Devonshire, Dorset, Monmouth, and Edward Russell, and five Tories—Caermarthen, Pembroke, Nottingham, Marlborough, and Lowther.

## Nine Provinces

- the totality of China in the mythical reign of Emperor Yu, for whom Nine Shepherds brought metal to cast the Nine Cauldrons of the Hsia dynasty.

## *Nine Queens*

- a caper film from Argentina (2000) in which two con artists attempt to swindle a stamp collector by selling him a sheet of counterfeit rare stamps (the nine queens). Memorable for its clever plotting and sharp direction.

## niner

- Australian slang for a woman nine months pregnant.

## nine ranks of mandarins

- mandarins were bureaucrats in imperial China, their grade indicated by the color of the button worn in their dress cap: 1, ruby; 2, coral; 3, sapphire; 4, an opaque blue stone; 5, crystal; 6, an opaque white shell; 7, wrought gold; 8, plain gold; and 9, silver.

## nine realms of Norse mythology

- which emanate from the roots of Yggdrasil, the Tree of Life: Helheim (domain of the dead), Niflheim (the realm of frost), Jotunheim (land of the giants), Nidavellir (land of the dwarfs), Midgard (middle-earth, realm of mankind),

Svartalfheim (domain of the dark elves), Alfheim (land of the light elves), Vanaheim (world of the Vanir), and Asgard (world of the Aesir).

## nine red-hot plowshares

- one of the English versions of trial by ordeal had the accused walk blindfolded across a floor on which nine red-hot plowshares had been placed randomly. If the accused could cross the floor without injury, he was judged innocent of the charges.

## Nine Skies

- a theory in Chinese mythology expressed by Chu Tzu, that the sky is composed of nine levels, stacked vertically, each separated from the next by a gate guarded by panthers and tigers.

## nine spheres

- (in the early Ptolemaic system of astronomy): Thus, in his "Arcades," Milton refers to "nine enfolded spheres," which are those of the Moon, Mercury, Venus, the Sun, Mars, Jupiter, Saturn, the Firmament, and the Crystalline Sphere. The Earth was assumed to be at the center of the arrangement.

## 9 Thermidor

- date in the French Revolutionary calendar (July 27, 1794) on which dissident Jacobins and some members on the Committee of Public Safety, fearing for their own security, arrested Robespierre along with his closest associates on the floor of the legislature. Robespierre and eighty-two followers were guillotined.

## *Nine to Five*

- three secretaries (Jane Fonda, Lily Tomlin, and Dolly Parton) join forces to exact revenge on their chauvinistic boss (Dabney Coleman) in this 1980 romp. This was Parton's first film.

## nine Valkyries

- in his operas *Die Walküre* and *Die Götterdämmerung*, Richard Wagner cites the Valkyries as nine daughters of Wotan and Erda who are named Brynhild, Helmwige, Ortlinde, Gerhilde, Waltraute, Siegrune, Rossweisse, Grimgerde, and Schwertleite.

## The Nine Worthies

- men of note usually grouped together in medieval literature, organized as follows: pagan— Hector, Alexander the Great, Julius Caesar; Jewish—Joshua, David, Judas Maccabeus; Christian—King Arthur, Charlemagne, Godfrey of Bouillon.

## Nine Worthies of London

- a 1592 account by Richard Johnson of nine prominent figures of London: Sir William Walworth, Sir Henry Pritchard, Sir William Sevenoke, Sir Thomas White, Sir John Bonham, Christofer Croker, Sir John Hawkwood, Sir Hugh Caverly, Sir Henry Maleverer.

## Nine Worthy Women

- according to the sixteenth-century woodcuts of Burgkmair: pagan—Lucretia, Veturia, Virginia; Jewish—Esther, Judith, Jahel; Christian—St. Helena, St. Brigita of Sweden, St. Elizabeth of Hungary.

## Nine Years' War

- (a) (aka, War of the Grand Alliance): from 1688 to 1697, with most of the German states united as the League of Augsburg joined by England, Portugal, Sweden, the United Provinces, and Spain, allied to fight French expansion. Despite several French victories, the War was inconclusively ended by the Treaty of Ryswick (September 30, 1697), which pretty much restored the European continent to its prewar condition.
- (b) (aka, Tyrone's Rebellion): the Gaelic Irish against the Elizabethan English government of Ireland, 1597–1601, led by Hugh O'Neill, Earl of Tyrone.

## Ninth Amendment to the Constitution

- states that the Constitution is not to be construed to deny other rights retained by the people.

## Ninth Commandment

- Thou shalt not bear false witness against thy neighbor.

## Ninth Crusade

- in 1271 Prince Edward (later King Edward I of England) led the last Crusade to the Holy Land. Within one year he had concluded a truce and soon returned to England. In 1268 Antioch was retaken by the Muslims; in 1289, Tripoli; and in 1291, Acre, removing the last vestiges of Christian presence in the area.

## Ninth House

- in astrology, deals with knowledge, both learned and disseminated.

## ninth president

- William Henry Harrison, March 4, 1841 to April 4, 1841.

## ninth state of the Union

- New Hampshire ratified the Constitution on June 21, 1788.

## ninth wedding anniversary gift

- pottery is customary.

## nones

- (a) in the ancient Roman calendar, the ninth day before the ides of a month; the seventh of March, May, July, or October, or the fifth day of the other months.
- (b) the ninth hour after sunrise; the time of day devoted to the observance of the fifth of the seven canonical hours (also called the nones).

## nonet

- a musical group composed of nine instruments or nine voices.

**novena**

- a consecutive nine-day period of prayers and devotions in the Catholic Church.

**number 9 billiard ball**

- the ball that is conventionally striped yellow.

**the Pierides**

- nine maidens who challenged the Muses to a singing contest and lost.

*Plan 9 from Outer Space*

- an unintentionally hilarious film about aliens who plan to conquer earth, considered by many critics as the worst film ever made. Bela Lugosi died during filming and was replaced by an actor who played the rest of movie with his face hidden. This 1959 fiasco—the inept creation of Ed Wood, the legendary auteur of the low-budget, low-quality movie—has become a cult classic.

**Schubert's Ninth Symphony**

- "Great" symphony (in C major).

**shortest overtime in NHL Playoff history**

- nine seconds, when Brian Skrusland scored in game two of a Stanley Cup playoff, winning for the Montreal Canadiens over the Calgary Flames, 3–2, on May 18, 1986.

**a stitch in time saves nine**

- a maxim that counsels one to handle problems early before they become difficult to manage.

**to the nines**

- (adv.) ornately, perfect in very detail; usually used to describe the way a person dresses (i.e., "He was dressed to the nines").

**whole nine yards**

- slang phrase meaning everything, all included.

**Zsa-Zsa Gabor's nine husbands**

- Burhan Belge (four years); hotelier Conrad Hilton (five years); actor George Sanders (five years); Herbert Hunter (three years); Joshua S. Cosden, Jr. (one year); Jack Ryan (one year); Michael O'Hara (five years); Felipe de Alba (one day); Frederick, Prinz von Anhalt (married in 1986). The first seven ended in divorce, de Alba was annulled, but she is still married to von Anhalt at this writing.

## 9.5

- Richter scale magnitude of the most powerful earthquake ever read, off the coast of south-central Chile on May 22, 1960.

## 010

- Decimal ASCII code for [line feed].

## 10

- age of the youngest person ever to receive a regular Academy Award: Tatum O'Neal, for her role in the 1973 *Paper Moon*, for which she won Best Supporting Actress. Shirley Temple had been awarded an Oscar at five years old in 1934, but it was an honorary one.
- atomic number of the chemical element neon, symbol Ne.
- baths in a cor (ancient Hebrew measure of liquid capacity).
- Beaufort number for wind force of 55 to 63 miles per hour, called "storm."
- dental chart designation for the left lateral incisor.
- gauge (number) of a wood screw with a shank diameter of 3/16 of an inch.
- gerahs in a beka (half a shekel), ancient Hebrew measure of weight.
- highest score attainable in a gymnastic event.
- highest score on the Apgar test of a newborn's general condition at birth, determined by assessing the baby's heart rate, respiration, muscle tone, reflex, and color.
- in Hebrew, the numerical value of the letter *yod*.

- in mineralogy, Mohs Hardness Scale maximal level, characterized by diamond.
- magnums in a Nebuchadnezzar (champagne measure).
- most consecutive strikeouts pitched (baseball record): by Tom Seaver, New York Mets, against the San Diego Padres, April 22, 1970.
- most field goals scored in NFL Pro Bowl games in a season (football record): by Morten Andersen, for the NFC, 1986–89, 1990, 1992, and 1996.
- most games scoring three or more goals in a season (hockey record): by Wayne Gretzky, twice: 1981–82 (6 three-goal, 3 four-goal, 1 five-goal) and 1983–84 (6 three-goal, 4 four-goal).
- most home runs by a team in a single game (baseball record): by the Toronto Blue Jays against the Baltimore Orioles, September 14, 1987.
- most kickoff returns in an NFL game (football record): twice, by Desmond Howard for the Oakland Raiders against the Seattle Seahawks, October 26, 1997, and by Richard Alston for the Cleveland Browns against the Cincinnati Bengals, November 28, 2004.
- most Olympic gold medals won by an American in track and field: Ray Ewry, from 1900 to 1908. Two of the medals were won at the Intercalated Games of 1906, which, because they fell outside the usual four-year Olympic cycle, are not now considered part of the official record. His medals were won in events no longer in the Olympic program: standing high jump, standing long jump, and and standing triple jump.
- most overtime wins in a hockey playoff year: by the Montreal Canadiens in 1993.
- most penalties served in one NHL game (hockey record): by Chris Nilan for the Boston Bruins against the Hartford Whalers, March 31, 1991.
- most punt returns to touchdowns, lifetime (football record): by Eric Metcalf for the Cleveland Browns, 1989–94; then six different teams, 1989–2002.
- most seasons as leading scorer in NBA (basketball record): Michael Jordan for the Chicago Bulls, 1987–93 and 1996–98.

- most steals in an NBA playoff game (basketball record): by Allen Iverson for the Philadelphia 49ers against the Orlando Magic, May 13, 1999.
- most team touchdowns scored in a game (football record): three times, by the Philadelphia, Eagles against the Cincinnati Bengals, November 6, 1934; the Los Angeles Rams against the Baltimore Colts, October 22, 1950; the Washington Redskins against the New York Giants, November 27, 1966.
- most World Series games won by a pitcher (baseball record): by Whitey Ford, New York Yankees, 1950–64. Ford pitched in twenty-two games in 11 series.
- numeric base of the decimal system.
- number of arms on a squid.
- number of counties in the state of New Hampshire.
- number of Earth hours in a day on the planet Jupiter.
- number of the French Département of Aube.
- number of presidential electoral votes apportioned to each of the states of Arizona, Maryland, Minnesota, and Wisconsin. (May be reapportioned for the 2012 election.)
- number of seasons Wayne Gretzky won the Art Ross Trophy for leading the league in scoring (hockey record): seven in a row for the Edmonton Oilers, 1980–81 and 1986–87; and three for the Los Angeles Kings, 1989–90, 1990–91, and 1993–94.
- number of seats in the House of Representatives allotted to the state of Massachusetts (as of the 2000 Census).
- omers in an ephah (ancient Hebrew measure of dry capacity).
- players on a lacrosse team.
- point value of the Q or the Z in the game of Scrabble.
- record number of points scored by a player in a professional hockey game: twice, by Jim Harrison with three goals and seven assists for the Alberta Oilers (later the Edmonton Oilers) in a WHA match, January 30, 1973; and by Darryl Sittler, with 6 goals and 4 assists for the Toronto Maple Leafs against the Boston Bruins in an NHL match, February 7, 1976.
- retired baseball uniform number of:
  ⇨ manager Sparky Anderson, retired by the Cincinnati Reds.
  ⇨ both outfielder Andre Dawson and first baseman/outfielder Rusty Staub, retired by the Montreal Expos.
  ⇨ manager Dick Howser, retired by the Kansas City Royals.

    ⇨ shortstop Phil Rizzuto, retired by the New York Yankees.
    ⇨ third baseman Ron Santo, retired by the Chicago Cubs.

- retired basketball jersey number of:
    ⇨ Maurice Cheeks, retired by the Philadelphia 76ers.
    ⇨ Walt Frazier, retired by the New York Knicks.
    ⇨ Tim Hardaway, retired by the Miami Heat.
    ⇨ Bob Love, retired by the Chicago Bulls.
    ⇨ Nate McMillan, retired by the Seattle SuperSonics.
    ⇨ Earl Monroe, retired by the Washington Wizards.
    ⇨ Jo Jo White, retired by the Boston Celtics.

- retired football jersey number of:
    ⇨ Steve Bartowski, retired by the Atlanta Falcons.
    ⇨ Fran Tarkenton, retired by the Minnesota Vikings.

- retired hockey shirt number of:
    ⇨ Alex Delvecchio, retired by the Detroit Red Wings.
    ⇨ Ron Francis, retired by the Carolina Hurricanes.
    ⇨ Dale Hawerchuk, retired by the Phoenix Coyotes.
    ⇨ Guy Lafleur, retired by the Montreal Canadiens.

- sides on a decagon.
- soccer jersey number of Pelé, with the Santos Futebol Clube, favored by soccer fans around the world.
- square chains in an acre.
- sum of the first three prime numbers $(2 + 3 + 5)$.
- sum of the first four integers $(1 + 2 + 3 + 4)$.
- ten feet:
    ⇨ height of the crossbar on the football goalposts.
    ⇨ height of the ring on a basketball backboard.
- ten pounds of water in a gallon.
- tenth day of the seventh month shall be a day of atonement (Leviticus 23:27).
- ten years: prison time served by Eugene V. Debs, Socialist Party leader and presidential candidate, for violating the Espionage Act in a speech in Canton, Ohio, in June 1918.
- Title of the United States Code dealing with the armed forces.
- value in the game of pinochle for the nine of trumps, called the deece.
- value of Roman numeral X.
- years in a decade.

### Beethoven's Tenth

- conductor Hans von Bulow's name for Brahms' first symphony (in C minor).

### Big Ten Conference

- an association of eleven schools sharing the same educational policies and goals and each featuring within its programs a focus on intercollegiate athletics. Founded in 1896, the member universities are Illinois, Indiana, Iowa, Michigan, Michigan State, Minnesota, Northwestern, Ohio State, Penn State, Purdue, and Wisconsin.

### Council of Ten

- a secret tribunal formed in Venice in 1310, ostensibly to safeguard the republic. Later expanded to seventeen—the doge and sixteen members—it soon became the supreme authority on both internal and external affairs.

### count to ten

- take time to cool down.

### DC-10

- a wide-cabin three-engine jetliner from Douglas Aircraft, in service since 1971. Various models are designed for air routes up to 6,000 miles, with passenger capacity up to 380 in all-economy seating.

### decathlon

- men's competition of ten events. Day 1: 100-m dash, long jump, shot put, high jump, and 400-m dash; Day 2: 110-m hurdles, discus, pole vault, javelin, and 1,500-m running event.

### dime

- ten-cent piece; U.S. coin in circulation since 1796.

### dime bag

- ten dollars' worth of narcotics (underworld slang).

### eagle

- ten dollars gold piece, minted by the U.S. from 1795 to 1933.

**etto**

- a measure of weight used in Italy equal to 10 grams.

## *Force 10 from Navarone*

- 1978 action film has Harrison Ford, Robert Shaw, Edward Fox, and Franco Nero behind enemy lines on a mission to blow up a bridge vital to the Nazis.

## Group of Ten

- eleven industrially advanced nations whose banks cooperate to regulate international finance: Belgium, Canada, France, Germany, Italy, Japan, Netherlands, Sweden, Switzerland, United Kingdom, United States.

## Group 10 elements (of the Periodic Table)

- nickel (28), palladium (46), platinum (78), darmstadtium (110).

**hang ten**

- (surfing slang): surfing; an allusion to the ten toes gripping the board.

## The Hollywood Ten

- a group of American screenwriters and directors who in 1947 were called before the House Un-American Activities Committee (HUAC) investigating communist influence in Hollywood. Citing the First Amendment, they rejected the Committee's constitutional authority and refused to testify. The ten were Alvah Bessie, Herbert Biberman, Lester Cole, Edward Dmytryk, Ring Lardner Jr., John Howard Lawson, Albert Maltz, Samuel Ornitz, Adrian Scott, and Dalton Trumbo. They were cited for contempt of Congress, and after several unsucessful appeals, they were imprisoned and blacklisted.

## I-10

- major east-west interstate highway fom Los Angeles to Phoenix, El Paso, San Antonio, Houston, Baton Rouge, New Orleans, Mobile, Tallahassee, and Jacksonville.

## KC-10

- military variant of the Douglas DC-10 jetliner, the Air Force KC-10 is an aircraft used for aerial refueling and cargo transport.

## *The Last Ten Days*

- a chilling depiction of what it might have been like in Hitler's bunker in the final days of the war, as the Third Reich crumbled. A 1956 German film by director G. W. Pabst.

## manipuraka chakra

- in yoga, the ten-petal lotus.

## minyan

- ten men, minimum number required to conduct a Jewish communal prayer service.

## number 10 billiard ball

- the ball that is conventionally striped blue.

## number 10 wood screw

- a screw with a screw shank diameter of 3/16 inches.

## perfect 10

- describes a woman who exhibits all the ideal qualities of beauty.

## "playmates ten"

from the nursery rhyme (a mnemonic poem to help children learn Roman numbers):

> X shall stand for playmates ten,
> V for five stalwart men,
> I for one, as I'm alive,
> C for hundred, and
> D for five (hundred),
> M for a thousand soldiers true, and
> All these figures, I've told you.

**take ten**

- take a short rest from work or from any other activity.

**10**

- Dudley Moore is a wealthy, successful songwriter, suffering the longings of middle age, when he spots the woman who he thinks can satisfy his unfilled desires. But he discovers he wants more out of a relationship than just sex, and returns to marry his longtime girlfriend, Julie Andrews. Blake Edwards directed this 1979 comedy that offers cogent observations on maleness.

## *10*, *Admiral Grove*

- childhood home of Beatle Ringo Starr, in Liverpool, England.

**ten Animals in Muslim Heaven**

*1.* the ram sacrificed by Abraham;
*2.* Solomon's ant;
*3.* the lapwing of Balkis;
*4.* the camel of the Prophet Saleh;
*5.* Balaam's ass;
*6.* the ox of Moses;
*7.* Jonah's whale;
*8.* the dog Katmir of the Seven Sleepers of Ephesus;
*9.* Muhammad's steed Al Borak;
*10.* Noah's dove.

**ten-armed Hindu goddess**

- Durga, Hindu goddess of fertility, who vanquished the buffalo-demon Mahish.

**ten Attic orators**

- the ten orators included in the Alexandrian Canon: Aeschines, Andocides, Antiphon, Demosthenes, Dinarchus, Hyperides, Isaens, Isocrates, Lysias, and Lycurgus.

**ten avatars (incarnations) of the Indian god Vishnu**

- Narasimha, the man-lion; Kurma, the tortoise; Matsya, the
  fish; Varaha, the boar; Rama, a prince; Vamana, the dwarf;
  Kalkin, a warrior; Parashurama, Rama with an ax; e; Krishna,
  cow-herd; and the Buddha.

**ten-carat**

- (adj.) large, outstanding, especially noticeable; usually used
  with a negative quality.

## Ten Commandments

- the principles of the supreme law as given to Moses by God
  on Mount Sinai:
  1. Thou shalt have no other gods before Me.
  2. Thou shalt make no graven images.
  3. Thou shalt not take the name of the Lord in vain.
  4. Thou shalt keep holy the Sabbath day.
  5. Thou shalt honor thy father and thy mother.
  6. Thou shalt not murder.
  7. Thou shalt not commit adultery.
  8. Thou shalt not steal.
  9. Thou shalt not bear false witness against thy neighbor.
  10. Thou shalt not covet thy neighbor's house or his wife
  or anything that is thy neighbor's.

The Ten Commandments appear in two places in the Old
Testament: Exodus 20:1–17 and Deuteronomy 5:6–21.

### *The Ten Commandments*

- the biblical story brought to the screen twice, both times
  by Cecil B. DeMille: First, a silent film of 1923 strangely
  composed of two parts, the first recounting episodes from
  the Bible, the second a contemporary melodrama, a morality
  tale of two sons, one reverent, the other corrupt. Second,
  a large-scale, ornate 1956 spectacular tracing the story of
  Moses from birth to eventual sanctuary in the Promised
  Land. Charlton Heston is a commanding Moses. This was
  DeMille's last film, a true epic.

## ten corn maidens

- in Zuni legend, ten corn maidens appear in the spring, bringing food and offerings for rituals and ceremonies.

## ten-day week

- calendar change instituted in France during the French Revolution in 1792. There were three weeks in a month; each day had ten hours, each hour one hundred minutes, each minute one hundred seconds.

## *Ten Days that Shook the World*

- John Reed's eye-witness account of the ten most critical days of the Russian Revolution, a significant record of how the Bolsheviks came into power. Reed's narrative was the basis of a 1928 Russian film by Sergei Eisenstein, known both as *October* and as the original book title.

## 10 December

- Wyoming Day, celebrating the granting of suffrage rights to women in the Wyoming Territory on this day in 1869.

## Ten Degree Channel

- passage between the Andaman and Nicobar Islands, south of Burma.

## ten dollars

- denomination of U.S. paper money that bears the portrait of Alexander Hamilton.

## 10, Downing Street

- London residence of the British prime minister.

## the ten epithets of the Buddha

1. Tathagata (one who has come from Thusness)
2. Arhat (one worthy of receiving alms)
3. Samyaksambuddha (fully enlightened)
4. Vidya-carana-sampanna (one having wisdom and practice)
5. Sugata (well-gone; one who has attained emancipation)
6. Lokavid (the knower of the world)

7. Anuttara, (the unsurpassed)
8. Purusa-damya-sarathi (the tamer of gods and men)
9. Shasta deva-manusyanam (the teacher of gods and men)
10. Buddha-lokanatha (the enlightened and world-honored one).

**Ten Evil Acts**

• As defined in the Buddhist Kusha Ron: three physical evils—killing, theft, and adultery; four verbal evils—lying, flattery, slander, and duplicity; and three mental evils—greed, anger, and stupidity.

**ten-four**

• CB and police band radio code for "Acknowledged; message received."

**tenner**

• slang for a ten-dollar bill.

**ten "fetters"**

• in Buddhism, the ten impediments that must be overcome in order to achieve correct understanding of the beyond. Five objective bonds, namely: the desire for life in form, the desire for formless life, spiritual pride, self-love, and the lingering remnant of ignorance of error; and five subjective bonds, namely: the belief that the body itself is important, uncertainty on the path to liberation, dependence upon outward rules or ceremonies, and likings and dislikings from the past that have become habits in the present. By freeing oneself from all ten, one may achieve that status of "Arhat."

**ten forces of nature**

• Native Americans believe that the Great Spirit has delegated to elemental gods and spirits control over the natural forces: the Sun, Earth, Summer, Winter, Lightning, Rain, and the Four Winds.

## ten-gallon hat

- a broad-brimmed hat with an exceptionally high crown, worn mostly in the western and southwestern U.S.; cowboy hat. Name said to derive from the Spanish word *galon*, a decorative braid that was wrapped several times around the crown above the brim.

## *Ten Great Religions*

- subtitled, "An Essay in Comparative Theology," James Freeman Clarke's 1871 study reflected the increasing awareness of religions other than Christianity, several from other parts of the world. The work was a popular success, producing twenty-one editions in fifteen years.

## ten gauge

- a measure of shotgun bore diameter = 19.67 mm.

## Ten Hills

- a district on the west side of Baltimore, Maryland.

## Ten Hours Act

- Parliamentary act of 1847 that limited working hours by women and adolescents in textile mills to ten hours per day, Monday through Friday, and eight hours on Saturday. New hours were to be phased in, a sixty-three-hour week starting July 1, 1947, then a fifty-eight-hour week starting May 1, 1848. In 1850, however, a subsequent act defined the working day as 6:00 AM to 6:00 PM, with one and a half hours for meals, which returned to a ten and half-hour work day, with a maximum of sixty hours per week.

## ten Indriyas

- (in yoga): The ten organs of the body by which man gains knowledge of the world: the five sensory organs, concerned with seeing, hearing, touching, tasting, and smelling; plus the five action organs, concerned with walking, handling, speaking, procreation, and evacuation.

## 10-iron

- golf club called a wedge.

## Ten Kingdoms of Southern China

- in the period before unification of the empire: Wu, Southern T'ang, Wuyue, Chu, Min, Southern Han, Jinnan, Former Shu, Later Shu, Northern Han.

## *Ten Little Indians*

- (a) a 1939 mystery novel by Agatha Christie, originally titled *And Then There Were None*. Adapted by the author for the stage, where it ran successfully in both England and the U.S.; in 1945 it became a film (with its original title), directed by Rene Clair, with Barry Fitzgerald, Roland Young, Walter Huston, and Judith Anderson. Other film versions appeared in 1965 (with Hugh O'Brian, Shirley Eaton, and the singer Fabian); in 1975 (with Oliver Reed, Elke Sommer, and Charles Aznavour); and in 1989 (starring Donald Pleasence), but the earliest one remains the best.
- (b) a children's song:

> One little, two little, three little Indians,
> Four little, five little, six little Indians,
> Seven little, eight little, nine little Indians,
> Ten little Indian boys.

## "ten lords a-leaping"

- the gift on the tenth day of Christmas in the carol "The Twelve Days of Christmas."

## Ten Mile Lake

- a body of water in Newfoundland, Canada, at the north end of the Long Range Mountains.

## Tenmile River

- a waterway east of Poughkeepsie, New York, near the Connecticut border.

## Ten Mile River

- a town in Sullivan County, New York, about 50 miles west of Middletown, near the Pennsylvania border.

## ten-million-gallon oil spill

- discharge into Prince William Sound in Alaska when the 900-foot oil tanker *Exxon Valdez* hit ground on May 24, 1989, polluting 1,500 miles of coast.

## *Ten Nights in a Bar-Room and What I Saw There*

- melodramatic tale by Timothy S. Arthur (1854) that portrays the destructive influences of a barroom on a town, evil liquor causing the degradation of a good man and the death of his daughter. Became a favorite story for temperance workers, even more so when adapted for the stage by William W. Pratt in 1858. Filmed as a silent film at least a half-dozen times—the 1926 production by the Colored Players Film Corporation, with an all-black cast—and with sound in 1931, starring William Farnum, with Peggy Lou Lynd as his doomed daughter. The word "bar-room" variously appeared as one or two words, with or without a hyphen.

## "a ten o'clock scholar"

- from the nursery rhyme:

  A diller, a dollar, a ten o'clock scholar,
  What makes you come so soon?
  You used to come at ten o'clock,
  But now you come at noon.

## 10 October

- October Historical Day, commemorating the birthday of Pierre Chouteau (1758), "the father" of Oklahoma.

## ten-ounce glass

- the standard size of a beer glass in Australia, having different names in the several states: "Middie" in New South Wales and Western Australia, "pot" in Victoria and Queensland, "schooner" in South Australia, and "handle" in the Northern Territory; in Tasmania, it's simply called a "ten."

## ten-penny nail

- a carpenter's common nail, 3 inches long (coded 10d).

**ten-percenter**

- an agent, who ordinarily gets 10 percent of his client's income.

**Ten Percent Plan of Reconstruction**

- President Lincoln's December 3, 1863, "Proclamation of Amnesty and Reconstruction to the Southern People," in which he offered full pardon to any Southerner, excluding some leaders, who would pledge to support the Constitution and "the union of the States there-under." Those who made the pledge in each state would be permitted to form a new state government on two conditions: that the new government would accept the elimination of slavery in accord with the Emancipation Proclamation of January 1963; and that those voting for the new state government would constitute at least 10 percent of the voters in the 1860 presidential election.

**tenpins**

- another name for the game of bowling, since ten wooden pins provide the target to be knocked down.

**ten plagues of Egypt**

- those wrought upon the Egyptians by God because Pharaoh would not free the Hebrews (Exodus 7:19 to 12:36): water turned to blood, swarms of frogs, epidemic of gnats, flies, murrain, boils, hail, locusts, darkness, death of the firstborn in each Egyptian family.

**10-point card**

- in the game of Canasta, any 8, 9, 10, jack, queen, or king.

**Ten Point Program**

- the political principles intended to direct the PLO, drafted by the Palestine National Council (the PLO's governing body) meeting in Cairo, Egypt, in June 1974.

**Ten Points**

- a program of reform in Hungary, later known as the March Laws, proposed by Francis Deák following the elections of 1847, which brought a new majority of Liberals to the Table of Deputies. The program—including proposals for popular representation, freedom of the press, right of public assembly, religious freedom, universal equitable taxation, the incorporation of Transylvania, and abolition of serfdom with compensation to the landlords—was eventually adopted by the Table of Deputies, effectively making Hungary all but independent of the Habsburg empire.

### 10-point type

- typographer's measure, known as "long primer."

### ten provinces of Canada

- Ontario, Quebec, Nova Scotia, New Brunswick, Manitoba, British Columbia, Prince Edward Island, Saskatchewan, Alberta, Newfoundland/Labrador. Canada also has three territories: Northwest Territory, Yukon, and Nunavut.

### *10 Rillington Place*

- Richard Attenborough in a disturbing 1971 film based on the true story of John Reginald Christie, who murdered several women and helped convict the innocent husband of one of them to be hanged. The case led to the abolition of capital punishment in Britain.

### ten sages

- in Hindu mythology, the Prajapatis, Lords of Creation, from whom humanity is descended: Marichi, Atri, Angiras, Pulastya, Pulaha, Kratu, Vasishtha, Daksha (or Prachetas), Bhrigu, and Narada.

### ten-spot

- (a) a ten-dollar bill.
- (b) (obsolete slang) a ten-year prison sentence.

### 10 Stigwood Avenue, Brooklyn Heights, New York

- residence of Dr. Heathcliff Huxtable and family, on the NBC-TV sitcom, *The Cosby Show*.

## The Ten Talents of Quakerism

1. God's spiritual light, lightening every man;
2. indwelling of the spirit with the discipline;
3. the headship of Christ in the church;
4. the priesthood of all true believers;
5. the freedom of the Gospel ministry;
6. spiritual equality of the sexes;
7. spiritual baptism and spiritual communion only—no sacraments;
8. the unlawfulness of war;
9. the unlawfulness of oaths;
10. the duty of brotherly love and simplicity of life.

## Ten-Ten

- 10-10 International Net, a volunteer organization of radio amateurs dedicated to preserving quality in amateur radio communications in the 10-meter amateur band (28.0–29.7 Mhz).

## Tenth Amendment to the Constitution

- clarifies the rights of states under the document.

## Tenth Commandment

- Thou shalt not covet thy neighbor's house or his wife or anything that is thy neighbor's.

## Tenth House

- in astrology, deals with social standing, career, and reputation.

## tenth president

- John Tyler, April 6, 1841 to March 3, 1845.

## tenth state of the Union

- Virginia ratified the Constitution on June 25, 1788.

## Ten Times One Is Ten

- an 1871 novelette by Edward Everett Hale that examines the influence of a dead man's ghost on his friends.

**tenth wedding anniversary gift**

- tin or aluminum is customary.

**10-V**

- slang for the worst, the lowest (in contrast to 1-A, the best).

**Ten Years' War**

- a Cuban revolt against Spain, 1868 to 1878, fought mostly as a bloody guerrilla war without major battles. Although the rebels were unsuccessful, the revolt presaged the Cuban war of independence, 1895, and the Spanish-American War, 1898, which eventually led to an independent Cuba.

**top ten**

- the ten most popular or highest-rated items in any category, usually for a limited period of time.

**top ten of Beaujolais**

- the ten villages ("crus") that produce the best wines in the region: Brouilly, Chénas, Chiroubles, Côte de Brouilly, Fleurie, Juliénas, Morgon, Régnié, Moulin-à-Vent, and Saint-Amour.

**The Wilmington Ten**

- Ben Chavis, a field organizer for the United Church of Christ's Commission for Racial Justice, and eight other black men and one white woman were convicted in 1972 of fire-bombing a white-owned grocery store in Wilmington, North Carolina. Four years later the witnesses against them recanted their testimony.

**10.5**

- percentage of the labor force in agriculture in 2003, compared with 52 percent in 1870.

**10.6**

- number of Earth hours in a day on the planet Saturn.

# 011

- Decimal ASCII code for [vertical tab].
- Dewey Decimal System designation for bibliographies.
- medical classification code for pulmonary tuberculosis.

# 11

- age at which the youngest pope took office. Benedict IX was installed in 1032.
- atomic number of the chemical element sodium, symbol Na.
- Beaufort number for wind force of 64 to 72 miles per hour, called "violent storm–severe storm."
- dental chart designation for the left canine tooth (cuspid).
- eleven days: length of the ancient Babylonian New Year Festival.
- eleven holes: longest sudden-death golf playoff: in the 1949 Motor City Open, Cary Middlecoff and Lloyd Mangrum were declared co-winners by mutual consent.
- 11 ounces: average weight of the human heart.
- eleven years: approximate periodicity in the sunspot activity cycle.
- most consecutive golf tournaments won on the PGA Tour: by Byron Nelson, from the Four-Ball Tournament at Miami Springs Golf Club, March 8–11, through the Canadian Open, August 2–4, 1945.
- most passes caught in a Super Bowl game (football record): four times, by Dan Ross for the Cincinnati Bengals against the San Francisco 49ers in SB-XVI, 1982; by Jerry Rice for the San Francisco 49ers against the Cincinnati Bengals in SB-XXIII, 1989; by Deion Branch for the New England Patriots against the Philadelphia Eagles, SB-XXXIX, 2005; and by Wes Walker for the New England Patriots, SB-XLII, 2008.
- most punt returns in an NFL game (football record): by Eddie Brown for the Washington Redskins against the Tampa Bay Buccaneers, October 9, 1977.
- most steals in an NBA game (basketball record): twice, by Larry Kenon, for the San Antonio Spurs against the Kansas City Kings, December 26, 1976; and by Kendall Gill for the New Jersey Nets against the Miami Heat, April 3, 1999.

- most touchdown passes in Super Bowl games, career (football record): by Joe Montana for the San Francisco 49ers: SB-XVI, 1982; SB-XIX, 1985; SB-XXIII, 1989; and SB-XXIV, 1990.
- natural, in craps.
- number of heads on Avalokitesvara, Lord of Compassion, patron deity of Tibet.
- number of known rings of the planet Uranus, the brightest known as the epsilon ring.
- number of Muhammad's wives. Or maybe twelve, or thirteen, or sixteen; commentators give various counts. But they all agree that the Prophet's first wife was Khadijah bint Khuwaylid, married in 595 when he was twenty-five, his only wife until she died in 619; and that his next four wives were name Sawda, Aisha, Hafsa, and Zainab. After them, accounts differ.
- number of presidential electoral votes apportioned to each of the states of Indiana, Missouri, Tennessee, and Washington. (May be reapportioned for the 2012 election.)
- number of seats in the House of Representatives allotted to the state of Virginia (as of the 2000 Census).
- number of tales in the Mabinogion, a collection of medieval Welsh tales, several of which were to become part of the Arthurian legend.
- number of the French Département of Aude.
- number of winners of horse racing's Triple Crown—the Kentucky Derby, the Preakness, and the Belmont Stakes: Sir Barton (1919), Gallant Fox (1930), Omaha (1935), War Admiral (1937), Whirlaway (1941), Count Fleet (1943), Assault (1946), Citation (1948), Secretariat (1973), Seattle Slew (1977), and Affirmed (1978).
- players on a football team, soccer team, or cricket team.
- Raquel Welch's uniform number in the roller-derby film, *Kansas City Bomber.*
- retired baseball uniform number of:
    ⇨ shortstop Luis Aparacio, retired by the Chicago White Sox.
    ⇨ shortstop/manager Jim Fregosi, retired by the Anaheim Angels.
    ⇨ pitcher Carl Hubbell, retired by the San Francisco Giants.

⇨ outfielder Paul Waner, retired by the Pittsburgh Pirates.
- retired basketball jersey number of:
  ⇨ Bob Davies, retired by the Sacramento Kings.
  ⇨ Elvin Hayes, retired by the Washington Wizards.
  ⇨ Isiah Thomas, retired by the Detroit Pistons.

- retired football jersey number of Phil Simms, retired by the New York Giants.
- retired hockey shirt number of:
  ⇨ Mike Gartner, retired by the Washington Capitals.
  ⇨ Wayne Maki, retired by the Vancouver Canucks.
  ⇨ Mark Messier, retired by both the New York Rangers and the Edmonton Oilers.
  ⇨ Gilbert Perreault, retired by the Buffalo Sabres.
  ⇨ Brian Sutter, retired by the St. Louis Blues.

- sides on a hendecagon.
- the smallest number requiring three syllables to verbalize.
- Title of the United States Code dealing with bankruptcy.
- value of Roman numeral XI.

### *Apollo 11*
- NASA mission placing the first men on the moon, Neil Armstrong and Buzz Aldrin. Launched July 16, 1969, landing module *Eagle* on the lunar surface on July 20, 4:17:40 EDT; the two astronauts spent two hours and thirty-one minutes out of the vehicle on the lunar surface.

### Chapter 11
- a section of the U.S. bankruptcy code allowing for a company to stay in business while a bankruptcy court oversees a reorganization of the company's debts and payment to its creditors.

### CV-11
- number of the U.S. Navy aircraft carrier, USS *Intrepid*. Converted to a museum at New York City in 1982.

### divisibility by 11
- to determine if a number of any length is divisible by 11: Add the alternate digits into two sums. If the two sums are

equal, the original number is divisible by 11. For example, the number 64893521: 6+8+3+2 = 19; 4+9+5+1 = 19; thus 64893521 is divisible by 11.

## The Eleven

- a panel of public officials (*hoi hendeka*) responsible for administering the state prisons of ancient Athens. Most criminals were brought before them, and they also were in charge of executions.

## 11 Avenue du Président Wilson

- address of the Musée d'Art Moderne de la Ville de Paris (Paris Museum of Modern Art).

## Eleven Basic Passions of the Soul

- a taxonomy by St. Thomas Aquinas, from his "Summa Theologica" (1.81.2), defining (a) passions of the concupiscible appetite: those of love, hate, desire, pleasure/joy, avoidance, and pain/sadness; and (b) passions of the irascible appetite: those of hope, despair, fear, daring, and anger.

## eleven Confederate states

- those seceding from the Union to form the Confederacy: South Carolina, Mississippi, Florida, Alabama, Georgia, Louisiana, Texas, Virginia, Arkansas, North Carolina, and Tennessee.

## the eleven Greek popes

- Evaristus (ca. 100–109), Telesphoros (ca. 125–136), Hyginus (ca 138–142), Anterus (235–236), Sixtus II (257–258), Eusebius (310), Zosimus (417–418), Theodore (642–649), John VI (701–705), John VII (705–707), Zacharius (741–752). Two are of disputed origins: Anacletus (ca. 79–91) and Eleutherius (174–189).

## 11 January

- Christian feast day commemorating the Baptism of Jesus.

## 11 June

- Kamehameha Day in Hawaii; celebrates King Kamehameha I, who conquered the other Sandwich Islands in 1810, becoming ruler of them all and establishing his dynasty.

## Eleven Mile Canyon Reservoir

- a six-mile-long water catchment area in northern Colorado.

## 11 November

- Veterans Day in the U.S., Remembrance Day in Canada; observes the ending of World War I.

## "eleven pipers piping"

- the gift on the eleventh day of Christmas in the carol "The Twelve Days of Christmas."

## Eleven Point River

- through the Ozark Hills in southern Missouri, the state's only federally designated Wild and Scenic River.

## 11-point type

- typographer's measure, known as "small pica."

## eleven rivers that flowed from Hvergelmir

- in Norse mythology, the rivers (Elivagar) emanating from the spring Hvergelmir, "the roaring kettle," in the ice-cold realm of Niflheim: Svöl ("cool"), Gunnthrá ("Battle-defiant"), Fjörn, Fimbulthul ("Loud-bubbling"), Slið ("Fearsome"), Hrið ("Storming"), Sylg, Ylg, Við ("Wide"), Leipt ("Fast-as-lightning") and Gjöll.

## Eleventh Amendment to the Constitution

- by which judicial powers are construed; ratified on February 7, 1795.

## Eleventh Commandment

- Thou shalt not be found out.

## eleventh hour

- last chance to attend to unfinished matters; often implies impending danger or emergency.

## Eleventh House

- in astrology, deals with friendships and humanitarian activities.

**eleventh president**

- James K. Polk, March 3, 1845 to March 3, 1849.

**eleventh state of the Union**

- New York ratified the Constitution on July 26, 1788.

**Eleventh Vow**

- in Buddhism, the Vow of Unfailing Attainment of Nirvana promises that those born in the Pure Land dwell in the Stage of Right Assurance and unfailingly reach Nirvana.

**eleventh wedding anniversary gift**

- steel is customary.

**11 to 1**

- the odds against tossing either a 4 or tossing a 10 in dice; or the odds in poker against filling an inside straight or a straight open at one end.

**11 West 53rd Street**

- New York City address of The Museum of Modern Art (MOMA), the single most important collection of modern art in the United States.

**Eleven Years' Tyranny**

- the rule of Charles I of England without a Parliament, from 1629 to 1640; also known as "Charles' Personal Rule."

**F-11**

- the Grumman Tiger, the U.S. Navy's carrier-based day fighter aircraft.

**Group 11 elements (of the Periodic Table)**

- the noble metals: copper (29), silver (47), gold (79), and roentgenium (111). This group is sometimes called "coinage metals," because they are relatively inert and do not corrode easily.

## MD-11

- planned as a successor to the DC-10 jetliner, the three-engine McDonnell Douglas MD-11 was designed to carry either passengers or freight. In service since early 1991, demand was light, and the last one built was delivered to Lufthansa in February 2001.

## most Oscars won by a film

- three times, most recently in 2003 by *The Lord of the Rings: The Return of the King*, which won in all categories for which it was nominated, including Best Picture. Others were *Titanic* in 1997 and *Ben-Hur* in 1959.

## number 11 billiard ball

- the ball that is conventionally striped red.

## PDD-11

- Presidential Decision Directive 11 deals with Moratorium on Nuclear Testing, issued by President Bill Clinton on July 3, 1993.

## Shostakovich's *Symphony No. 11*

- "The Year 1905" symphony (in G minor).

## Winning Eleven

- soccer video game, popular world-wide, originating in Japan. The American version is called "Winning Eleven International." To the rest of the world, it is "Pro Evolution Soccer."

## *11.1*

- percentage of the U.S. population who are foreign-born, according to the 2000 census.

## *11.8*

- the proportion of the Earth's land surface occupied by South America.

## *11.9*

- number of Earth years in Jupiter's circuit around the sun.

## *012*

- Decimal ASCII code for [form feed].

## *12*

- atomic number of the chemical element magnesium, symbol Mg.
- base of the number system used by the Romans.
- Beaufort number for wind force of at least 73 miles per hour, called "hurricane force."
- bottles in a Salmanazar (champagne measure).
- days of Christmas.
- dental chart designation for the left first premolar (first bicuspid).
- dozen in a gross.
- gauge (number) of a wood screw with a shank diameter of 7/32 of an inch.
- half-tones in a musical scale.
- hours on a clock.
- inches in a foot.
- largest number pronounced in a single syllable.
- logs in a hin (ancient Hebrew measure of liquid capacity).
- months in a year.
- most career touchdowns scored off pass interceptions (football record): by Rod Woodson for four different teams, 1987–2003.
- most home runs by both teams in a single game (baseball record): twice, by the Detroit Tigers (7) against the Chicago White Sox (5), May 28, 1995, and by the Detroit Tigers (6) against the Chicago White Sox (6), July 2, 2002.
- most in-the-park home runs in a season (baseball record): by Sam Crawford for the Cincinnati Reds, 1901.
- most NBA seasons scoring 2,000 or more points (basketball record): by Karl Malone for the Utah Jazz, 1987–97 and 1999–2000.
- most NFL Pro Bowl games played (football record): by Randall McDaniel, for the NFC, 1990–2001.

- most nominations for Best Actress Oscar, Katharine Hepburn, the only four-time winner: for *Morning Glory* (1933), *Guess Who's Coming to Dinner* (1967), *The Lion in Winter* (1968), and *On Golden Pond* (1981).

- most Oscar nominations for a male actor: Jack Nicholson, four for Best Supporting and eight for Best Actor, with three wins: Best Actor for *One Flew Over the Cuckoo's Nest* (1975) and *As Good as it Gets* (1997); and Best Supporting Actor for *Terms of Endearment* (1983).

- most runs batted in, in a nine-inning game: by both Jim Bottomley (St. Louis Cardinals) on September 16, 1924, and Mark Whiten (St. Louis Cardinals) on September 7, 1993.

- most runs batted in, in a World Series (baseball record): by Bobby Richardson for the New York Yankees against Pittsburgh Pirates in the seven-game 1960 Series, for which Richardson has a slugging percentage of .667.

- most three-point field goals in an NBA game (basketball record): twice, by Kobe Bryant for the Los Angeles Lakers against the Seattle SuperSonics, January 7, 2003; and Donyell Marshall for the Toronto Raptors against the Philadelphia 76ers, March 13, 2005.

- most touchdown passes by both teams in an NFL game (football record): New Orleans Saints against the St. Louis Cardinals (six by each team), November 2, 1969.

- number comprising a dozen.

- number of countries in the European Union introducing the euro as the new currency in 2002.

- number of buttons or dialing slots on a telephone keyboard (1 through 9, 0, #, and *).

- number of decimal places in the inverse multiple designated by the International System prefix pico-, written "p-".

- number of gates into the heavenly Jerusalem (Revelation 21:12).

- number of husbands of Martha Jane Canary (1852–1903), eleven of whom died prematurely; thus, her nickname, "Calamity Jane."

- number of laps in the four-horse-chariot races of the ancient Olympic games.

- number of letters in the Hawaiian alphabet: a, e, h, i, k, l, m, n, o, p, u, w.

- number of men (pieces) on each side in the game of checkers.

- number of popes named Pius.
- number of precious stones in the foundation wall of the heavenly Jerusalem; namely, jasper, sapphire, chalcedony, emerald, sardonyx, sardius, chrysolyte, beryl, topaz, chrysoprasus, jacinth, amethyst (Revelation 21:19–20).
- number of presidential electoral votes apportioned to the state of Massachusetts. (May be reapportioned for the 2012 election.)
- number of stars on the flag of the European Union; signifying the founding twelve members, adopted by the Council of Europe on December 8, 1955.
- number of syllables in an alexandrine line of poetry.
- number of the French Département of Aveyron.
- number of years that Ty Cobb (with the Detroit Tigers) had baseball's highest batting average: 1907–15 and 1917–19.
- number of 0s in a trillion (American) or a billion (British).
- number of 0s in the multiple designated by the International System prefix tera-, written "T-".
- ounces troy in a pound troy.
- pints of blood in the average human body.
- players on a Canadian football team.
- playing pieces on each side in the Moorish game Alquerque.
- points in a pica (typographer's measure).
- quarterback sacked most times in an NFL game (football record): three times—Bert Jones for the Baltimore Colts against the St. Louis Cardinals, October 26, 1980; Warren Moon for the Houston Oilers against the Dallas Cowboys, September 29, 1985; and Donovan McNabb for the Philadelphia Eagles against the New York Giants, September 30, 2007.
- record number of U.S. Open tennis doubles titles won by a women's team: Louise Brough and Margaret du Pont, 1942–50, 1955–57.
- record number of women's Wimbledon tennis doubles titles: won by Elizabeth Ryan, 1914, 1919–23, 1925–27, 1930, and 1933–34.
- retired baseball uniform number of third baseman Wade Boggs, retired by the Tampa Bay Devil Rays.
- retired basketball jersey number of:

⇨ Dick Barnett, retired by the New York Knicks.
⇨ John Stockton, retired by the Utah Jazz.
⇨ Maurice Stokes, retired by the Sacramento Kings.

- retired football jersey number of:
  ⇨ John Brodie, retired by the San Francisco 49ers.
  ⇨ Bob Griese, retired by the Miami Dolphins.
  ⇨ Jim Kelly, retired by the Buffalo Bills.
  ⇨ Joe Namath, retired by the New York Jets.
  ⇨ The Fans, "the 12th man," retired by the Seattle Seahawks.

- retired hockey shirt number of:
  ⇨ Sid Abel, retired by the Detroit Red Wings.
  ⇨ Dickie Moore and Yvan Cournoyer, both retired by the Montreal Canadiens.
  ⇨ Stan Smyl, retired by the Vancouver Canucks.

- sides on a dodecagon.
- sum of the first six Fibonacci numbers $(0 + 1 + 1 + 2 + 3 + 5)$.
- Title of the United States Code dealing with banks and banking.
- 12 feet: width of the goal cage on a regulation field hockey playing field.
- 12 seconds: length of the Wright Brothers' successful power-driven flight at Kitty Hawk, North Carolina, on December 17, 1903.
- 12 volts: capacity of most car batteries.
- value of Roman numeral XII.

### Adam 12

- police drama starring Martin Milner, on NBC-TV from September 1968 to August 1975. Produced by Jack Webb, whose *Dragnet* had popularized the genre.

### the Apostles

- the prime disciples of Jesus Christ: Peter, Andrew, James (the son of Zebedee), John, Thomas, James (the son of Alphaeus), Jude (or Thaddeus), Philip, Matthew, Simon, Bartholemew, and Judas Iscariot (replaced by Matthias). Number was symbolic of the twelve tribes of Israel.

## A-12

- the Curtis Shrike, the first single-wing plane to be used in large numbers by the U.S. Army Air Corps in the early 1930s. It was also pressed into service delivering mail during the airmail emergency in 1934.

## "boxcars"

- in craps, when both dice turn up, each with six spots showing.

## CV-12

- number of the replacement aircraft carrier, USS Hornet, commissioned by the U.S. Navy on November 29, 1943. The CV-12 was built to avenge the CV-8, also named Hornet, which was sunk off the Santa Cruz Islands in October 1942.

## dime a dozen

- (adj.) extremely common; readily available.

## *Dirty Dozen*

- a particularly violent war mission film of 1967, in which a group of misfits is trained to blow up a resort used by high-ranking Nazi officers. Directed by Robert Aldrich, with Lee Marvin, Ernest Borgnine, Charles Bronson, John Cassavetes, Robert Ryan, Telly Savalas, Donald Sutherland, and several others well-known actors.

## duodecimo

- (See *12mo*)

## fastest score in an All-Star hockey game

- twelve seconds after the start of an All-Star hockey game in 2008, by Rick Nash for the West.

## Group 12 elements (of the Periodic Table)

- zinc (30), cadmium (48), mercury (80), ununbium (112).

## heraldic crosses

- the twelve crosses of heraldry (as per Brewer's *Dictionary of Phrase and Fable*):
  *1.* ordinary cross;
  *2.* cross humetté or couped;

3. cross urdé or pointed;
4. cross potent;
5. cross crosslet;
6. cross botonné or treflé;
7. cross moline;
8. cross potence;
9. cross fleury;
10. cross paté;
11. Maltese cross (or eight-pointed cross);
12. cross cleché or fitché.

## longest winning streak in NBA Playoff games—twelve games

- this basketball record was set by the San Antonio Spurs in 1999.

## most decorated Olympic runner

- Paavo Nurmi, the "Flying Finn," won twelve medals, the most ever in track and field—nine gold and three silver—between 1920 and 1928.

## most decorated winter Olympic athlete

- Norway's nine-time world-champion skier Bjorn Daehlie won twelve medals for cross-country skiing—eight gold and four silver—between 1992 and 1998.

## most Tonys won by a Broadway show

*The Producers,* written by comedian/writer Mel Brooks, won twelve Tonys on May 7, 2004, surpassing the previous record of ten Tonys to *Hello, Dolly!* in 1964.

## number 12 billiard ball

- the ball that is conventionally striped purple.

## number 12 wood screw

- a screw with a screw shank diameter of 7/32 inches.

## Olympians

- the divine family of Greek gods who ruled from Mount Olympus: Zeus, the supreme god; his brother Poseidon and sisters Demeter and Hestia; Hera, Zeus's wife; Ares, their son; Zeus's other children, Athena, Apollo, Aphrodite, Hermes, and Artemis; and Hera's son Haphaestus. Two important deities are missing from this list, Hades and Persephone; although members of Zeus's family, they reside in the underworld, not on Mount Olympus.

## Quorum of the 12

- in the Church of Jesus Christ of Latter-day Saints (Mormons), the highest ruling council in the administration of the Church, just under the First Presidency. Unanimous decisions by the Quorum of Apostles have authority equal to those of the First Presidency.

## Shostakovich's *Symphony No. 12*

- "To the Memory of Lenin—1917" symphony (in D minor).

## Titans

- the elder gods of Greek mythology, all brother-sister pairs born to Gaea, the Earth goddess, and Uranus, her son and husband; namely Cronus and Rhea, Oceanus and Tethys, Hyperion and Theia, Creus and Mnemosyne, Iapetus and Themis, Coeus and Phoebe. Overthrown by their descendants, the Olympians, in the battle called Titanomachy.

## Twelfth Amendment to the Constitution

- establishes separate ballots for the election of the president and vice president of the U.S., ratified September 25, 1804.

## Twelfth House

- in astrology, deals with the subconscious mind and deep-seated habits.

## twelfth night

- eve of Epiphany; end of the Christmas season.

## *Twelfth Night*

- Shakespeare's comedy on the twins, Sebastian and Viola, separated after a shipwreck off the coast of Illyria, and how they find one another. Film adaptation, transposed to the 1890s, directed by Trevor Nunn in 1996.

**twelfth night cake**

- a traditional spice cake with candied fruits and almond paste filling. Sometimes a coin is added to the topping and used to choose a "king" or "queen" of the festival marking the twelfth day of Christmas.

**twelfth president**

- Zachary Taylor, March 4, 1849 to July 9, 1850.

**twelfth state of the Union**

- North Carolina ratified the Constitution on November 21, 1789.

**twelfth wedding anniversary gift**

- silk or linens are appropriate.

*Twelve Angry Men*

- 1957 film, directed by Sidney Lumet, based on a 1954 Reginald Rose teleplay on *Studio One*, portrays the interpersonal dynamics of a jury deliberating a verdict. Henry Fonda leads an all-star cast, including Lee J. Cobb, E. G. Marshall, Jack Klugman, Ed Begley, etc. A 1997 film, directed by William Friedkin, was produced for cable TV with Jack Lemmon, George C. Scott, Hume Cronyn, Ozzie Davis, and other well-known actors. In yet another life, a Broadway stage adaptation opened in October 2004 and ran for 228 performances until May 2005. Directed by Scott Ellis, the stage version featured Philip Bosco, Boyd Gaines, and Tom Aldredge.

**Twelve Apostles of Ireland**

- students from the monastery of Clonard who, starting in the sixth century, established schools in Ireland, and later, on the Continent.

**12 April**

- Halifax Resolution Day in North Carolina, commemorating their adoption in 1776, which contributed to the acceptance of the Declaration of Independence.

## 12, Arnold Grove

- childhood home of Beatle George Harrison, in Liverpool, England.

## Twelve Articles

- demands made by the peasants in Swabia and Franconia, in 1524, against the social and economic inequalities of feudalism in Germany, including the right to choose their own pastors, abolition of some tithes, abolition of serfdom, the right to hunt and fish freely, impartiality of the courts, and abolition of the death tax.

## 12 bars

- format of a standard blues musical progression, unlike the pop song, which was structured in eight-bar sections.

## twelve causes of misery

- in Buddhism:
  1. ignorance;
  2. activation of the habit-molds in the three worlds of mental, emotional, and physical;
  3. recognition of objects;
  4. the setting up of names and forms;
  5. awakening of the six organs of sense;
  6. contact with things;
  7. feelings of pleasure or pain therefrom;
  8. desire to enjoy or to avoid;
  9. attachment and activity;
  10. coming to birth;
  11. condition and class of the body;
  12. the series of decay, death, grief, lamentation, pain, and dejection or mental affliction.

    (E. Woods, *Yoga Dictionary*)

## Twelve Chairs

- a humorous novel of 1928, by Ilya Arnoldovich Ilf and Yevgeni Petrovich Katayev, about master crook Ostap Bender's odyssey to recover a cache of money hidden in one of a dozen identical chairs. Filmed in 1970 by Mel Brooks, starring Ron Moody, Frank Langella, and Dom DeLuise.

**twelve-day pope**
- the shortest reign of a pope, Urban VII, September 15 to 27, 1590.

**Twelve Days of Christmas**
- on which "My true love gave to me: a partridge in a pear tree, two turtledoves, three French hens, four calling birds, five golden rings, six geese a-laying, seven swans a-swimming, eight maids a-milking, nine ladies dancing, ten lords a-leaping, eleven pipers piping, twelve drummers drumming."

**Twelve Deva Guardians (Juniten)**
- protector of the twelve directions in Esoteric Buddhism in Japan: Bonten (guards the upward direction), Taishakuten (guards the east), Suiten (guards the west), Bishamonten (guards the north), Enmaten (the south), Katen (the southeast), Rasetsuten (the southwest), Futen (the northwest), Ishanaten (the northeast), Jiten (the downward direction), Nitten (the Sun Deva), and Gatten (the Moon Deva).

**"twelve drummers drumming"**
- the gift on the twelfth day of Christmas in the carol "The Twelve Days of Christmas."

**12 February**
- Abraham Lincoln's birthday, 1809, in a log cabin in Hodgenville, Hardin County (now Larue County), Kentucky. Now observed on Presidents' Day, the third Monday in February.

**twelve gauge**
- a measure of shotgun bore diameter = 18.52 mm.

**Twelve Houses of Astrology**
- House of . . .
    1. Self (ruled by the sign of Ares and the planet Mars);
    2. Possessions (ruled by the sign of Taurus and the planet Venus);
    3. Communication (ruled by the sign of Gemini and the planet Mercury);

4. Family and Home (ruled by the sign of Cancer and the Moon);
5. Pleasure (ruled by the sign of Leo and the Sun);
6. Health (ruled by the sign of Virgo and the planet Mercury);
7. Partnerships (ruled by the sign of Libra and the planet Venus);
8. Sex (ruled by the sign of Scorpio and the planet Mars and the dwarf planet Pluto);
9. Philosophy (ruled by the sign of Sagittarius and the planet Jupiter);
10. Social Status (ruled by the sign of Capricorn and the planet Saturn);
11. Friendships (ruled by the sign of Aquarius and the planets Saturn and Uranus);
12. Subconscious (ruled by the sign of Jupiter and the planet Neptune).

## Twelve Labors of Hercules

- extraordinary tasks undertaken as demanded by King Eurystheus:
  1. killing the lion of Nemea;
  2. destroying the nine-headed Hydra;
  3. capturing the boar of Erymanthus;
  4. capturing the hind of Mount Cerynea;
  5. conquering the man-eating birds of Stymphalus;
  6. obtaining the girdle of Queen Hippolyte of the Amazons;
  7. cleaning the Augean stables;
  8. capturing the mad bull of Crete;
  9. capturing the man-eating horses of Thrace;
  10. capturing the oxen of Geryones;
  11. obtaining the golden apples of the Hesperides;
  12. bringing back the dog Cerberus from the lower world.

## *Twelve Men*

- Theodore Dreiser's 1919 studies of twelve actual persons.

## Twelve Mile Lake

- a lake in south-central Saskatchewan Province, Canada.

## Twelvemile Summit

- a peak in east-central Alaska, about 100 miles northeast of Fairbanks.

## 12mo

- twelvemo or duodecimo; the page size of a book from printer's sheets that are folded into twelve leaves, each about 5 x 7 -¾ inches.

## *Twelve Monkeys*

- director Terry Gilliam's 1995 science fiction film in which Bruce Willis travels back in time to locate a potent antivirus that will counteract a virus that makes animals rule the world.

## 12 Norse gods of Scandinavian mythology

- authorities agree that the Aesir, the great gods of the Norse pantheon who lived in heavenly Asgard, were twelve in number, plus Odin, the "all-father." But the list varies by authority. For example:
- according to Snorri Sturluson, the thirteenth-century Icelandic historian, the principal dieties of Scandinavians, all men, were: Odin, along with Thor, Njord, Frey, Tyr, Heimdall, Bragi, Vidar, Vali, Ull, Hœnir, Forsetti, and Loki.
- A. S. Murray's *The Manual of Mythology* concurs with the *Encyclopedia Americana*, stating that the Aesir most likely consisted of Thor, Baldr, Freyr, Tyr, Bragi, Hödr, Heimdall, Vithar, Vali, Ullr, Ve, and Forseti—and, of course, Odin.
- Benet's *Reader's Encyclopedia* includes Odin in the twelve, and adds Thor and Tiu (Odin's sons), and Balder, Bragi, Vidar, Hoder, Hermoder, Hoenir, Odnir, Loki, and Vali.

## *Twelve O'Clock High*

- tense 1949 war picture, directed by Henry King, dealing with the stresses of wartime command. Stars Gregory Peck, Gary Merrill, and Dean Jagger, who won the Oscar for Best Supporting Actor.

## 12 October

- Columbus Day, observed the second Monday in October.

## 12, Orme Square

- London residence of composer Richard Wagner.

## The Twelve Paladins

- the warriors in Charlemagne's court; namely: Roland, Charlemagne's nephew; Astolpho, cousin of Roland; Ferumbras, a Saracen who became a Christian; Florismart, a faithful follower of Roland; Ganelon, the traitor; Maugris, the sorceror, cousin of Rinaldo; Ogier, the Dane; Namo; Oliver, son of Regnier comte de Gennes and a rival to Roland; Otuel, another converted Saracen; Rinaldo, son of duke Aymon; and one other, name not certain, but likely Guy de Bourgogne. (List from the Italian epics of Torquato Tasso and Ludovico Ariosto.)

## twelve-penny nail

- a carpenter's common nail, 3 -1/4 inches long (coded 12d).

## the Twelve Pins

- mountains in Western Connach County, Republic of Ireland.

## 12-point type

- typographer's measure, known as "pica."

## Twelvers

- (also known as Ithna 'Ashariyah): the largest and most conservative branch of the Shi'ite sect of Islam; those who accept as the seventh imam Musa al-Kazim, the younger son of Ja'far ibn Muhammad, the sixth imam, and acknowledge Musa's successors through the twelfth imam.

## 12 September

- Defenders' Day in Maryland, commemorating the Battle of Fort McHenry in the War of 1812.

## twelve ships "with crimson-painted bows"

- those commanded by Odysseus in the forces aligned against Troy (so says *The Iliad*).

## twelve signs of the Zodiac

- Aries the Ram, Taurus the Bull, Gemini the Twins, Cancer the Crab, Leo the Lion, Virgo the Virgin, Libra the Balance, Scorpio the Scorpion, Sagittarius the Archer, Capricorn the Goat, Aquarius the Water-Bearer, and Pisces the Fishes.

## twelve spokes (kalas) in the Jain wheel of time

- six facing upward (*utsarpini*) and six facing downward (*avasarpini*), each representing a period or age.

## 12-step

- a program designed to assist in recovery from addictive or compulsive behavior, especially one employing spiritual guidance and admission of one's own incapability for self-healing. Specifically, the Twelve Steps as guiding concepts of Alcoholics Anonymous.

## Twelve Tables

- the first code of Roman law, dating from about 450 BC. Remained the basis of Roman law until the "Corpus Iurus" under Emperor Justinian in about AD 530.

## 12-tone scale

- a modern system of musical composition in which the twelve tones of an octave are organized in a structured manner. Devised by composer Arnold Schoenberg early in the twentieth century.

## Twelve Tribes of Israel

- the sons of Jacob: Reuben, Simeon, Levi, Judah, Zebulun, Issachar, Dan, Gad, Asher, Naphtali, Joseph, Benjamin (Genesis 49).

## Twelve Wise Masters

- the original corporation of the Meistersingers, guilds of workingmen and craftsmen in Germany in the fourteenth century (continued on through the fifteenth and sixteenth centuries) for the cultivation of poetry and the composition and performance of music. The original members were: Master Altschwert; Hans Blotz, barber; Sebastian Brandt, jurist; Hans Folz, surgeon; Konrad Harder; Heinrik von Mueglen; Master Muscablüt, tailor; Thomas Murner; Master Barthel Regenbogen, blacksmith; Hans Rosenblüt, armorial painter; Wilhelm Weber; and Hans Sachs, cobbler. Sachs was, in reality, not one of the founders, but was so outstanding that he is included among the wise meistersingers.

## twelve-year molar

- the maxillary second molar, the second permanent tooth, appearing at age twelve.

## Twelve Years' Truce

- in 1609, after the death of Philip II, Spain's power in Europe went into decline and a truce of twelve years was signed with the North Provinces of the Netherlands. When the truce expired in 1621, Europe returned to the endless wars between the Catholics and the Protestants.

## world-record sixty-minute rainfall

- 12 inches fell at Holt, Missouri, on June 22, 1946.

## *12.4*

- percentage of the U.S. population age sixty-five years and over, according to the 2000 Census. The percentage had increased to 12.7 in 2008.

## *013*

- Decimal ASCII code for [carriage return].

## *13*

- age at which a Jewish boy celebrates his bar mitzvah and thus becomes an adult member of the religious community.
- age of Huckleberry Finn in the Mark Twain 1885 novel.
- age of the youngest actress ever nominated for a Best Actress category Oscar, Keisha Castle-Hughes, for her role as Pai, the intrepid Maori girl in *Whale Rider*, 2004. (Tatum O'Neal

had won at age ten for *Paper Moon*, but hers was for Best Supporting Actress.)

- atomic number of the chemical element aluminum, symbol Al.
- average number of annual vacation days in the U.S., one of the lowest among industrialized nations.
- composer Richard Wagner, with thirteen letters in his name, was born in 1913, and composed thirteen operas.
- dental chart designation for the left second premolar (second bicuspid).
- in modern Wicca, the maximum size of a coven.
- most birdies in an LPGA 18-hole golf round: by Annika Sorenstam, in the second round of the Standard Register PING Tournament at Moon Valley CC, Phoenix, Arizona, March 16, 2001.
- most career touchdowns scored on kick returns (NFL football record): by Brian Mitchell for the Washington Redskins, 1990–99, the Philadelphia Eagles, 2000–02, the New York Giants, 2003–04 (nine punt returns and four kickoff returns).
- most games passing 400 or more yards in an NFL career (football record): by Dan Marino for the Miami Dolphins, 1983–99.
- most goals scored by a team in an NHL playoff game (hockey record): by the Edmonton Oilers against the Los Angeles Kings (3), April 9, 1987.
- most points scored in a Stanley Cup final (hockey record): by Wayne Gretzky for the Edmonton Oilers against the Boston Bruins, 1988.
- most goals scored in a World Cup soccer final: by Juste Fontaine (France) in the 1958 Championship in Sweden.
- most seasons leading the football league in touchdowns scored (football record): by the Chicago Bears, 1932, 1934–35, 1939, 1941–44, 1946–48, 1956, 1965.
- most strikeouts taken in a single world series (baseball record): by Ryan Howard of the Philadelphia Phillies, 2009, against the New York Yankees.
- most strokes under par in an LPGA 18-hole golf round: by Annika Sorenstam, in the second round of the Standard Register PING at Moon Valley CC in Phoenix, Arizona, March 16, 2001.

- most strokes under par in a PGA 18-hole golf round: three times, by Al Geiberger in the second round of the 1977 Memphis Classic (1 eagle, 11 birdies); by Chip Beck in the third round of the 1991 Las Vegas Invitational (thirteen birdies); and by David Duval in the final round in winning the 1999 Bob Hope Chrysler Classic (one eagle, eleven birdies).
- number of books in Euclid's *Elements*.
- number of books in St. Augustine's *Confessions*.
- number of cards in each of the four suits of a poker deck: 2 through 10, jack, queen, king, and ace.
- number of dwarfs taking the journey in J. R. R. Tolkien's *The Hobbit*: Thorin, Dwalin, Balin, Kili, Fili, Dori, Nori, Ori, Oin, Gloin, Bifur, Bofur, and Bombur.
- number of heavens in the Mayan religion (along with nine underworlds).
- number of lunar months in a year.
- number of popes named Leo; number of popes named Innocent.
- number of presidential electoral votes apportioned to the state of Virginia. (May be reapportioned for the 2012 election.)
- number of satellites of the planet Neptune; Triton, the largest, is the coldest body yet visited in the solar system, with surface temperature at about 391 degrees Fahrenheit. The most recent, four were discovered in 2002 and one in 2003.
- number of seats in the House of Representatives allotted to each of the states of Georgia, New Jersey, and North Carolina (as of the 2000 Census)
- number of steps on the pyramid on the back of a U.S. one-dollar bill.
- number of the French Département of Bouches-du-Rhône.
- number of Valois rulers of France, from Philip VI in 1328; John II (the Good), 1350; Charles V the Wise), 1364; Charles VI (the Mad), 1380; Charles VII (the Victorious), 1422; Louis XI, 1461; Charles VIII, 1483; Louis XII, 1498; Francis I, 1515; Henry II, 1547; Francis II, 1559; Charles IX, 1560; Henry III, 1574–89. (The Bourbon Dynasty followed.)

- number of volumes in Marcel Proust's monumental masterpiece, *Remembrance of Things Past*, his first volume published in 1913.
- number of original American colonies.
- players on a rugby league team (but fifteen in rugby union), seven backs and six forwards.
- record number of career goals scored in All-Star hockey games (hockey record): by Wayne Gretzky in eighteen games played, between 1980 and 1999.
- record number of men's Wimbledon tennis titles: won by Laurie Doherty of the UK: five singles, 1902–06, and eight men's doubles, 1897–1901, 1903–05.
- record number of women's French Open tennis titles: won by Margaret Smith-Court; five singles, 1962, 1964, 1969–70, 1973; four doubles, 1964–66, 1973; four mixed doubles 1963–65, 1969.
- record number of women's U.S. Open tennis doubles titles: won by Margaret du Pont, 1941–50, 1955–57.
- retired baseball uniform number of shortstop Dave Concepcion, retired by the Cincinnati Reds.
- retired basketball jersey number of:
  - ⇨ Wilt Chamberlain, retired by the Los Angeles Lakers, Golden State Warriors, and Philadelphia 76ers, the only basketball player to have his number retired by three teams.
  - ⇨ Bobby Phills, retired by the New Orleans Hornets.
  - ⇨ James Silas, retired by the San Antonio Spurs.
  - ⇨ Dave Twardzik, retired by the Portland Trail Blazers.

- retired football jersey number of:
  - ⇨ Dan Marino, retired by the Miami Dolphins.
  - ⇨ Don Maynard, retired by the New York Jets.

- stripes on the American flag—seven red and six white— each representing one of the original thirteen colonies.
- 13 feet:
  - ⇨ deepest part of Venice's Grand Canal, near the Rialto Bridge.
  - ⇨ separation between crossbar uprights in the Olympic high jump.

- 13 miles: length of Manhattan Island in New York City.
- Title of the United States Code dealing with census.
- TV channel (WNET) of Public Broadcasting System in New York City.
- value of Roman numeral XIII.
- widely accepted unlucky number. Many buildings and hotels have no thirteenth floor. However, 13 has some interesting characteristics. For example, 13 reversed is 31; the square of 13 is 169, the square of 31 is 961, also a reversal. The sum of digits in 169 is 16. The sum of the digits in 13 is 4, the square root of 16. If the number 13 is unlucky to some, it certainly was not to the designers of American money: on the back of the dollar bill is an eagle with thirteen stars above it, thirteen horizontal stripes on its shield, thirteen war arrows in one talon, and in the other, a peace branch with thirteen olive leaves. The pyramid in the Great Seal has thirteen levels, and the phrase e pluribus unum contains thirteen letters.

### Apollo 13

- ill-fated NASA moon mission launched April 11, 1970, which was aborted after an oxygen tank ruptured fifty-six hours into the flight. The crew—James A Lovell, Fred W. Haise Jr., and John L. Swigert Jr.—in constant contact with ground control, accomplished the formidable task of bringing the command module back to Earth. The events were dramatized in a 1995 film, appropriately named *Apollo 13*, directed by Ron Howard, with Tom Hanks, Bill Paxton, and Kevin Bacon portraying the three astronauts.

### Assault on Precinct 13

- a riveting 1976 thriller, directed by John Carpenter, in which an LA police station is held under siege by a multiracial youth gang. A modernized update of Howard Hawks's classic western, *Rio Bravo*.

### baker's dozen

- many bakers still sell thirteen as a dozen; this dates from a 1266 British law stipulating that exactly eighty loaves of bread were to be baked from a standard sack of flour, and prescribing stiff penalties for any baker who sold loaves weighing less than a set weight. To avoid punishment, bakers added an extra loaf to every twelve they sold, just in case the composite of a dozen was underweight.

## Chapter 13

- a section of the U.S. bankruptcy code providing for the reorganization of an individual's debt by a court-approved payment schedule, managed by a court-appointed trustee.

## file 13

- a wastebasket, in World War II armed forces slang (also "file 17").

## *Friday the 13th*

- a bus is struck by lightning in this 1934 British film, leading the audience to speculate who will survive among the six passengers whose stories had been told previously. Not to be confused with a 1980 teenage gore stinkeroo of the same title.

## Group 13 elements (of the Periodic Table)

- the boron group: boron (5), aluminum (13), gallium (31), indium (49), thalium (81), and ununtrium (113).

## *In a Year of Thirteen Moons*

- disturbing 1980 German film by Rainer Werner Fassbinder (who wrote, directed, photographed, and edited) about the repercussions of a man's sex-change operation. Grim portrayal of alienation and loneliness.

## Louis Treize

- pertaining to the late-Renaissance style in architecture, furnishings decoration typical of the reign of King Louis XIII of France in the early seventeenth century.

## number 13 billiard ball

- the ball that is conventionally striped orange.

## Shostakovich's *Symphony No. 13*

- "Babi Yar" symphony (in B-flat minor).

## Stalag 13

- setting for the CBS-TV sitcom *Hogan's Heroes.*

## *The Thirteen*

- a well-crafted Russian soldier-squad-lost-in-the-desert film, directed by Mikhail Romm. Made in 1937, the film was reworked for American audiences as Sahara in 1943, starring Humphrey Bogart.

## 13 April

- birthday of Thomas Jefferson, 1743, third president of the United States.

## Thirteen Articles of Faith

- which define the beliefs of the Church of Jesus Christ of Latter-day Saints (the Mormons).

## Thirteen Classics

- works in the Confucian tradition, thought to be the Táng Dynasty canon, comprising the so-called Nine Classics (q.v.) plus *Book of Filial Piety, Erya, Analects,* and *Mencius.* At times, only Twelve Classics are cited, *Mencius* being excluded.

## Thirteen Colonies

- a term used to designate the British colonies of North America that joined together in the American Revolution and became the United States. They were New Hampshire, Rhode Island, Massachusetts, Connecticut, New York, New Jersey, Pennsylvania, Delaware, Maryland, Virginia, North and South Carolina, and Georgia.

## *Thirteen Days*

- tense 2000 film, directed by Roger Donaldson, dealing with the two-week Cuban missile crisis in October, 1962, at which time President Kennedy had to deal with the buildup of Soviet offensive weapons that could trigger a nuclear war. Kevin Costner plays Kenny O'Donnell, adviser to JFK (Bruce Greenwood).

## *13 Going on 30*

- 2004 comedy-fantasy, directed by Gary Winick, with Jennifer Garner. Having trouble fitting in with her school mates, she wishes she were a grown-up. When she awakes the next day, her wish has come true. But she discovers being an adult isn't all she expected.

## 13-gun salute

- honors the arrival or departure of major generals or rear admirals (upper half).

## 13 Lords of the Day

- gods of the dawn in the Toltec tradition (along with nine gods of the night).

## 13 May

- Jamestown Day, celebrated in Virginia on the anniversary of the settlement of Jamestown in 1607.

## 13 November

- the date that Felix Unger's wife threw him out of his home on the ABC-TV series, *The Odd Couple.*

## Thirteen Principles of Faith

- Moses Maimonides's personal summary of the basic tenets of Judaism, as formulated in his commentary on the Mishna. Included are several doctrines concerning the nature of God, the law, and Moses, and they affirm the future coming of the Messiah.

## *13 Rue Madeleine*

- World War II spy movie (1946) starring James Cagney, Annabella, Richard Conte, and Walter Abel. Directed by Henry Hathaway.

## thirteens

- slang for obsolete Irish shillings, which, in the beginning of the nineteenth century, were worth about thirteen pence each.

**Thirteenth Amendment to the Constitution**
- abolishes slavery; ratified December 6, 1865.

*The Thirteenth Chair*
- highly successful 1916 Broadway play by Bayard Veiller, in which a séance is staged to catch a murderer. Converted to film three times: first, as a silent film in 1919; with sound in 1930, directed by Tod Browning, with Margaret Wycherly as Madame. LaGrange, the phony mystic, and Bela Lugosi as Inspector Delzante; then again in 1937, with Dame May Witty as LaGrange and Lewis Stone as the inspector.

**thirteenth president**
- Millard Fillmore, July 7, 1850 to March 3, 1853.

**thirteenth state of the Union**
- Rhode Island ratified the Constitution on May 29, 1790.

**thirteenth wedding anniversary gift**
- lace is customary.

**"Thirteen Ways of Looking at a Blackbird"**
- a 1917 haiku-inspired poem by Wallace Stevens (1879–1955).

**thirteen years of exile**
- describes a period in the life of Confucius, ending in 483 BC, during which he wandered from state to state, searching for a position from which he could practice his ideas on peace and good government.

## 014

- Decimal ASCII code for [shift out].
- Dewey Decimal System designation for bibliographies of anonymous and pseudonymous works.

## 14

- atomic number of the chemical element silicon, symbol Si.

- badge number of Agent April Dancer, played by Stefanie Powers, on the NBC-TV spy spoof, *The Girl from U.N.C.L.E.* (1966–67).
- badge number of Detective Mark Saber, played by Tom Conway, on the TV series, *Mark Saber.*
- days and nights in a fortnight; two weeks.
- dental chart designation for the left first molar.
- English size women's coat equivalent to American size 12.
- fourteen innings, longest complete game pitched to a World Series victory (baseball record): by Babe Ruth, for Boston Red Sox against the Brooklyn Robins, 1916.
- gauge (number) of a wood screw with a shank diameter of 15/64 of an inch.
- lines in a sonnet.
- lunch-counter code for a special order.
- maximum number of golf clubs permitted in competition play.
- most consecutive pitches fouled off (baseball record): by Los Angeles Dodger Alex Cora against Chicago pitcher Wilson Alvarez, May 12, 2004. On the eighteenth pitch, Cora hit a two-run homer.
- most gold medals won by a country in the Winter Olympics: by Canada in 2010 at the XXIst games.
- most gold medals won by any Olympian: swimmer Michael Phelps, six in 2004 and eight in 2008.
- most pass interceptions in a season (football record): by Richard "Night Train" Lane for the Los Angeles Rams, 1952.
- most seasons making 1,000 or more yards receiving passes (football record): – by Jerry Rice, 1986–96, 1998, 2001–02.
- most stolen bases in World Series games (baseball record): twice, by Eddie Collins for Philadelphia Athletics, 1910–14, and Chicago Cubs, 1917 and 1919; and by Lou Brock for St. Louis Cardinals, 1964–68.
- most turnovers in an NBA game (basketball record): by John Drew for Atlanta Hawks against New Jersey Nets, March 1, 1978.
- number of counties in each of the states of Massachusetts and Vermont.
- number of generations from Abraham to David, acording to the Gospel of St. Matthew (Matthew 1:17).

- number of museums comprising the Smithsonian Institution, plus the National Zoo. Established in 1846, the Smithsonian is the world's largest museum complex, holding some 100 million artifacts and specimens. Includes the National Portrait Gallery, the National Air and Space Museum, the National Museum of African Art, and also the Cooper-Hewitt Museum in New York City.
- number of nations, represented by 241 athletes, competing in the first modern Olympics, in Athens, Greece, in 1896.
- number of popes named Clement.
- number of propositions in Book II of Euclid's *Elements*, on geometric algebra.
- number of the French Département of Calvados.
- pH of normal sodium hydroxide, the most alkaline substance. Upper limit of the pH scale.
- pounds avoirdupois in a stone (British unit of weight).
- principal classes of gods in the early texts of Buddhism—six in the lower realms, eight in the higher.
- retired baseball uniform number of:
  - ⇨ shortstop Ernie Banks, retired by the Chicago Cubs.
  - ⇨ third baseman Ken Boyer, retired by the St. Louis Cardinals.
  - ⇨ pitcher Jim Bunning, retired by the Philadelphia Phillies.
  - ⇨ outfielder Larry Doby, retired by the Cleveland Indians.
  - ⇨ manager Gil Hodges, retired by the New York Mets.
  - ⇨ first baseman Kent Hrbek, retired by the Minnesota Twins.
  - ⇨ outfielder Jim Rice, retired by the Boston Red Sox.

- retired basketball jersey number of:
  - ⇨ Bob Cousy, retired by the Boston Celtics.
  - ⇨ Jeff Hornacek, retired by the Utah Jazz.
  - ⇨ Jon McGlocklin, retired by the Milwaukee Bucks.
  - ⇨ Lionel Hollins, retired by the Portland Trail Blazers.
  - ⇨ Tom Meschery, retired by the Golden State Warriors.
  - ⇨ Oscar Robertson, retired by the Sacramento Kings. Robertson's number 1 with the Milwaukee Bucks has also been retired.

- retired football jersey number of:
  - ⇨ Dan Fouts, retired by the San Diego Chargers.

⇨ Otto Graham, retired by Cleveland Browns.

⇨ Don Hutson, retired by the Green Bay Packers.

⇨ Y. A. Tittle, retired by the New York Giants.

- retired hockey shirt number of Rene Robert, retired by the Buffalo Sabres.
- sum of the first three square numbers $(1 + 4 + 9)$.
- Title of the United States Code dealing with the Coast Guard.
- value of Roman numeral XIV.

## carbon 14

- a radioactive isotope of carbon with mass number 14 and half-life of 5,730 years, used widely in archaeology in determining the age of organic material.

## fastest touchdown in a Super Bowl game

- in SB-XLI, on February 14, 2007, Devin Hester of the Chicago Bears returned the opening kickoff 92 yards, scoring a touchdown in the first fourteen seconds of the game, the only player to take a Super Bowl kickoff back for a TD. Nonetheless, the Bears lost to the Indianapolis Colts, 29–17.

## F-14

- U.S. Navy aircraft, the Grumman Tomcat, a carrier-based interceptor.

## fourteen

- a solitaire card game in which kings count 13, queens 12, jacks 11, and aces 1. The deck is divided into twelve piles, cards face up, and the player removes pairs of top cards totaling 14. The object is to empty the table.

## 14 April

- (a) Pan-American Day.
- (b) John Hanson Day in Maryland, born in 1721.

## 14 February

- Valentine's Day.

## fourteener

- a poetic line of fourteen syllables, usually seven iambic feet.

## Fourteeners

- collective name for the fifty-four peaks over 14,000 feet in the Colorado Rockies, so called by recreational climbers. The tallest is Mount Elbert in the Sawatch Range, elevation 14,433 feet.

### fourteen human wives (or consorts) of Zeus

- (in Greek mythology): Alcmene, Antiope, Callisto, Aegina, Europa, Io, Laodamia, Niobe, Pluto, Taygete, Danae, Leda, Electra, Semele.

## 14 June

- Flag Day; on this date in 1777, Congress adopted the Stars and Stripes as the official emblem of the United States.

## 14 July

- Bastille Day, French holiday commemorating the storming of the Bastille in 1789.

## Fourteenmile Point

- a knob of land on Lake Superior in northern Michigan.

## 14-point type

- typographer's measure, known as "English."

## Fourteen Points

- President Wilson's goals prescribing "the only possible program" for attaining peace following World War I, presented to Congress on January 8, 1918.

### fourteen punctuation marks in English

- Period (.), comma (,), colon (:), semicolon (;), dash (—), hyphen (-), apostrophe (',‘), question mark (?), exclamation point (!), quotes ("‚"), parentheses ( (,) ), brackets ([,]), braces ({,}), ellipsis ( . . . ). Several other signs are used [e.g., ampersand (&), number (#), solidus (/)], but these are not truly punctuation.

**14 September**
- Holyrood Day, Christian feast day commemorating the Exaltation of the Holy Cross.

**Fourteen Stations of the Cross**
- depict what happened to Jesus from the time he was condemned by Pilate until he died and was interred.

**Fourteen Streams**
- a town in the Republic of South Africa, about 80 miles north of Kimberley.

**Fourteenth Amendment to the Constitution**
- prohibits the abridgement of citizenship rights to any person born or naturalized in the U.S., ratified July 28, 1868.

**fourteenth president**
- Franklin Pierce, March 4, 1853 to March 3, 1857.

**Fourteenth State of the Union**
- Vermont, admitted March 4, 1791.

**fourteenth wedding anniversary gift**
- ivory is customary.

**Gang of Fourteen**
- a coalition of fourteen moderate U.S. senators, seven from each party, meeting in the spring of 2005 to work out a compromise that would forestall a Democratic filibuster and the retaliatory Republican cloture vote (the "nuclear option") relating to congressional confirmation of three conservative appellate court judges nominated by President George W. Bush. The seven Republicans were Lincoln Chafee (RI), Susan Collins (ME), Mike DeWine (OH), Lindsey Graham (SC), John McCain (AZ), Olympia Snowe (ME), and John Warner (VA); the seven Democrats were Robert Byrd (WV), Daniel Inouye (HI),

Mary Landrieu (LA), Joe Lieberman (CT), Ben Nelson (NE), Mark Pryor (AR), and Ken Salaz (CO). The group agreed that Democrats would not filibuster in "all but extraordinary circumstances," and the GOP would not call for cloture. The three nominees (Janice Rogers Brown, Priscilla Owen, and William Pryor) were all confirmed by full Senate vote.

## Group 14 elements (of the Periodic Table)
- carbon (6), silicon (14), germanium (32), tin (50), lead (82), ununquadium (114).

## Louis Quatorze
- pertaining to the style of architecture, furnishings, and decoration, heavily ornamented, overly grand, reaching its height in France in the late seventeenth century.

## Louis XIV
- the Sun King of France, who reigned for seventy-two years from 1643 to 1715, the longest sitting ruler of France or any other major European monarchy.

## number 14 billiard ball
- the ball that is conventionally striped green.

## number 14 wood screw
- a screw with a screw shank diameter of 15/64 inches.

## 015
- Decimal ASCII code for [shift in].
- Dewey Decimal System designation for bibliographies from specific places.

## 15
- Article of the New York State Vehicle and Traffic Law dealing with registration of motorcycles.
- atomic number of the chemical element phosphorus, symbol P.

- dental chart designation for the left second molar.
- 15 feet:
  ⇨ distance from blue line to center line on a regulation ice hockey rink.
  ⇨ distance from the basket to the free-throw line on a regulation basketball court.

- 15 feet below sea level, the altitude of the world's lowest international airport, Schiphol Airport, outside Amsterdam, The Netherlands.
- in the card game cribbage, a pair of cards totaling 15 in pip values, or the score of two points for such a combination.
- largest margin of victory in a major golf championship, by Tiger Woods at the 2000 U.S. Open at Pebble Beach, California, shooting a 12-under-par 272.
- magic constant of a 3 x 3 magic square:

  | 4 | 9 | 2 |
  |---|---|---|
  | 3 | 5 | 7 |
  | 8 | 1 | 6 |

- members of a hurling team.
- minimal SPF sunscreen recommended for protection against UV index of 2 or less.
- minutes in a football game quarter.
- most fumbles by both teams in an NFL Pro Bowl game (football record): by the NFC (10); by the AFC (5), in 1974.
- most goals in soccer World Cup Finals: by Brazilian striker Ronaldo Luis Nazário de Lima, on June 27, 2006, against Ghana.
- most NHL seasons scoring at least one hundred points (hockey record): by Wayne Gretzky, for Edmonton Oilers, Los Angeles Kings, St. Louis Blues, and New York Rangers, 1979–99.
- most women's golf majors career wins: by Patty Berg: one U.S. Women's Open (1946), seven Titleholders Championships (1937–39, 1948, 1953, 1955, 1957), and seven Western Opens (1941, 1943, 1948, 1951, 1955, 1957–58).
- number of Capetian rulers of France, from Hugh Capet in 987; then Robert II (the Pious), 996; Henry I, 1031; Philip I, 1060; Louis VI (the Fat), 1108; Louis VII (the Young), 1137; Philip II Augustus, 1180; Louis VIII (the Lion), 1223; Louis IX

(St. Louis), 1226; Philip III (the Bold), 1270; Philip IV (the Fair), 1285; Louis X (the Stubborn), 1314; John I, 1316; Philip V (the Tall), 1316; Charles IV (the Fair), 1322–28. (The Valois Dynasty followed.)

- number of counties in the state of Arizona.
- number of days in each of the twenty-four cycles of the Chinese calendar.
- number of decimal places in the inverse multiple designated by the International System prefix femto-, written "f-".
- number of marbles of each color in the game Chinese checkers.
- number of markers, or stones, on each side in the game of backgammon.
- number of presidential electoral votes apportioned to each of the states of Georgia, New Jersey, and North Carolina. (May be reapportioned for the 2012 election.)
- number of presidents of France under the Third Republic: Louis Jules Trochu (provisional), 1870; Adolphe Thiers, 1871; Patrice MacMahon, 1873; Jules Grévy, 1879; Sadi Carnot, 1887; Jean Casimir-Périer, 1894; Félix Faure, 1895; Emile Loubet, 1899; Armand Falliéres, 1906; Raymond Poincaré, 1913; Paul Deschanel, 1920; Alexandre Millerand, 1920; Gaston Doumergue, 1924; Paul Doumer, 1931; Albert Lebrun, 1932–40. [The Vichy Government followed.]
- number of seats in the House of Representatives allotted to the state of Michigan (as of the 2000 Census).
- number of stars and stripes on the American flag from 1794 until 1818, when the thirteen stripes were restored to represent the original thirteen colonies.
- number of the French Département of Cantal.
- number of 0s in a quadrillion.
- number of 0s in the multiple designated by the International System prefix peta-, symbol "P-".
- the ozone layer starts at about 15 kilometers above sea level.
- players on a rugby union team (but thirteen in rugby league), eight forwards and seven backs.
- record number of Tony nominations for *The Producers*, written by writer/comedian Mel Brooks, winning twelve of them.

- retired baseball uniform number of catcher Thurman Munson, retired by the New York Yankees.
- retired basketball jersey number of:
  - ⇨ Brad Davis, retired by the Dallas Mavericks.
  - ⇨ Hal Greer, retired by the Philadelphia 76ers.
  - ⇨ Tom Heinsohn, retired by the Boston Celtics.
  - ⇨ Vinnie Johnson, retired by the Detroit Pistons.
  - ⇨ Both Earl Monroe and Dick McGuire, retired by the New York Knicks.
  - ⇨ Larry Steele, retired by the Portland Trail Blazers.

- retired football jersey number of:
  - ⇨ Bart Starr, retired by the Green Bay Packers.
  - ⇨ Steve Van Buren, retired by the Philadelphia Eagles.

- retired hockey shirt number of Milt Schmidt, retired by the Boston Bruins.
- seahs in a lethech (ancient Hebrew measure of dry capacity).
- sum of the first five integers $(1 + 2 + 3 + 4 + 5)$.
- sum of the first three squared numbers $(1 + 4 + 9)$.
- Title of the United States Code dealing with commerce and trade.
- value of Roman numeral XV.

## F-15

- U.S. warplane, the McDonnell Douglas Eagle, a twin-engine, all-weather interceptor-fighter. Also used by Japan and Israel.

## Fifteen

- (a) a program created by chef Jamie Oliver in London to train disadvantaged young people for careers in the restaurant industry. As of this writing, graduates are active in food establishments in London, at the two-level Fifteen Restaurant, and others in Cornwall, Melbourne, and Amsterdam.
- (b) a version of the game blackjack for two players, in which only one card is dealt to each, face down, and an ace counts only as one. Additional cards are dealt as requested. The object is to score closer to 15 than the opponent without

going over. If a player exceeds 15, he doesn't declare it until after both players have drawn their cards. Also known as Quince.

## "The Fifteen"

- the First Jacobite Rising, mostly in Scotland, against George I in 1715. The Jacobites, led by the Earl of Mar, sought to regain the throne for James Edward Stuart. They were defeated at Preston and surrendered on November 14, 1715.

## 15 April

- deadline each year for filing a federal income tax return.

## 15 August

- Christian feast day commemorating the Assumption of the Virgin Mary.

## 15 February

- anniversary of the bombing of the USS *Maine* in Havana harbor in 1898. Celebrated in Maine as Battleship Day, and in Massachusetts as Maine Day.

## 15 December

- Bill of Rights Day; the Bill of Rights was adopted on this day, 1791.

### *Fifteen Decisive Battles of the World*

- Sir Edward Creasy's 1851 historical work in which he discusses those battles he believes have been most influential in producing significant change in Western political history; the battles of: Marathon (490 BC), Syracuse (413 BC), Arbela (331 BC), Metaurus (207 BC), Teutoberg Forest (AD 9), Chalons (AD 451), Tours (AD 732), Hastings (1066), Orleans (1429), Blenheim (1704), defeat of the Spanish Armada (1858), Pultowa (1709), Saratoga (1777), Vlamy (1792), and Waterloo (1815).

## 15-gun salute

- honors the arrival of U.S. or foreign envoys or ministers accredited to the U.S., or a lieutenant general or vice admiral.

## 15 January 1929

- birthday of Martin Luther King, Jr., celebrated on the third Monday in January.

## fifteen kilotons of TNT

- the explosive power of the first atom bomb dropped on Hiroshima, August 6, 1945.

## 15 May 1940

- the first MacDonald's restaurant opens in San Bernardino, California.

## "Fifteen men on a dead man's chest"

- "Fifteen men on a dead man's chest; Yo-ho-ho, and a bottle of rum"—sung by seaman Bill Bones in the 1883 novel, *Treasure Island*, by Robert Louis Stevenson.

## Fifteen Mile Creek

- a waterway in northern Wyoming.

## fifteen million dollars

- amount paid to France in 1803 for the Louisiana Purchase, which added 828,000 square miles to the country, almost doubling its size.

## 15 O's of St. Bridget

- prayers of St. Bridget of Sweden intended to bring the person closer to Christ in his passion and death, so called because all fifteen begin with the words "O Jesus."

## 15 percent tip

- the customary amount for adequate service in a restaurant to a waiter or waitress serving two people. In recent years, the amount has been edging up to 18 percent.

## 15 Place Vendome

- address of the Ritz Hotel in Paris, site of the famous Hemingway Bar, where, according to legend, Papa chose to celebrate the Allied liberation of Paris.

## 15 September

- Battle of Britain Day in the UK, commemorating the RAF's service in World War II. On that date in 1940, the Luftwaffe sent over 1,000 sorties against London, which were beaten back by pilots of the RAF with heavy losses to the Nazi air force.

## Fifteenth Amendment to the Constitution

- states that suffrage shall not be denied or abridged because of race, color, or previous condition of servitude, ratified March 30, 1870.

## 15th day of the seventh month

- shall be the feast of tabernacles for seven days (Leviticus 23:34).

## fifteenth president

- James Buchanan, March 4, 1957 to March 3, 1861.

## fifteenth state of the Union

- Kentucky, admitted June 1, 1792.

## fifteenth wedding anniversary gift

- crystal or glass is customary.

## freshman fifteen

- slang term for the weight gain of fifteen pounds expected during freshman year of college attendance, especially if living on campus in residence halls or dormitories.

## Group 15 elements (of the Periodic Table)

- pnictogens: nitrogen (7), phosphorus (15), arsenic (33), antimony (51), bismuth (83), ununpentium (115).

## Ides of March

- Julius Caesar assassinated at the Roman Forum in 44 BC.

## I-15

- major north-south interstate highway from San Diego to Las Vegas, Salt Lake City, Pocatello, Helena, Great Falls, and to Sweetgrass (Montana) on the Canadian border.

## number 15 billiard ball

- the ball that is conventionally striped maroon.

## Proposition 15

- of Book I of Euclid's *Elements*; contains his proof of the equality of vertical angles created by the intersection of two lines.

## Quinceañera

- in some Latin American countries, a coming-of-age celebration of a girl's fifteenth birthday, denoting her budding womanhood. Equivalent to the Sweet Sixteen celebration among Anglos.

## Shakespearean Sonnet XV

- begins, "When I consider everything that grows . . ."

## size of bases on a baseball field

- standard is 15 inches by 15 inches.

## Special Field Order No. 15

- issued by Union general William Tecumseh Sherman on January 16, 1865, giving freed slaves a tract of land and a surplus Union animal so they could become self-sufficient farmers. Whence the phrase, "Forty acres and a mule."

## world's longest vehicular tunnel

- the Lærdal tunnel, connecting Lærdal and Aurland in western Norway, is 15.2 miles long. Completed in 2000, the tunnel is a section of the E16 highway from Oslo to Bergen.

## 15.5

- 15-½ inches, maximum length of the blade on a goalkeeper's hockey stick, measured from the heel to end of the blade (as against 12-½ inches for other players' sticks).

## 016

- Dewey Decimal System designation for bibliographies from specific subjects.

## 16

- age of Judy Garland when she starred in the film *The Wizard of Oz*.
- atomic number of the chemical element sulphur, symbol S.
- bottles in a Balthazar (champagne measure).
- dental chart designation for the left third molar (wisdom tooth).
- drams in an ounce (avoirdupois).
- fluid ounces in a pint.
- the fourth power of 2.
- gauge (number) of a wood screw with a shank diameter of 17/64 of an inch.
- in Japanese myth, sinners are relegated to one of the sixteen districts of the hellish region called Jigoku.
- in Tibetan Buddhism, the number of Arhats ("worthy ones") who have attained a state of spiritual perfection.
- minimum number of members for a federal grand jury (twenty-three is maximum).
- most balks by a pitcher in a season (baseball record): by Dave Stewart, Oakland Athletics, 1988.
- most consecutive regular games won in an NFL season (football record): by the New England Patriots, 2007, the entire season.
- most games lost in relief in a season (baseball record): by Gene Garber, Atlanta Braves, in 1979 (6-16 for the season).
- most goals by a team in an NHL game (hockey record): by the Montreal Canadiens against the Quebec Bulldogs on March 3, 1920. Montreal won, 16–3.

- most punts in an NFL game (football record): by Leo Araguz for the Oakland Raiders against the San Diego Chargers, October 11, 1998.
- number of counties in each of the states of Maine and Nevada.
- number of dams in a gold mohur (obsolete Indian coin).
- number of Earth hours in a day on the planet Neptune (actually, sixteen hours, six minutes).
- number of eyes on the all-seeing Yoruba God, Fa.
- number of nations cooperating in the International Space Station project: U.S., Russia, UK, Belgium, Brazil, Canada, Denmark, France, Germany, Italy, Japan, Netherlands, Norway, Spain, Sweden, and Switzerland.
- number of peerages in the House of Lords dating from before the accession of Henry VIII in 1485.
- number of pieces on each side in a chess game: king, queen, two bishops, two knights, two rooks, and eight pawns.
- number of popes named Benedict, the most recent being Benedict XVI, who had been German Cardinal Joseph Ratzinger, chosen pope on April 19, 2005, on fourth ballot.
- number of popes named Gregory.
- number of the French Département of Charente.
- the number of topics that must be understood in order to discover truth and attain the Supreme Felicity, according to the opening verse of the Nyayasutra of India: 1) means of right knowledge (*pramana*), 2) object of right knowledge (*prameya*), 3) doubt (*samsaya*), 4) purpose (*prayojana*), 5) familiar example (*drstanta*), 6) established tenet (*siddhanta*), 7) members of a syllogism (*avayava*), 8) confutation (*tarka*), 9) ascertainment (*nirnaya*), 10) discussion (*vada*), 11) controversy (*jalpa*), 12) cavil (*vitanda*), 13) fallacy (*hetvabhasa*), 14) equivocation (*chala*), 15) futility (*jati*), 16) disagreement in principle (*nigrahasthana*). (T. Bernard, *Hindu Philosophy*, Philosophical Library, 1947).
- ounces in a pound.
- quarts in a bucket (dry measure)
- record number of kills by an American fighter pilot in the Korean War, U.S. Air Force Captain Joseph M. McConnell, flying an F-86 Sabre.

- record number of men's Grand Slam singles tennis titles: won by Swiss Roger Federer: six Wimbledons (2003–07, 2009), five U.S. Opens (2004–08), four Australian Opens (2004, 2006–07, 2010), and one French Open (2009).
- record number of shutouts pitched in a baseball season: twice, by George Bradley, St. Louis Brown Stockings, 1876; and by Grover Cleveland Alexander, Philadelphia Phillies, 1916.
- record number of touchdowns scored by both teams in an NFL game (football record): by the Washington Redskins (10) against the New York Giants (6), November 27, 1966.
- record number of U.S. Open tennis titles won by a man: Bill Tilden, including seven men's singles (1920–25 and 1929), five doubles (1918, 1921–23, 1927), and four mixed doubles (1913–14, 1922–23). Tilden was the first American to win at Wimbledon.
- record number of walks pitched in a nine-inning baseball game: three times, by Bill George, New York Giants, on May 30, 1887; by George Van Haltren, Chicago Cubs, on June 27, 1887; and by Bruno Haas, Philadelphia Athletics, on June 23, 1915.
- retired baseball uniform number of:
  - ⇨ pitcher Whitey Ford, retired by the New York Yankees.
  - ⇨ pitcher Ted Lyons, retired by the Chicago White Sox.
  - ⇨ pitcher Hal Newhouser, retired by the Detroit Tigers.
- retired basketball jersey number of:
  - ⇨ Al Attles, retired by the Golden State Warriors.
  - ⇨ Bob Lanier, retired by both the Detroit Pistons and the Milwaukee Bucs.
  - ⇨ Satch Sanders, retired by the Boston Celtics.
- retired football jersey number of:
  - ⇨ Len Dawson, retired by the Kansas City Chiefs.
  - ⇨ Frank Gifford, retired by the New York Giants.
  - ⇨ Joe Montana, retired by the San Francisco 49ers.
- retired hockey shirt number of:
  - ⇨ Bobby Clarke, retired by the Philadelphia Flyers.
  - ⇨ Marcel Dionne, retired by the Los Angeles Kings.
  - ⇨ Michel Goulet, retired by the Quebec Nordiques.
  - ⇨ Brett Hull, retired by the St. Louis Blues.
  - ⇨ Pat LaFontaine, retired by the Buffalo Sabres.

⇨ Trevor Linden, retired by the Vancouver Canucks.

⇨ Henri Richard, retired by the Montreal Canadiens.

- shots in a pint (spirits measure).
- sixteen days: average gestation period of a hamster.
- 16 feet square: minimum-size ring for amateur or Olympic boxing matches.
- sixteen granite columns support the porch providing entrance to the Pantheon in Rome.
- 16 meters: length of a beach doubles volleyball court.
- 16 pounds:
  ⇨ weight of a regulation bowling ball.
  ⇨ weight of a shot putt or hammer in men's athletics (women use one that weighs 8 pounds, 13 ounces).
- sixteen years:
  ⇨ age of Uzziah when he became king of Judah (2 Chronicles 26:1).
  ⇨ reign of Jehoash, son of Jehoahaz, over Israel in Samaria (2 Kings 13:10).

- the square of 4.
- square rods in a square chain.
- tablespoons in a cup.
- Title of the United States Code dealing with conservation.
- value of Roman numeral XVI.

## Benedict XVI

- the present pope, formerly Joseph Ratzinger of Germany, announced by Cardinal Medina from St. Peter's balcony at 6:43 PM on April 19, 2005. He is the first German pope since the eleventh century, and at seventy-eight, the oldest to be elected pope since Clement XII (also seventy-eight) in 1730.

## Block 16

- the red-light district in Las Vegas in the early twentieth century. The district was shut down during World War II to "protect" young army trainees at a nearby camp.

## F-16

- U.S. warplane, the Fighting Falcon, produced by General Dynamics. In service with the U.S. Air Force and the air forces of several other countries.

## Group 16 elements (of the Periodic Table)

- oxygen (8), sulfur (16), selenium (34), tellurium (52) polonium (84), the synthetic ununhexium (116).

## in sixteens

- describes a printing process whereby each printed sheet contains sixteen book pages.

## League of Sixteen

- group formed in Paris in the late 1580s purposing to depose the weak King Henry III.

## most decorated male Olympian

- American swimmer Michael Phelps won six gold medals and two bronze in 2004, and eight gold in 2008. Phelps set a new world record in seven of the eight events he entered in 2008.

## M16

- U.S. Army's standard assault rifle, in use since 1967.

## number 16 wood screw

- a screw with a screw shank diameter of 17/64 inches.

## semiquaver

- sixteenth note, in musical notation.

## 16 April 1912

- American Harriet Quimby is the first woman pilot to fly across the English Channel.

## 16 August

- Bennington Battle Day, a legal holiday in Vermont commemorating the 1777 defeat of the British by the Green Mountain Boys.

### *Sixteen Candles*

- just shy of sixteen, Molly Ringwald suffers her first serious crush, and friend Anthony Michael Hall helps by setting her up with her crushee in this 1984 growing-pains film, the first directed by John Hughes.

### *16 Days of Glory*

- film coverage of the 1984 Olympic Games in Los Angeles. Directed by Bud Greenspan, narrated by David Perry.

### 16 Elwood Avenue, Devonshire, England

- address of the Fawlty Towers Hotel, Basil Fawlty and wife, Sybil, proprietors, on the BBC TV comedy series, *Fawlty Towers.*

### sixteen gauge

- a measure of shotgun bore diameter = 16.81 mm.

### sixteen herbs having magical powers

- those described in the first book of *The Secrets of Albertus Magnus.*

### 16 June 1904

- Bloomsday, the day in Dublin covered by James Joyce's novel, *Ulysses.*

### Sixteen Kingdoms

- a group of short-lived sovereignties in northern China (AD 304 to 439) that took hold after the Jin Dynasty was forced south by the incoming barbarians from the north. The sixteen states were Han Zhao (304–329), Former Liang (314–376), Later Zhao (319–351), Later Liang (386–403), Southern Liang (397–414), Northern Liang (397–439), Southern Yan (398–410), Former Qin (351–394), Former Yan (370–384), Later Yan (384–397), Later Qin (384–417), Western Qin (385–431), Western Liang (400–421), Northern Yan (409–436), Xia (407–431), and Cheng Han (304–347). The period saw frequent wars and was not stabilized until the formation of the Northern Dynasties.

## 16mo

- sixteenmo or sextodecimo: page size of a book from printer's sheets that are folded into sixteen leaves, each about 4 x 6 inches to 4 -1/4 x 6 -¾ inches.

## sixteen-penny nail

- a carpenter's common nail, 3-½ inches long (coded 16d).

## 16-point type

- typographer's measure, known as "Columbian."

## 16 September

- Cherokee Strip Day in Oklahoma; commemorates the 1893 opening of the previously protected Cherokee Strip to settlers. A controversial holiday because of the death of many Cherokees as a consequence.

## Sixteen-String Jack

- an eighteenth-century English highwayman, John Rann, known for his foppish manner of dress. The nickname referred to the many ribbons or "strings" he wore on his costumes.

## Sixteenth Amendment to the Constitution

- authorizes the federal income tax, adopted February 25, 1913.

## Sixteenth Arrondissement

- section of Paris containing the Arc de Triomphe.

## sixteenth president

- Abraham Lincoln, March 4, 1861 to April 15, 1865.

## sixteenth state of the Union

- Tennessee, admitted June 1, 1796.

## "16 Tons"

- hit recording of 1955 by Tennessee Ernie Ford proclaiming the grief of the coal miner's life. Other recordings by Merle Travis earlier, and Frankie Laine later.

## Sweet 16

- a rite of passage for a young lady, celebrating her sixteenth birthday.

# *17*

- age at which one may view R-rated films without an accompanying parent (according to MPAA standards).
- atomic number of the chemical element chlorine, symbol Cl.
- a corpse, in Australian underworld slang.
- most blocked shots in an NBA game (basketball record): by Elmore Smith for the Los Angeles Lakers against the Portland Trail Blazers, October 28, 1973.
- most consecutive losing seasons by a baseball team: Pittsburgh Pirates, 1993–2009.
- most field goals scored in an NBA All-Star game (basketball record): three times, by Wilt Chamberlain, 1962; Michael Jordan, 1988; and Kevin Garnett, 2003.
- most NBA championships won (basketball record): by the Boston Celtics: in 1957 and the eight consecutive years from 1959 through 1966, coached by Red Auerbach; in 1968–69, coached by Bill Russell; in 1974 and 1976, under Tommy Heinsohn; in 1981, under coach Bill Fitch; in 1984 and 1986, coached by K. C. Jones, and in 2008, coached by Doc Rivers.
- most runs scored by a team in one inning in the "modern era" (since 1900) (baseball record): by the Boston Red Sox against the Detroit Tigers, 0, June 18, 1953, in the seventh.
- most steer-roping titles: held by Guy Allen, known in rodeo circles as "The Legend."
- most strikeouts pitched in a World Series game (baseball record): by Bob Gibson for the St. Louis Cardinals, against the Detroit Tigers, in game one of the series, October 2, 1868.
- most team penalties in a Pro Football Hall of Fame game: by the Dallas Cowboys, against the Cleveland Browns, 1999.
- a number considered unlucky in Italy. Seventeen (XVII) is an anagram, and the sum of the letters in the Latin word "VIXI," which means "I have lived," indicating no longer

alive, thus dead. In Italy, some buildings have no seventeenth floor, and some hotels have no room 17.

- number of Japanese ships, including four aircraft carriers, sunk by American naval forces in the battle of Midway, June 4–6, 1942.
- number of pharaohs in the Early Dynastic Period of Egypt.
- number of presidential electoral votes apportioned to the state of Michigan. (May be reapportioned for the 2012 election.)
- number of Saxon rulers of England: the first, Egbert, in 802; then Ethelwulf, 839; Ethelbald, 858; Ethelbert, 860; Ethelred, 866; Alfred (the Great), 871; Edward (the Elder), 899; Athelstan, 924; Edmund I, 939; Edred, 946; Edwy, 955; Edgar, 959; Edward (the Martyr), 975; Ethelred II (the Unready), 978 and 1014; and Edmund II (Ironside), 1016. Then came a twenty-six-year period of Danish kings until Saxon Restoration under Edward III (the Confessor), 1042; and Harold II, 1066. (The Norman Dynasty followed.)
- number of string quartets written by Beethoven.
- number of the French Département of Charente-Maritime.
- number of trees visible from Dostoyevsky's prison window.
- police emergency phone number in France.
- record number of medals won by a country in a Winter Olympics: by the USA at the XXI Winter Olympic Games in Vancouver, 2010; nine gold, fifteen silver, thirteen bronze. Most gold (fourteen) were won by Canada.
- retired baseball uniform number of pitcher Dizzy Dean, retired by the St. Louis Cardinals.
- retired basketball jersey number of:
  ⇨ John Havlicek, retired by the Boston Celtics.
  ⇨ Ted Turner, retired by the Atlanta Hawks

- retired hockey shirt number of Jari Kurri, retired by the Edmonton Oilers.
- Section of the California Penal Code defining felonies.
- seventeenth day of the seventh month, when the ark came to rest on the mountains of Ararat (Genesis 8:4).
- seventeen years:
  ⇨ age of Boris Becker when he won singles tennis at Wimbledon, the youngest ever to do so.

⇨ appearance cycle of the periodical cicada (*Magicicada septendecim*), also known as the seventeen-year locust, known for its noisy scratching, screeching, and singing, especially by the male.

⇨ longest tenure as Speaker of the House, Texas Democrat Sam Rayburn, 1940–61.

⇨ time Joseph lived in the land of Egypt (Genesis 47:28).

- sum of the first four prime numbers $(2 + 3 + 5 + 7)$.
- Title of the United States Code dealing with copyrights.
- value of Roman numeral XVII.

## B-17
- the Boeing Flying Fortress, a long-range U.S. bomber in World War II.

## Beethoven's *Sonata No. 17*
- the "Tempest" sonata (in D minor).

## file 17
- a wastebasket, in World War II armed forces slang (also "file 13").

## Group 17 elements (of the Periodic Table)
- halogens: fluorine (9), chlorine (17), bromine (35), iodine (53), astatine (85).

## haiku
- Japanese poetic form of three lines and a total of seventeen syllables, in the pattern of five, seven, five.

## longest winning streak in the National Hockey League— seventeen games
- by the Pittsburgh Penguins, beginning with their win over the Boston Bruins (3–2), March 9, 1993, and ending with a tied game with the New Jersey Devils (6–6), April 10.

## Pier 17

- site of the South Street Seaport on the East River in New York City. Center of the area that had been the Fulton Fish Market, the location is now a major tourist attraction.

## *Seventeen*

- (a) a magazine primarily for teenage girls, founded in 1945.
- (b) Booth Tarkington's hilarious 1916 novel about a small-town adolescent, William Sylvanus Baxter (Silly Billy), suffering the pangs of a first summer love, aggravated by his disruptive younger sister, Jane.

## 17 April

- Verrazano Day in New York; commemorates the discovery of New York harbor by Giovanni da Verrazano in 1524.

## Seventeen Article Constitution

- Japan's first formal Constitution, formulated during the reign of Prince Shotoku in the year 604, never officially repealed. Non-militaristic, and avoiding economic matters, the Constitution prescribes rules for the conduct of public officials and focuses on maintaining harmony and seeking group consensus.

## 17, Cherry Tree Lane

- London home of the Banks family in the 1964 movie *Mary Poppins,* which starred Julie Andrews as the magical nanny.

## 17, Gough Street

- London residence of literary lion Samuel Johnson in the mid-eighteenth century.

## 17-gun salute

- honors the arrival or departure of generals or admirals, or the arrival of assistant secretaries of the military services.

## 17 July

- Pig's Day, celebrated on some U.S. college campuses since the 1960s in an attempt to make mathematics, and the study of mathematics, fun.

## 17 June

- Bunker Hill Day in Boston and Suffolk County, Massachusetts, celebrating the battle in 1775.

## 17 March

- (a) St. Patrick's Day.
- (b) Evacuation Day in Boston and Suffolk County, Massachusetts, commemorating the withdrawal of British troops from Boston in 1776.

## 17 September

- Citizenship Day; formerly called Constitution Day.

## Seventeen Standard Histories

- an early Chinese work published during the Sung Seventeenth Dynasty, a period known for its high-quality Chinese printing. Printing of this work began in 994 and was completed in 1061.

## Seventeenth Amendment to the Constitution

- provides for popular election of U.S. senators, ratified on May 31, 1913.

## 17th parallel

- the dividing line between North and South Vietnam, established by peace negotiations in 1954 following the French defeat by the Viet Minh at Dien Bien Phu, the State of Vietnam (later the Republic of Vietnam) to the South and the DRV (Democratic Republic of Vietnam) communists to the North.

## seventeenth president

- Andrew Johnson, April 15, 1865 to March 3, 1869.

**seventeenth state of the Union**

• Ohio, admitted March 1, 1803.

**17 to 1**

• the odds against tossing a 3 or tossing an 11 in dice.

**Seventeen Years' War**

• (a) the Vietnam War, also known as the Second Indochinese War, dating from 1956—when President Eisenhower began sending military advisors and intelligence operatives to support the South Vietnamese government of Ngo Dinh Diem against Communist guerrilla attacks from the north— to the January 1973 peace accord between the U.S. and the DRV. The Paris Peace Agreement, however, did not end the conflict in Vietnam, as the South Vietnam government continued fighting the northern Communists until April 30, 1975, when North Vietnamese troops entered Saigon, capturing the presidential palace. Saigon was then renamed Ho Chi Minh City.

• (b) the first Sudanese civil war, 1955 to 1972, the southern non-Arab populations rebelling against the northern Arab-controlled government, the English-schooled south seeking regional self-determination. Hostilities varied in intensity until the Addis Ababa accords, signed on March 27, 1972, which guaranteed autonomy for the southern region, including acceptance of English as the principal language of the south while establishing Arabic as Sudan's official language.

**17-0**

• football record for the Miami Dolphins in 1972, the only NFL team to achieve a perfect season since the AFL-NFL merger in 1970. With Don Shula coaching and Bob Griese as quarterback, the Dolphins won all fourteen games in their regular schedule, plus three more post-season, beating the Washington Redskins 14–7 in Super Bowl VII. (The New England Patriots went into the SB-XLII with an 18-0 record, the only undefeated 16-game regular season, and two postseason wins, but lost the Bowl game to the New York Giants 17–14.)

## *Stalag 17*

- Billy Wilder directed, William Holden starred (for which he won the Best Actor Oscar), in this quintessential World War II POW film, released in 1953. Comic relief was supplied by Robert Strauss and Harvey Lembeck (reprising their roles in the Broadway play on which the film is based), and menace, by Otto Preminger as the camp commander.

## *17 1/4*

- number of Earth hours in a day on the planet Uranus.

## *18*

- atomic number of the chemical element argon, symbol Ar.
- eighteen days: average incubation period of a pigeon.
- 18 feet square: minimum size ring for professional boxing matches.
- 18 inches: diameter of the rim on a regulation basketball basket.
- eighteen innings: the longest playoff game in baseball history—the Atlanta Braves against the Houston Astros, October 9, 2005.
- 18 meters (= 59 feet), length of a regulation volleyball court.
- eighteen months: average life of a one-dollar bill.
- fewest walks pitched in a season (baseball record): by Babe Adams, Pittsburgh Pirates, in 263 innings, 1920.
- gauge (number) of a wood screw with a shank diameter of 19/64 of an inch.
- Imperial gallons in a kilderkin (British unit of capacity).
- inches usually in a cubit (biblical measure of length), but it varied.
- kabs in a bath (ancient Hebrew measure of liquid capacity).
- kabs in an ephah (ancient Hebrew measure of dry capacity).
- most career home runs hit in World Series games (baseball record): by Mickey Mantle for the New York Yankees, 1951–64.
- most goals scored by both teams in an NHL playoff game (hockey record): Los Angeles Kings, 10, Edmonton Oilers, 8, April 7, 1982.

- most golf major tournaments won in a career, by Jack Nicklaus: six Masters (1963, 1965–66, 1972, 1975, 1986), five PGA Championships (1963, 1971, 1973, 1975, 1980), four U.S. Opens (1962, 1967, 1972, 1980), and three British Opens (1966, 1970, 1978).
- most individual points scored in an NFL Pro Bowl game (football record): four times, by John Brockington for the NFC, 1973; Mike Alstott for the NFC, 2000; Jimmy Smith for the AFC, 2000; Shaun Alexander for the NFC, 2004.
- most individual points scored in a Super Bowl game (football record): five times, by Roger Craig for the San Francisco 49ers against the Miami Dolphins, SB-XIX, 1985; by Jerry Rice for the San Francisco 49ers against the Denver Broncos, SB-XXIV, 1990, and against the San Diego Chargers, SB-XXIX, 1995; by Ricky Watters for the San Francisco 49ers; and by Terrell Davis for the Denver Broncos against the Green Bay Packers, SB-XXXII, 1998.
- most runs scored by a team in one inning (baseball record): by the Chicago White Stockings against the Detroit Wolverines, ' 0, September 6, 1883, in the seventh.
- number of Academy Award ceremonies emceed by Bob Hope between 1939 and 1977.
- number of animals having magical powers as described in the third book of *The Secrets of Albertus Magnus*.
- number of chapters in the Bhagavad Gita, the Vedantic "Song of the Lord."
- number of decimal places in the inverse multiple designated by the International System prefix atto-, written "a-".
- number of holes on a regulation golf course.
- number of minor demons listed in *The Grimorium Verum*, the true black book of sorcerers; namely, Clauneck, Musisin, Bechaud, Frimost, Klepoth, Khil, Mersilde, Clisthert, Sirchade, Segal, Hicpacth, Humots, Frucissière, Guland, Surgat, Morail, Frutimière, and Huictiigaras.
- number of months in the solar calendar of the Mayan (called Haab) and of the Aztecs (called Xihuitl), each such month composed of twenty days, the sum increased by five "unlucky days" (Uayeb), adding up to a year of 365 days. The eighteen Mayan months are known, in sequence, as: Pop,

Uo, Zip, Zotz, Tzec, Xuc, Yaxkin, Mol, Chen, Yax, Zac, Ceh, Mac, Kankin, Maun, Pax, Kayab, and Cumku.

- number of seats in the House of Representatives allotted to the state of Oklahoma (as of the 2000 Census).
- number of the French Département of Cher.
- number of times actress Susan Lucci was nominated for an Emmy but failed to win for her role of Erica Kane on the daytime-TV soap, *All My Children.* In 1999, on her nineteenth nomination, she finally received the coveted award.
- number of 0s in a quintillion (American) or a trillion (British).
- number of 0s in the multiple designated by the International System prefix exa-, written "E-".
- the number worn on Adam Sandler's football jersey in the 2005 film remake of *The Longest Yard.*
- players on the field in an Australian football team (plus four on the interchange bench).
- record number of All-Star basketball games played: by Kareem Abdul-Jabbar, every year from 1970 to 1989, with the exception of 1978.
- record number of baseball games won in a season by a relief pitcher: by Roy Face, Pittsburgh Pirates, 1959.
- record number of golf match wins in PGA Tournaments in a season (of thirty entered): by Byron Nelson, 1945.
- retired baseball uniform number of:
  ⇨ pitcher Mel Harder, retired by the Cleveland Indians.
  ⇨ outfielder Ted Kluszewski, retired by the Cincinnati Reds.

- retired basketball jersey number of:
  ⇨ Dave Cowens, retired by the Boston Celtics.
  ⇨ Jim Loscutoff, retired by the Boston Celtics.

- retired football jersey number of Frank Tripucka, retired by the Denver Broncos.
- retired hockey shirt number of:
  ⇨ Danny Gare, retired by the Buffalo Sabres.
  ⇨ Dennis Savard, retired by the Chicago Blackhawks.
  ⇨ Serge Savard, retired by the Montreal Canadiens.
  ⇨ Dave Taylor, retired by the Los Angeles Kings.

- Surah 18 of the Quran tells the legend of Alexander the Great.
- Title of the United States Code dealing with crimes and criminal procedure.
- used as a code for Adolf Hitler in neo-Nazi circles, A and H being, respectively, the first and eighth letters of the alphabet.
- value of Roman numeral XVIII.

**chai**

- 18 in Hebrew, a number believed to be lucky among Jewish people.

**18 June, 1815**

- Napoleon's forces defeated by the English under the Duke of Wellington at Waterloo, Belgium. Napoleon abdicated June 22.

**18, Kensington Square**

- London residence of philosopher John Stuart Mill.

**18mo**

- eighteenmo, the page size of a book from printer's sheets that are folded into eighteen leaves, each about 4 × 6 -1/4 inches.

**eighteenpence**

- rhyming slang for common sense.

**18 percent tip**

- customary amount for adequate service in a restaurant to a waiter or waitress serving four or more people. In recent years, the amount has been edging up to 20 percent.

**18-point type**

- typographer's measure, known as "great primer."

**Eighteenth Amendment to the Constitution**

- institutes prohibition, ratified on January 29, 1919.

## Eighteenth Arrondissement

- section of Paris containing Sacre Coeur.

## 18th of April

- date in Henry Wadsworth Longfellow's 1861 poem, "Paul Revere's Ride":

> Listen my children and you shall hear
> Of the midnight ride of Paul Revere,
> On the eighteenth of April, in seventy-five.
> Hardly a man is now alive
> Who remembers that famous date and year . . .

## eighteenth president

- Ulysses S. Grant, March 4, 1869 to March 3, 1877.

## eighteenth etate of the Union

- Louisiana, admitted April 30, 1812.

## 18-wheeler

- a tractor-trailer rig, containing eighteen wheels.

## F/A-18

- military aircraft, the McDonnell Douglas Hornet, a carrier-based fighter and attack airplane. In use with the U.S. Navy, Marines, Canada, Australia, Spain, Finland, and others.

## Group 18 elements (of the Periodic Table)

- the noble gases: helium (2), neon (10), argon (18), krypton (36), xenon (54), radon (86). These were once called inert gases.

## most decorated Olympian

- Soviet gymnast Larissa Latynina won eighteen Olympic medals, the most in any sport: nine gold, five silver, and four bronze, between 1956 and 1964.

## Shakespearean Sonnet XVIII

- begins, "Shall I compare thee to a summer's day?"

# 18.2

- 18.2 percent, proportion of the Earth's land surface occupied by Asia.

# 19

- Article of the New York State Labor Law that deals with minimum wage.
- atomic number of the chemical element potassium, symbol K.
- impossible score in the card game cribbage.
- in the game of squash, the ball must hit the front wall no less than 19 inches above the floor.
- lunch-counter code for banana split.
- most runs scored in an inning by both teams (baseball record): the Cleveland Indians (13) against the Boston Red Sox (6), in the eighth inning on April 10, 1977, in the eighth.
- most years appearing in NBA playoff games (basketball record): by Karl Malone for the Utah Jazz, 1985–2002, and the Los Angeles Lakers, 2003.
- 19 x 19 lines on the board for the game Go.
- nineteen days: average gestation period of a house mouse.
- nineteen dollars: the per capita national debt in 1790, the beginning of the national government.
- 19 feet: height of the Daniel Chester French sculpture of Abraham Lincoln in the Lincoln Memorial in Washington, D.C.
- 19 mm: diameter of a U.S. penny.
- nineteen months: age of Charles Lindbergh Jr. when he was kidnapped on January 22, 1932.
- nineteen months of nineteen days each: the structure of the Bahai calendar.
- nineteen years: length of the Metonic Cycle used to calculate the date of Easter.
- number of angels guarding Hell, according to verses 27 through 31, chapter 74 in the Quran.

- number of seats in the House of Representatives allotted to each of the states of Illinois and Pennsylvania (as of the 2000 Census).
- number of ships sunk or damaged in the Japanese attack on the Pearl Harbor Naval Base on December 7, 1941, bringing the United States into World War II.
- number of the French Département of Corrèze.
- record number of consecutive baseball games lost by a pitcher: by Jack Nabors for Philadelphia Phillies, April 28 to September 28, 1916.
- record number of goals scored in a hockey playoff year: twice, by Reggie Leach for the Philadelphia Flyers, 1976; and by Jari Kurri for the Edmonton Oilers, 1985.
- retired baseball uniform number of:
  - ⇨ pitcher Bob Feller, retired by the Cleveland Indians.
  - ⇨ first baseman Jim Gilliam, retired by the Los Angeles Dodgers.
  - ⇨ outfielder Tony Gwynn, retired by the San Diego Padres.
  - ⇨ pitcher Billy Pierce, retired by the Chicago White Sox.
  - ⇨ shortstop/outfielder Robin Yount, retired by the Milwaukee Brewers.

- retired basketball jersey number of:
  - ⇨ Don Nelson, retired by the Boston Celtics.
  - ⇨ Willis Reed, retired by the New York Knicks.
  - ⇨ Lenny Wilkens, retired by the Seattle SuperSonics.

- retired football jersey number of:
  - ⇨ Lance Alworth, retired by the San Diego Chargers.
  - ⇨ Johnny Unitas, retired by the Indianapolis Colts.
- retired hockey shirt number of:
  - ⇨ John McKenzie, retired by the Hartford Whalers.
  - ⇨ Bill Masterton, retired by the Minnesota North Stars.
  - ⇨ Larry Robinson, retired by the Montreal Canadiens.
  - ⇨ Bryan Trottier, retired by the New York Islanders.
  - ⇨ Steve Yzerman, retired by the Detroit Red Wings.

- Title of the United States Code dealing with custom duties.
- value of Roman numeral XIX.

## C19A

- designation for the largest iceberg in the world, measuring approximately 2,264 square miles. Located in the Antarctic, approximately 158.8° W longitude, 62.3° latitude, C19A became the world's largest when iceberg B15, at 4,400 square miles, broke up in a powerful storm late in 2003.

## John Milton's Sonnet XIX

- "On His Blindness" (1655), which ends with the well-known line, "They also serve who only stand and wait."

## K-19

- designation of the first Soviet nuclear submarine, which suffered a major cooling system malfunction on its maiden voyage in July 1961. The story is told in a 2002 film, *K-19: The Widowmaker*, starring Harrison Ford and Liam Neeson.

## The Long Nineteenth Century

- historian's term for the period comprising the 125 years between the French Revolution, in 1789, and the beginning of World War I, in 1914.

## M-19

- 220-pound cluster of thirty-six M69 firebombs, developed to burn out Japanese cities during World War II.

## "Nineteen"

- 1985 hit song by Paul Hardcastle in the U.K., with a distinct antiwar message, "nineteen" denoting the (supposed) average age of soldiers in the Vietnam War.

## 19 August, 1929

- *Amos 'n' Andy* goes network on NBC radio. The show originally aired on WMAQ Chicago, starting March 19, 1928.

## 19, Curzon Street

- London residence of Benjamin Disraeli, once prime minister of England.

## 19-gun salute

- honors the arrival of the vice president of the U.S., a cabinet member, chief justice of the U.S., U.S. or foreign ambassador, state governor, or chief of staff of a military service.

## 19 January

- birthday of Robert E. Lee (in 1807), observed in Southern states.

## 19 June

- "Juneteenth," aka, Emancipation Day in Texas; anniversary of emancipation of slaves in that state in 1865, considered to be the day when the last slaves were freed. Not universally observed in all of Texas.

## Nineteen Propositions

- Parliament's proposal to Charles I in 1642, intended to diminish the power of the crown. The King's rejection led to the English civil war, following which Charles was convicted of treason and later beheaded.

## Nineteenth Amendment to the Constitution

- extends suffrage rights to women, ratified August 26, 1920.

## 19th hole

- euphemism for the bar at a golf clubhouse.

## nineteenth president

- Rutherford B. Hayes, March 4, 1877 to March 3, 1881.

## nineteenth state of the Union

- Indiana, admitted December 11, 1916.

## Product 19

- a breakfast cereal from Kellogg's.

## 020

- Dewey Decimal System designation for writings on library and information science.
- medical classification code for plague.

## 20

- age of majority in Japanese society. A person who is exactly twenty years of age is said to be *hatachi*.
- atomic number of the chemical element calcium, symbol Ca.
- base of the number system used by the Mayans.
- bottles in a Nebuchadnezzar (champagne measure).
- fluid ounces in a pint (British measure).
- gerahs in a shekel (ancient Hebrew currency and measure of weight).
- grains in a scruple (apothecary' measure).
- in Hebrew, the numerical value of the letter *kaph*.
- international telephone calling code for Egypt.
- long hundredweight in a long ton (= 2,240 pounds).
- members needed for a quorum in the Canadian House of Commons.
- minimum number of residents in a thorp, in the game Dungeons & Dragons.
- most career pinch-hit home runs (baseball record): by Cliff Johnson, with six different teams in both leagues, 1974–86.
- number comprising a score.
- number of amino acids contributing to the standard genetic code: Alanine, Arginine, Asparagine, Aspartic acid, Cysteine, Glutamic acid, Glutamine, Glycine, Histidine, Isoleucine, Leucine, Lysine, Methionine, Phenylalanine, Proline, Serine, Threonine, Tyrosisne, Tryptophane, Valine.
- number of Arrondissements (districts) comprising the city of Paris.
- number of children sired by composer Johann Sebastian Bach.
- number of days in each of the eighteen months in the calendar used by the Maya and other Mesoamericans.
- number of first moves possible in the game of chess.
- number of groupings of "The Sayings of Confucius."

- number of human baby teeth.
- number of lifetime Grammy Awards won by Aretha Franklin, the "Queen of Soul," since her initial win in 1967 for Best R&B Female Vocal Performance with "Respect." Her most recent Grammy was in 2008 for "Never Gonna Break My Faith," Best Gospel-Soul Duo with Mary J. Blige. She won Grammys each year from 1969 to 1975, and sang at the presidential inaugurations of both Jimmy Carter and Bill Clinton.
- number of presidential electoral votes apportioned to the state of Ohio. (May be reapportioned for the 2012 election.)
- pennyweights in an ounce troy.
- quires in a ream (paper measure).
- record number of strikeouts pitched in a major league nine-inning baseball game, done several times: twice by Roger Clemens with the Boston Red Sox, against the Seattle Mariners, April 29, 1986, and against the Detroit Tigers, September 18, 1996; and by Kerry Wood with the Chicago Cubs against the Houston Astros, May 6, 1998. Randy Johnson (Arizona Diamondbacks) pitched twenty strikeouts against the Cincinnati Reds on May 8, 2001; the game went into extra innings, but Johnson pitched only nine.
- record number of Wimbledon tennis titles won by a woman—twice: by Billie Jean King, six singles (1966–68, 1972–73, 1975), ten doubles (1961–62, 1965, 1967–68, 1970–73, 1979), and four mixed doubles (1967, 1971, 1973–74); and by Martina Navratilova, nine singles (1978–79, 1982–87, 1990), seven doubles (1976, 1979, 1981–84, 1986), and four mixed doubles (1985, 1993, 1995, 2003).
- retired baseball uniform number of:
  - ⇨ outfielder Lou Brock, retired by the St. Louis Cardinals.
  - ⇨ outfielder Luis Gonzalez, retired by the Arizona Diamondbacks.
  - ⇨ outfielder/first baseman Monte Irvin, retired by the San Francisco Giants.
  - ⇨ outfielder Frank Robinson, retired by both the Cincinnati Reds and the Baltimore Orioles.
  - ⇨ third baseman Mike Schmidt, retired by the Philadelphia Phillies.

    ⇨ pitcher Don Sutton, retired by the Los Angeles Dodgers.

    ⇨ third baseman Pie Traynor, retired by the Pittsburgh Pirates.

    ⇨ second baseman Frank White, retired by the Kansas City Royals.

- retired basketball jersey number of Maurice Lucas, retired by the Portland Trail Blazers.
- retired football jersey number of:
  - ⇨ Gino Cappelletti, retired by the New England Patriots.
  - ⇨ Barry Sanders, retired by the Detroit Lions.
- retired hockey shirt number of Luc Robitaille, retired by the Los Angeles Kings.
- shillings in a pound (obsolete coins used throughout the United Kingdom).
- short hundredweight in a short ton (= 2,000 pounds).
- stivers in a guilder (obsolete Dutch coin).
- sum of the first seven Fibonacci numbers (0 + 1 + 1 + 2 + 3 + 5 + 8).
- Title of the United States Code dealing with education.
- total number of perfect games pitched in major league baseball, the most recent by Philadelphia Phillies' Roy Halladay against the Florida Marlins at Sun Life Stadium on May 29, 2010, the second perfect game in this month. Final score, 1-0.
- 20 feet: width of a regulation badminton court.
- 20 inches: width of the gohonzon, the prayer scroll given to each new convert to the Nichiren Shoshu, a nontraditional form of Buddhism.
- 20 meters:
  - ⇨ height of the Sphinx.
  - ⇨ length of the exchange zone for passing the baton in a relay race.

- twenty minutes: length of a regulation hockey period. A game consists of three such periods, with fifteen-minute intermissions.
- 20 percent: the proportion of the Earth's land surface occupied by Africa.
- 20 tons: size of meteor that fell near Blackstone, Virginia, on May 12, 1922.

- 20 yards: depth of each end zone on a Canadian football field.
- twenty years: most consecutive years in the hockey playoffs— by Larry Robinson, Montreal Canadiens, 1973–92.
- value in the game of pinochle for a meld of king and queen in any suit but trumps, called a marriage, or common marriage.
- value of Roman numeral XX.

**double eagle**

- twenty-dollar gold piece, minted by the U.S. from 1849 to 1933.

**I-20**

- major east-west interstate highway from Pecos to Fort Worth, Shreveport, Jackson, Birmingham, Atlanta, Columbia, and Florence (South Carolina).

**longest winning streak in professional hockey home games—twenty wins**

- this NHL record was achieved twice: by the Boston Bruins, 1929–30, and by the Philadelphia Flyers, 1975–76.

*On the Twentieth Century*

- 1978 Broadway musical comedy with music by Cy Colemen, book and lyrics by Betty Comden and Adolph Green, based on the 1934 film, *Twentieth Century*.

**Proposition 20**

- of Book IX of Euclid's *Elements*; contains his proof of the infinitude of primes.

**The Roaring Twenties**

- the 1920s, perceived as a raucous period of fast cars, loose women, speakeasies, jazz, affluence, and unbridled self-indulgence.

*The Roaring Twenties*

- a 1939 fast-paced gangster film, directed by Raoul Walsh, that explores the excesses of the Prohibition era. Memorable ending of James Cagney dying on the steps of a church on New Year's Eve, his head cradled in the lap of Gladys George, who responds to a young cop's query, "Who was he?" with an understated "He used to be a big shot."

## Twentieth Amendment to the Constitution

- the "lame duck" amendment; provides that Congress would convene each year on January 3, and that the terms of the president and vice president would begin on January 20; ratified on February 6, 1933.

## *The 20th Century*

- a CBS weekly historical TV documentary, narrated by CBS News correspondent Walter Cronkite on Sunday evenings, from October 20, 1957 to January 4, 1970.

## *Twentieth Century*

- (a) 1932 stage comedy by Ben Hecht and Charles MacArthur, revived in December 1950 and again in March 2004. Adapted to:
- (b) 1934 Howard Hawks comedy film showing John Barrymore at the top of his form, with a screenplay by the original playwrights. The film served as the basis for the 1978 Broadway musical, *On the Twentieth Century*.

## Twentieth Century Limited

- a train of the New York Central Railroad running between Chicago and New York City, noted for its luxury. Service was inaugurated June 15, 1902, and discontinued December 3, 1967, when New York Central merged with the Pennsylvania Railroad to form the Penn Central.

## *Twentieth Century Magazine*

- Boston monthly journal, in print from 1909 to 1913, well-known in the muckraker movement.

## twentieth president

- James Garfield, March 4, 1881 to September 9, 1881.

**twentieth state of the Union**

- Mississippi, admitted December 10, 1817.

**twentieth wedding anniversary gift**

- china is customary.

**twenty day names in the Mesoamerican calendar**

- alligator, wind, horse, lizard, serpent, death, deer, rabbit, water, dog, monkey, grass, reed, jaguar, eagle, vulture, movement, flint, rain, and flower.

**twenty dollars**

- (a) the value "in controversy" in civil suits above which the right of trial by jury is preserved, as per the Seventh Amendment to the U.S. Constitution.
- (b) denomination of U.S. paper money that bears the portrait of Andrew Jackson.

**twenty fingers, twenty toes**

- refers to the poem:

  > Has twenty nails, upon each hand,
  > Five, and twenty on hands and feet:
  > All this is true, without deceit.

**20, Forthlin Road**

- teenage home of Beatle Paul McCartney in Liverpool. England. British National Trust labels it "the birthplace of the Beatles," where they wrote and rehearsed their earliest songs.

**twenty gauge**

- a measure of shotgun bore diameter = 15.90 mm.

**20 January**

- Inauguration Day in the U.S., on which day, every four years, the newly elected President of the United States is sworn into office.

## 20 July 1969

- important date in cultural and scientific history—the first time a human set foot on another planet: Six hours after landing at 4:17 PM (EDT), astronaut Neil A. Armstrong took that "small step" out of the lumar module *Eagle* onto the surface of the moon. He was soon joined by Buzz Aldrin, following which the two spent two hours and twenty-one minutes on the lunar surface. The lunar module stayed on the moon for a total of twenty-one hours.

## 20, Maresfield Gardens

- address of Sigmund Freud's residence in London. Now the Freud Museum.

## 20 May

- Mecklenburg Day in North Carolina; commemorates the signing of the Mecklenburg Declaration of Independence in 1775.

## *Twenty Million Sweethearts*

- engaging 1934 film satire on radio, with Pat O'Brien, Dick Powell, and Ginger Rogers. Remade in 1949 as *My Dream Is Yours*, with Doris Day.

## *Twenty Mule Team*

- entertaining and action-filled 1940 Western film featuring Wallace Beery and Leo Carillo, supported by Marjorie Rambeau and Anne Baxter.

## 20 Mule Team Borax

- a trademarked cleaning product, popular in the 1930s through the 1970s due to its sponsorship of *Death Valley Days*, a Western anthology, first on radio's Blue network in the 1930s and '40s, and later on television, from 1952 through 1975.

## twenty-penny nail

- a carpenter's common nail, 4 inches long (coded 20d).

**twenty pieces of silver**
- price paid for Joseph when his brothers sold him to the Ishmaelites (Genesis 37:28).

**20-point card**
- in the game of Canasta, any ace or deuce.

**20-point type**
- typographer's measure, known as "paragon."

**Twenty Questions**
- a parlor game in which one player selects a person, place, or thing which the other players try to identify by asking no more than twenty questions, each of which is answered only by a "yes" or "no." Which led to . . .

*Twenty Questions*
- a popular quiz show on which the panel asked the questions, after being told whether the unknown was animal, vegetable, or mineral. The show originated on Mutual Radio in 1946, and moved to TV in November 1949. Its final telecast was May 3, 1955.

**twenty pence per mile**
- reimbursement rate for use of bicycle for travel by a member of the British Parliament (2007–08 legislative session).

*Twenty Thousand Years in Sing Sing*
- Spencer Tracy is a kingping gangster in this 1933 prison movie, directed by Michael Curtiz and based on a book by Lewis E. Lawes, warden of the renowned prison when the film was made, with many scenes shot within its walls. Bette Davis plays Tracy's moll.

*Twenty Thousand Leagues Under the Sea*
- Jules Verne's prophetic 1870 novel tracing the travels of the sociophobic Captain Nemo, who roams the world's oceans aboard the *Nautilus*, a submarine he designed and built. A film adaptation appeared in 1954 from Disney Studios,

directed by Richard Fleischer, with James Mason as Captain Nemo and Kirk Douglas, Paul Lukas, and Peter Lorre as his captive traveling companions.

## 20 to 1

- the odds against being dealt two pair in five-card poker.

## 20/20

- measure of visual acuity; indicates normal vision at 20 feet.

## *20/20*

- ABC-TV's news magazine, intended to compete with CBS's successful *60 Minutes*. The initial telecast on June 6, 1978, co-anchored by Robert Hughes and Harold Hayes, took a beating from the critics. The show was redesigned, and Hugh Downs was brought in as anchor, a role he filled until 1999. Several co-anchors included Barbara Walters (1984–2004), John Miller (2002–2003), and later, John Stossel and Elizabeth Vargas. Dozens of "co-anchors" and correspondents, many well-known from other fields, filled out the roster. As of this writing, the show is still on the air.

## 20-20 hindsight

- the ability to predict an event after it has occurred.

## *Twenty Years After*

- Alexandre Dumas' 1845 exciting sequel to *The Three Musketeers*.

## *Twenty Years at Hull House*

- Jane Addams's 1910 autobiographical account of how she established and directed one of the nation's most successful settlement houses.

## U-20

- German submarine, commanded by Kapitanleutnant Walther Schwieger, that sank the *Lusitania* on May 7, 1915, with 1,198 persons killed of the 1,924 aboard.

# 21

- atomic number of the chemical element scandium, symbol Sc.
- card game also known as blackjack.
- fewest points scored by a hockey team in one season: Washington Capitals, 1974–75.
- had been the minimal voting age in most states until July 1971, when the twenty-sixth Constitutional Amendment was ratified, standardizing the voting age to eighteen.
- lunch-counter code for a glass of lemonade or limeade.
- most goals, both teams, in a regular season NHL game (hockey record): twice, the Montreal Canadiens, 14, the Toronto St. Pats, 7, January 10, 1920; the Edmonton Oilers, 12, and the Chicago Blackhawks, 9, December 11, 1985.
- most NHL playoff years played (hockey record): by Ray Bourque for the Boston Bruins, 1980–96 and 1998–99; and the Colorado Avalanche, 2000–01.
- most pass receptions in an NFL game (football record): by Brandon Marshall for the Denver Broncos against the Indianapolis Colts, December 13, 2009. But the Colts won, 28–16.
- most screen writers ever credited for a film: *Forever and a Day,* from RKO in 1943. Co-writers included C. S. Forester, John van Druten, Christopher Isherwood, James Hilton, Norman Corwin, and Gene Lockhart.
- most seasons as an NBA player (basketball record): by Robert Parish, for the Golden State Warriors, 1976–80; for the Boston Celtics, 1980–94; for the New Orleans Hornets, 1994–96; and for the Chicago Bulls, 1996–97 season.
- most three-point field goals in a game by one team (basketball record): by the Toronto Raptors against the Philadelphia 76ers, March 13, 2005.
- most U.S. Open tennis tournaments played: by Martina Navratilova, 1973–93.
- number of buildings comprising Rockefeller Center in New York City.
- number of consonants in the English alphabet, along with five vowels.
- number of counties in the state of New Jersey.

- number of days for the Graf Zeppelin, the first lighter-than-air airship, to fly around the world in 1929, leaving and returning to Lakehurst, New Jersey, covering 20,373 miles.
- number of decimal places in the inverse multiple designated by the International System prefix zepto-, written "z-".
- number of pips, or spots, on one cubic die $(1 + 2 + 3 + 4 + 5 + 6)$.
- number of presidential electoral votes apportioned to the states of Illinois and Pennsylvania. (May be reapportioned for the 2012 election.)
- number of the French Département of Côte-d'Or.
- number of zeroes in a sextillion.
- number of zeroes in the multiple designated by the International System prefix zetta-, written "Z-".
- record number of scoreless innings pitched in one baseball game: by Joe Oeschger, with the Boston Braves, May 1, 1920 (the 6th to 26th innings).
- record number of strikeouts pitched in a major league baseball game: by Tom Cheney for the Washington Senators against the Baltimore Orioles, September 12, 1962, in sixteen innings. Cheney threw a total of 228 pitches, striking out thirteen in the first nine innings, eight more in overtime.
- record number of tennis titles won in a season: by Margaret Smith Court, 1970, including all four Grand Slam Tournaments.
- record number of Tony Awards won: by producer/director Hal Prince, in over forty years in the theater, the most recent in 2006 for Lifetime Achievement.
- retired baseball uniform number of:
  ⇨ outfielder Roberto Clemente, retired by the Pittsburgh Pirates.
  ⇨ pitcher Bob Lemon, retired by the Cleveland Indians.
  ⇨ pitcher Warren Spahn, retired by the Atlanta Braves.

- retired basketball jersey number of:
  ⇨ Dave Bing, retired by the Detroit Pistons.
  ⇨ Vlade Divac, retired by the Sacramento Kings.
  ⇨ Bill Sharman, retired by the Boston Celtics.
  ⇨ Dominique Wilkins, retired by the Atlanta Hawks.

- retired hockey shirt number of:

⇨ Michel Briere, retired by the Pittsburgh Penguins.

⇨ Stan Mikita, retired by the Chicago Blackhawks.

- shillings in a guinea (obsolete British coin).
- sum of the first six integers $(1 + 2 + 3 + 4 + 5 + 6)$.
- sum of the spots on a die.
- Title of the United States Code dealing with food and drugs.
- twenty-one days: average gestation period of a field mouse, a rat, or a chicken.
- 21 feet:
  ⇨ distance from the service line to the net on a regulation tennis court.
  ⇨ width of a regulation squash court.

- 21 percent of the air we breathe is oxygen.
- twenty-one seconds: fastest three goals scored (hat trick) in one game (hockey record): by Bill Mosienko for the Chicago Blackhawks against the New York Rangers in the last game of the season, March 23, 1952. Chicago won 7–6.
- twenty-one stories:, the height of the *Queen Mary 2*, the world's largest operating passenger ship.
- U.S. dry quarts in an ephah (ancient Hebrew measure of dry capacity).
- value of Roman numeral XXI.
- winning score in ping pong (table tennis).

### *Bat*21*

- drawing on a real event, this 1988 film, directed by Peter Markle, has Captain Bartholomew Clark (Danny Glover) charged with rescuing Air Force Lieutenant Colonel Iceal Hambleton (Gene Hackman), an expert on missile weaponry, from enemy territory before the Vietcong can capture him. Based on Hambleton's 1980 book, *Bat 21*.

### Beethoven's *Sonata No. 21*

- the "Waldstein" (in C major).

### Century 21

- a real estate company.

## Century 21 Exposition

- alternate name for the 1962 World's Fair in Seattle, Washington.

## Channel 21

- in New York area, WLIW, the Public Broadcasting TV Station in Garden City, New York.

## LPD-21

- registry of the USS *New York*, an amphibious transport dock built in a New Orleans shipyard using 7.5 tons of recaptured steel from the collapsed World Trade Center. The ship made her first entry into New York Harbor in November 2, 2009.

## Twenty-First Amendment to the Constitution

- repeals Prohibition, ratified December 5, 1933.

## twenty-first president

- Chester A. Arthur, September 20, 1981 to March 3, 1885.

## twenty-first state of the Union

- Illinois, admitted December 3, 1818.

## *Twenty-One*

- a quiz show, emceed by Jack Barry, that debuted on NBC-TV on September 12, 1956. It fell victim to the quiz show scandal of 1958, when its producers admitted the game had been rigged, and that they had often given answers to the contestants. The most popular winner was college professor Charles Van Doren, who eventually admitted his complicity and was relieved of his academic position.

## 21 April

- San Jacinto Day; Texas holiday commemorating the Battle of San Jacinto in 1836, in which Sam Houston and 900 Texans demolished Mexican forces under Santa Ana, retaliating for the loss at the Alamo.

### The Twenty-One Balloons

- book for preteens by William Pene du Bois, winner of 1947 Newbery Medal, that depicts an island society in which wealth governs all activities, portraying the destructive effects of greed and excess. The balloons of the title refer to those planned as vehicles of escape from the potential danger of the active volcano on the island.

### 21 Club

- an exclusive restaurant on West 52nd Street in New York City, known for its signature collection of painted cast-iron jockeys on the steps above the entrance. Originally a glamorous speakeasy, it boasts ten private dining rooms, one in the Prohibition-era wine cellar.

### 21 December

- Forefathers' Day, commemorating the landing of the Pilgrims on Plymouth Rock in 1620. Observed since 1769 in New England.

### Twenty-One Demands

- ultimatum secretly given to China by Japan in 1915, by which Japan attempted to gain increased control over Chinese affairs.

### 21-gun salute

- honors the arrival or departure of the president of the U.S. Also used for a former president, a president-elect, a sovereign or chief of state of a foreign country, or a member of a reigning royal family.

### 21 January

- Lee-Jackson Day, the birthdays of Robert E. Lee (1807) and Stonewall Jackson (1824).

## 21 Jump Street

- TV series about an undercover police unit consisting of young-looking officers specializing in youth crime. Ran on the Fox Network 1987–90, then was syndicated in 1990–91. Cast featured Johnny Depp as Officer Tom Hanson, Peter DeLuise as Officer Doug Penhall, and Holly Robinson as Officer Judy Hoffs. Show was created by Stephen J. Cannell.

## 21 November

- Christian feast day commemorating the Presentation of the Virgin Mary.

## 21 Up

- the third installment (1977) of the Michael Apted documentary on British TV following the lives of a group of British children in seven-year intervals.

## Twenty-One Years' War

- the second Sudanese civil war, 1983 through the end of 2004. Since the end of the first civil war, in 1972, an uneasy peace prevailed between the non-Arab south and the Islamic north, through several governments in Khartoum. In the interim, leaders of the south, many schooled in England, were increasingly excluded from serving in the governance of their own country, and, in 1983, when President Nimiery declared his intent to impose Islamic law on the entire country, the civil war resumed. A treaty was finally signed on January 9, 2005, giving the south autonomy for six years, following which the people of south Sudan will be able to vote in a referendum on secession.

## 022

- medical classification code for anthrax disease.

## 22

- Article of the New York State Tax Law covering personal income tax.

- atomic number of the chemical element titanium, symbol Ti.
- badge number of Agent Mark Slate, played by Noel Harrison, on the NBC-TV spy spoof, *The Girl from U.N.C.L.E.* (1966–67).
- badge number of Inspector Nash Bridges, played by Don Johnson, on the CBS-TV police series, *Nash Bridges.*
- most pass completions in an NFL Pro Bowl game (football record): by Peyton Manning for the AFC, 2004.
- most Ryder Cup golf matches won by an American, Arnold Palmer (of 32 played), 1961 to 1974 (six singles, nine foursome, seven four-ball). In 1975 Palmer returned as non-playing captain.
- most shutouts in an NFL season (hockey record): by the Montreal Canadiens, 1928–29; all won by goalie George Hainsworth.
- most years in NHL playoffs (individual hockey record): by Chris Chelios for the Montreal Canadiens, 1984–90; the Chicago Blackhawks, 1990–97; and the Detroit Red Wings, 1999–2004, 2006–07.
- number of balls played in a snooker match: one each of black, white, green, yellow, blue, brown, pink; and fifteen red.
- number of bones in the human skull.
- number of letters in the Hebrew alphabet.
- number of seasons Babe Ruth played baseball, for three different teams: 1914–19 with the Boston Red Sox, 1920–34 with the New York Yankees, and 1935 with the Boston Braves.
- number of the French Département of Côte-d-Armor.
- the number worn on Burt Reynolds's football jersey in the 1974 movie, *The Longest Yard.*
- retired baseball uniform number of pitcher Jim Palmer, retired by the Baltimore Orioles.
- retired basketball jersey number of:
  - ⇨ Elgin Baylor, retired by the Los Angeles Lakers.
  - ⇨ Rolando Blackman, retired by the Dallas Mavericks.
  - ⇨ Dave DeBusschere, retired by the New York Knicks.
  - ⇨ Clyde Drexler, retired by both the Houston Rockets and the Portland Trail Blazers.
  - ⇨ Ed Macauley, retired by the Boston Celtics.

    ⇨ Larry Nance, retired by the Cleveland Cavaliers.

- retired football jersey number of:
  - ⇨ Bobby Layne, retired by the Detroit Lions.
  - ⇨ Buddy Young, retired by the Indianapolis Colts.
- retired hockey shirt number of Mike Bossy, retired by the New York Islanders.
- roulette number Rick Blaine (Humprey Bogart) tells a refugee to play twice in order to win enough money for he and his wife to get out of the city and back to the U.S. in the film *Casablanca*.
- slang for a .22 caliber gun.
- Title of the United States Code dealing with foreign relations and intercourse.
- twenty-two days: period of space flight of Russian dogs Veterok and Ugolek orbiting the Earth in Soviet *Kosmos 110*, launched on February 22, 1966.
- 22 miles: approximate thickness of the Earth's outer layer, the crust.
- 22 ounces: maximum weight of a regulation basketball (minumum is 20 ounces).
- twenty-two years: longest streak of NBA playoff appearances (basketball record): by the Syracuse Nationals / Philadelphia 76ers, 1949–71.
- value of Roman numeral XXII.
- wickets on a cricket pitch are 22 yards apart.
- yards in a chain.


**catch-22**

- an enigmatic problem for which any resolution leads to inconsistent and perverse outcomes; an insoluble dilemma. Describes the military regulations in...

*Catch-22*

- the best-selling 1961 novel by Joseph Heller; converted to a 1965 film directed by Mike Nichols and starring Alan Arkin as the panic-stricken Captain Yossarian.

## CVL-22

- number of the U.S. Navy light aircraft carrier *Independence*. Used in atom bomb tests at Bikini Atoll in July 1946; sunk as target ship on January 1951.

## Haydn's *Symphony No. 22*

- The "Philosopher" symphony (in E-flat major).

## Shakespearean Sonnet XXII

- begins, "My glass shall not persuade me I am old, So long as youth and thou are of one date . . ."

## Twenty-Second Amendment to the Constitution

- stipulates that no person shall serve more than two terms as President of the United States, declared ratified on February 27, 1951.

## twenty-second president

- Grover Cleveland, March 4, 1885 to March 3, 1889.

## twenty-second state of the Union

- Alabama, admitted December 14, 1819.

## 22-carat gold

- the purest used in most jewelry; twenty-two parts gold and two parts alloy to provide extra strength.

## twenty-two common traits of heroes

- scale of twenty-two characteristics describing the archetypal hero, created by Lord Raglan in 1936 in *The Hero*.

## 22 February

- George Washington's birthday, 1732, in Wakefield, Virginia; observed on Presidents' Day, the third Monday in February.

## 22 November, 1906

- International Radio Telegraphic Convention adopts "SOS" as international distress signal.

## 22 November, 1963

- date of John Kennedy's assassination in Dallas, Texas.

## 22-point type

- typographer's measure, known as "double small pica."

### twenty-two trumps

- the major arcana in the traditional tarot deck, number 0 to 21, and corresponding to the twenty-two letters in the Hebrew alphabet.

## 23

- atomic number of the chemical element vanadium, symbol V.
- the first prime number in which both digits are prime and make another prime when combined.
- the letter "W" is the twenty-third in the alphabet. Numerologists point out that it has two points down and three points up.
- most All-Star games played (hockey record): by Gordie Howe, between 1948 and 1980.
- most blocked shots in an NBA game (basketball record): by Toronto Raptors against Atlanta Hawks, March 23, 2001.
- most completed passes in a Pro Bowl game (football record): by Peyton Manning for the AFC, 2004.
- most fumbles in an NFL season (football record): twice, by Kerry Collins for the New York Giants, 2001, and Daunte Culpepper, Minnesota Vikings, 2002.
- most Ryder Cup golf matches won by an individual (of forty-six played): Nick Faldo of the UK, 1977 to 2004 (six singles, ten foursome, seven four-ball).
- most team three-point field goals in an NBA game (basketball record): by the Orlando Magic against the Sacramento Kings, January 13, 2009.
- most touchdown pass receptions in an NFL season (football record): by Randy Moss for the New England Patriots, 2007.
- most touchdowns scored in a Canadian Football League season: by Mike Stegall for the Winnepeg Blue Bombers, 2002.

- the Nissan automobile reportedly takes its name from "ni," which means two in Japanese, and "san," which means three. Thus, *Nissan* means "23."
- number of chromosomes contributed to the offspring by each parent.
- number of counties in each of the states of Maryland and Wyoming.
- number of letters in the classical Latin alphabet.
- number of men on each side in the game Peralikatuma, played in Ceylon.
- number of plants having magical powers as described in the first book of *The Secrets of Albertus Magnus.*
- number of popes named John.
- number of rooms in Elvis Presley's Graceland mansion in East Memphis.
- number of the French Département of Creuse.
- record number of career grand-slam home runs (baseball record): by Lou Gehrig, New York Yankees, 1923–39.
- record number of consecutive team losses in the NBA (basketball record): by the Cleveland Cavaliers, March 19 to November 5, 1982; the last nineteen games of the 1981–82 season and the first five games of the 1982–83 season.
- record number of Stanley Cup wins by a hockey team: the Montreal Canadiens, the first in 1924, the most recent in 1993.
- retired baseball uniform number of:
    - ⇨ Outfielder Willie Horton, retired by the Detroit Tigers.
    - ⇨ first baseman Don Mattingly, retired by the New York Yankees.
    - ⇨ second baseman Ryne Sandberg, retired by the Chicago Cubs.

- retired basketball jersey number of:
    - ⇨ Lou Hudson, retired by the Atlanta Hawks.
    - ⇨ Michael Jordan, retired by both the Chicago Bulls and the Miami Heat.
    - ⇨ Calvin Murphy, retired by the Houston Rockets.
    - ⇨ Frank Ramsey, retired by the Boston Celtics.
    - ⇨ John Williamson, retired by the New Jersey Nets.

- retired hockey shirt number of:
    - ⇨ Bob Nystrom, retired by the New York Islanders.

⇨ Bob Gainey, retired by the Montreal Canadiens.

- twenty-three games: the longest suspension in hockey history for an on-ice IPS action, against Marty McSorley (Boston Bruins) for slashing Donald Brashear (Vancouver Canucks) on the head with his stick during a game in Vancouver on February 21, 2000. McSorley was banned for the remainder of the playing year.
- telegrapher's code for "line break."
- there were exactly twenty-three characters (numbers and letters) on the face of all U.S. coins (excluding mint marks, where they appeared), until the new state quarters, which have fifty-six.
- Title of the United States Code dealing with highways.
- value of Roman numeral XXIII.

### Beethoven's *Sonata No. 23*

- the "Appassionata" (in F minor).

### longest winning streak in the National Football League—twenty-three games

- Indianapolis Colts regular-season games, beginning with their 18–15 win over the New England Patriots, November 2, 2008, and ending with their loss to the New York Jets, 29–15, December 27, 2009.

### "On His Having Arrived at the Age of Twenty-Three"

- a sonnet by John Milton in which he expresses his concern that time is passing too quickly and that his talent may wane as he grows older.

### Sanhedrin

- among the ancient Jews, a local tribunal of twenty-three members having authority over minor civil and criminal actions in law. Every city had such a Lesser Sanhedrin, which functioned subordinate to the Great Sanhedrin [q.v. under 71]. According to tradition, both were founded by Moses.

**Twenty-Third Amendment to the Constitution**
- gives citizens of the District of Columbia the right to vote in presidential elections, adopted March 29, 1961.

**twenty-third president**
- Benjamin Harrison, March 4, 1889 to March 3, 1893.

**twenty-third state of the union**
- Maine, admitted March 15, 1820.

**23 April**
- the date of William Shakespeare's birth in 1556 and death in 1616.

**23 February**
- Iwo Jima Day; commemorates the raising of the American flag on Mount Suribachi on Iwo Jima by the Marines in World War II.

**23, Fitzroy Road**
- London residence of poet W. B. Yeats.

**23 March 1775**
- Patrick Henry tells the 2nd Virginia Convention meeting in Richmond, "Give me liberty or give me death."

**23 October 1932**
- Fred Allen's radio debut on CBS, on a show then called *The Linit Bath Club*. It went through several other names, usually identifying the sponsor, before ending up simply as *The Fred Allen Show*.

***23 Paces to Baker Street***
- a 1956 suspense film starring Van Johnson as a blind writer who overhears a kidnapping plot. Directed by Henry Hathaway, the cast includes Vera Miles and Cecil Parker.

## "twenty-three skidoo!"

• (archaic): phrase with variable meaning, sometimes expressing surprise or pleasure, sometimes skepticism, sometime rejection or negation, as in "Beat it!" Originated about 1900 but now associated with the 1920s.

## 23, Tedworth Square

• Mark Twain's residence in London in the mid-1890s.

## *024*

• Decimal ASCII code for [cancel].

## *24*

• the Atlantic Ocean covers 24 percent of the Earth's surface.
• atomic number of the chemical element chromium, symbol Cr.
• carats in pure gold.
• grains in a pennyweight (troy).
• hours in a day.
• mathematical value of 4 factorial (4!) = (4 x 3 x 2 x 1).
• most eagles in golf's Masters Tournaments, by Jack Nicklaus, in play 1959–2000.
• most field goals scored in an NBA playoff game (basketball record): three times, by Wilt Chamberlain for the Philadelphia Warriors against the Syracuse Nationals, March 14, 1960; by John Havlicek for the Boston Celtics against the Atlanta Hawks, April 1, 1973; and by Michael Jordan for the Chicago Bulls against the Cleveland Cavaliers, May 1, 1988.
• most game-winning goals scored in NHL playoffs, career (hockey record): by Wayne Gretzky for four different teams, 1979–99.
• most World Series games pitched: by Mariano Rivera, of the New York Yankees, 1996, 1998, 1999, 2000, 2001, 2003, and 2009.
• most Stanley Cups won (hockey record): by the Montreal Canadiens, the first in 1916, then in 1924, 1930–31, 1944, 1946, 1953, 1956–60, 1965–66, 1968–69, 1971, 1973, 1976–79, 1986, and 1993.

- number of books in Homer's epic poem (in dactylic hexameter), *The Iliad*; the same in *The Odyssey*.
- number of cycles (each of fifteen days) in the Chinese solar year.
- number of decimal places in the inverse multiple designated by the International System prefix yocto-, written "y-".
- number of different stories in Chaucer's *Canterbury Tales*.
- number of great sages (*tirthamkaras*) in the present Jain era, the instructors of humanity who have learned to transcend human ignorance and misery and to achieve the peace of Nirvana.
- number of letters in both the modern and classical Greek alphabet.
- number of pieces in a checkers game, twelve on each side.
- number of points on a backgammon board.
- number of ribs in the human body.
- number of the French Département of Dordogne.
- number of uniform sheets of paper in a quire.
- number of 0s in a septillion (American) or a quadrillion (British).
- record number of consecutive baseball games won by a pitcher: by Carl Hubbell, New York Giants, July 17, 1936 to May 27, 1937; sixteen in 1936, eight in 1937.
- record number of women's Grand Slam singles tennis titles: won by Australian Margaret Smith Court—five U.S. Opens (1962, 1965, 1969–70, 1973), three Wimbledons (1963, 1965, 1970), five French Opens (1962, 1964, 1969–70, 1973), and eleven Australian Opens (1960–66, 1969–71, 1973).
- retired baseball uniform number of:
  - ⇨ manager Walt Alston, retired by the Los Angeles Dodgers.
  - ⇨ outfielder Rickey Henderson, retired by the Oakland Athletics.
  - ⇨ manager Whitey Herzog, retired by the St. Louis Cardinals
  - ⇨ outfielder Willie Mays, retired by the San Francisco Giants.
  - ⇨ first baseman Tony Perez, retired by the Cincinnati Reds.
  - ⇨ outfielder Jimmy Wynn, retired by the Houston Astros.

- retired basketball jersey number of:
  ↪ Rick Barry, retired by the Golden State Warriors.
  ↪ Bill Bradley, retired by the New York Knicks.
  ↪ Tom Chambers, retired by the Phoenix Suns.
  ↪ Spencer Haywood, retired by the Seattle Super Sonics.
  ↪ Bobby Jones, retired by the Philadelphia 76ers.
  ↪ Sam Jones, retired by the Boston Celtics.
  ↪ Moses Malone, retired by the Houston Rockets.

- retired football jersey number of Lenny Moore, retired by the Indianapolis Colts.
- retired hockey shirt number of:
  ↪ Bernie Federko, retired by the St. Louis Blues.
  ↪ Terry O'Reilly, retired by the Boston Bruins.

- Title of the United States Code dealing with hospitals and asylums.
- total number of major and minor keys in Western tonal music.
- twenty-four dollars: the estimated value of the beads that Peter Minuit used as payment for Manhattan Island.
- twenty-four elders sat around the throne in heaven, clothed in white gold crowns on their heads (Revelation 4:4).
- 24 feet square: maximum size ring for professional boxing matches.
- twenty-four frames per second: speed of commercial film projector used in most movie theaters.
- 24 miles: length of the longest bridge in the world, the Second Lake Ponchartrain Causeway, in Mandeville, Louisiana, opened in 1956. (The Bang Na Expressway, a viaduct in Bangkok, is actually longer, at 33.55 miles, but is not in contention for longest bridge because only a small portion of it crosses water.)
- twenty-four seasons: lifetime of the CBS-TV variety show, *The Ed Sullivan Show.*
- value of Roman numeral XXIV.

## A-24

- a U.S. Army Air Force dive-bomber, the Douglas Banshee, which saw service in the Pacific early in World War II. Under the designation SBD, it was also used by the Navy, called the Dauntless.

## B-24

- the Consolidated Liberator, used by the Air Force early in World War II, the airplane that was produced in larger numbers (more than 18,000) than any other American military aircraft. The B-24A was used mostly as a long-range transport; the B-24D bombed the Ploesti oil fields in Romania in June 1942.

## "four and twenty blackbirds"

- in the nursery rhyme:

  > Sing a song of sixpence,
  > A pocketful of rye,
  > Four and twenty blackbirds
  > Baked in a pie.
  > When the pie was opened
  > The birds began to sing.
  > Wasn't that a dainty dish
  > To set before the king?

## Group of 24

- (G-24): a chapter of the G-77 established in 1971 to coordinate monetary and financial concerns of developing countries and to represent them in international dealings.

## *Hill 24 Doesn't Answer*

- 1955 Israeli film portraying the fight for modern Israel in the stories of four soldiers stationed on a hill outside Jerusalem in the 1948 war.

## longest losing streak in professional basketball— twenty-four games

- by Cleveland Cavaliers over two seasons: from the loss to the Milwaukee Bucks on March 19, 1982 (1981–82 season) through the first five games in 1982–83 to win over the Golden State Warriors on November 10, 1982.

## Rule 24

- deals with personal fouls in the water polo rulebook.

## *24*

- a suspenseful TV spy series, debuting on the Fox Network on November 6, 2001, starring Kiefer Sutherland as agent Jack Bauer of the CTU (Counterterrorism Unit), with the final episode airing on May 24, 2010.

## 24 August

- Saint Bartholomew's Day; commemorates the massacre of Huguenots on the order of King Charles IX of France in the year 1572.

## 24 July

- Pioneer Day, a legal holiday in Utah, commemorating the first Mormon settlement near Salt Lake in 1847.

## 24mo

- twenty-fourmo or vegismoquarto, the page size of a book from printer's sheets that are folded into twenty-four equal sections, each about 3-½ x 6 inches.

## 24 October

- United Nations Day.

## 24 Parganas

- a district of West Bengal, India.

## 24-point type

- typographer's measure, known as "double pica."

## twenty-four proprietors

- a group of investors, mostly Scots, who formed a board of proprietors to administer land in East New Jersey sold by the estate of Sir George Carteret on his death in 1682. Sir George, along with Lord John Berkeley, had been given a lease to this territory by James, Duke of York, who had been granted patents for this land by his brother, King Charles II of England.

## twenty-four-second clock

- refers to the rule in basketball that a team must attempt a field goal within twenty-four seconds after gaining possession of the ball.

## 24-7

- abbreviation for twenty-four hours a day, seven days a week, meaning continuously. Usually used to describe a business or service that never closes.

## 24 Sussex Drive, Ottawa, Ontario

- residence of the Prime Minister of Canada.

## Twenty-Fourth Amendment to the Constitution

- abolishes the poll tax, ratified January 23, 1964.

## twenty-fourth president

- Grover Cleveland, March 4, 1893 to March 3, 1897.

## twenty-fourth state of the Union

- Missouri, admitted August 10, 1821.

## 24-0

- score of the first Army–Navy football game, won by Army on November 29, 1890, at West Point Military Academy.

## 24.6

- number of Earth hours in a day on the planet Mars.

## 25

- atomic number of the chemical element manganese, symbol Mn.
- average number of annual vacation days in Japan and Korea, compared to 13 in the U.S.
- fewest passes completed by a team in an NFL season (football record): by the Cincinnati Bengals, 1933.
- links in a rod (surveyor's measure).

- most birdies in a U.S. Masters Tournament: by Phil Mickelson, 2001, at the Augusta National Golf Club in Augusta, Georgia.
- most penalty minutes served in All-Star games, career (hockey record): by Gordie Howe, in twenty-three games played.
- most runs batted in by a pinch hitter in a season (baseball record): shared by Jerry Lynch, Cincinnati Reds, 1961; Rusty Staub, New York Mets, 1983; and Joe Cronin, Boston Red Sox, 1943.
- most Ryder Cup team golf wins: by the U.S., through 2008, to Europe's (and Great Britain's) ten. Next match is to be played in Wales in October, 2010.
- number of days for the sun to rotate 360 degrees.
- number of primes smaller than 100: 2, 3 ,5, 7, 11, 13, 17, 19, 23, 29, 31, 37, 41, 43, 47, 53, 59, 61, 67, 71, 73, 79, 83, 89, 97.
- number of seats in the House of Representatives allotted to the state of Florida (as of the 2000 Census).
- number of successful defenses of his heavyweight boxing title by Joe Louis, 1937–48, after winning the title from Jim Braddock on June 22, 1937.
- number of the French Département of Doubs.
- percentage of meals served at an average home consisting of frozen or prepared food from a market or restaurant, as per a *Newsweek* survey in 2004.
- record number of points scored in hockey All-Star games, career: by Wayne Gretzky (thirteen goals and twelve assists in eight games played).
- record number of seasons played in the Canadian Football League, by Lui Passaglia for the BC Lions, 1976–2000.
- record number of U.S. Open tennis titles won by a woman: Margaret du Pont, three singles (1948–50), thirteen doubles (1941–50, 1955–57), and nine mixed doubles (1943–46, 1950, 1956, 1958–60).
- retired baseball uniform number of outfielder Jose Cruz, retired by the Houston Astros.
- retired basketball jersey number of:
    ⇨ Gail Goodrich, retired by the Los Angeles Lakers.
    ⇨ Gus Johnson, retired by the Washington Wizards.
    ⇨ K. C. Jones, retired by the Boston Celtics.
    ⇨ Bill Melchionni, retired by the New Jersey Nets.
    ⇨ Mark Price, retired by the Cleveland Cavaliers.

- retired hockey shirt number of Thomas Steen, retired by the Winnipeg Jets.
- the smallest square number that is the sum of two other square numbers: $25 = 5^2 + 3^2 + 4^2$.
- Title of the United States Code dealing with Native Americans.
- 25 feet: width of a doubles squash court.
- twenty-five years:
    ⇨ minimum age for election to the U.S. House of Representatives; in addition, the person must have been a citizen for at least seven years.
    ⇨ the spread between Cher's two best-selling singles: "Dark Lady" in 1974 and "Believe" in 1999.

- value of Roman numeral XXV.

## A-25

- a Curtiss-built army dive-bomber of the early 1940s called the Helldiver. Although it never saw combat by the United States, it was used by the Australian air force and the Royal navy.

## B-25

- the Mitchell medium bomber, produced by North American Aircraft, widely used in the Pacific and Mediterranean in World War II.

## *Buck Rogers in the 25th Century*

- a popular sci-fi comic strip created in 1929 which later became a radio show (on air from 1931 to 1939), and then appeared as a TV series from September 1979 to April 1981.

## I-25

- major north-south interstate highway from Las Cruces to Albuquerque, Colorado Springs, Denver, Cheyenne, Casper, and to Buffalo (Wyoming).

## M25

- the designation for the orbital motorway around London.

## names on the Winchester Round Table

- the "Round Table," hanging in the great hall at Winchester Castle (arguably the present location of what has been called Camelot) has inscribed on it the names of twenty-five knights, supposedly those of King Arthur's court: King Arthur, Sir Galahad, Sir Lancelot du Lac, Sir Gawain, Sir Percivale, Sir Lionell, Sir Bors de Ganis, Sir Kay, Sir Tristram de Lyones, Sir Gareth, Sir Bedivere, Sir Bleoberis, La Cote Male Taile, Sir Lucan, Sir Palomedes, Sir Lamorak, Sir Safer, Sir Pelleas, Sir Ector de Maris, Sir Dagonet, Sir Degore, Sir Brunor le Noir, le Bel Desconneu, Sir Alymere, and Sir Mordred. Dating tests have indicated that the table actually was made some seven centuries after the time Arthur is believed to have lived.

## quarter

- twenty-five-cent piece; quarter of a dollar; U.S. coin in circulation since 1796.

## Twenty-Fifth Amendment to the Constitution

- establishes provisions for presidential succession and disability, ratified February 10, 1967.

## twenty-fifth anniversary gift

- silver is customary.

## twenty-fifth president

- William McKinley, March 4, 1897 to September 14, 1901.

## twenty-fifth state of the Union

- Arkansas, admitted June 15, 1836.

## "25"

- the national card game of Ireland, played frequently in pubs.

## Twenty-Five Articles of Religion

- John Wesley's 1784 adaptation of the Anglican Church's Thirty-nine Articles provide the basic doctrine of the Methodist Episcopal Church. Several of the articles are intended to distinguish Wesley's beliefs, as well as the Anglicans', from those of Roman Catholicism.

## 25, Brook Street

- London residence of composer George Frideric Handel.

## 25 December

- Christmas Day, celebrating the birth of Christ.

## 25 July

- Commonwealth Constitution Day, an official holiday in Puerto Rico, commemorating the effective date of the Puerto Rican constitution in 1952.

## 25 March

- (a) Maryland Day, observed in that state commemorating the landing of colonists in 1634.
- (b) Christian feast day commemorating the Annunciation of the Virgin Mary.

## Twenty-Five Mile Creek State Park

- a marine camping park on the south shore of Lake Chelan in north-central Washington.

## *$25,000 Pyramid*

- daytime-TV game show hosted by Dick Clark; debuted on CBS on March 26, 1973, moved to ABC in May 1974, then back to CBS in September 1982. Went off the air in July 1988, but produced several syndicated versions.

## 25 with an L

- rap slang for a jail sentence of twenty-five years to life.

## VC25A

- designation for Air Force One, the presidential airplane. Actually, there are two Air Force Ones, both 747-220B, with the tail numbers 28000 and 29000.

## 25.5

- twenty-five and a half minutes, mean travel time to work in the U.S., according to the 2000 Census.

## 026

- Decimal ASCII code for [substitute].
- medical classification code for rat-bite fever.

## 26

- age of Jackie Robinson when he was signed by Brooklyn Dodgers' general manager Branch Rickey in 1945, effectively ending baseball's Jim Crow policy. Robinson played with the Dodgers' farm team, the Montreal Royals, until his major league debut on April 15, 1947.
- atomic number of the chemical element iron, symbol Fe.
- average number of annual vacation days in Canada, compared to thirteen in the U.S.
- letters in the English alphabet.
- members of the Senate in the first Congress (1789–91), seventeen Federalist and nine Democrat-Republican.
- most games played by a team in an NHL playoff year (hockey record): by the Philadelphia Flyers, 1987.
- most Grammy Awards won by a female artist: Alison Krauss, bluegrass-country singer and fiddler. Her most recent in 2009, five Grammys for her collaboration with Robert Plant on Album of the Year, *Raising Sand,* and several individual tracks on it. She has won as a solo artist, collaborator, producer, and with her band, Union Station. She has also won several CMA awards.
- most innings pitched by a pitcher in a single game (baseball record): twice, by Leon Cadore, with the Brooklyn Robins,

and by Joe Oeschger, with the Boston Braves, both on May 1, 1920.

- most individual Oscars won: by Walt Disney, in sixty-four nominations (including cartoons and documentaries).
- most seasons played in the NFL (football record): George Blanda, 1949–75; for the Chicago Bears, 1949 and 1950–58, for Baltimore Colts, 1950, for the Houston Oilers, 1960–66, and for the Oakland Raiders, 1967–75.
- most seasons played in the NHL (hockey record): by Gordie Howe with the Detroit Red Wings, 1946–71 and the Hartford Whalers, 1979–80 season. Howe also played six years in the WHA, with the Houston Astros, 1973–77, and the New England Whalers, 1977–79.
- most touchdowns scored by a team in Super Bowls (football record): by the Dallas Cowboys in V, VI, X, XII, XIII, XXVII, XXVIII, XXXX.
- number of counties in the republic of Ireland: Carlow, Cavan, Clare, Cork, Donegal, Dublin, Galway, Kerry, Kildare, Kilkenny, Laoighis, Leitrim, Limerick, Longford, Louth, Mayo, Meath, Monaghan, Offaly, Roscommon, Slingo, Tipperary, Waterford, Westmeath, Wexford, and Wicklow.
- number of popes who either resigned or were deposed, from Pontian, in 235, to Pius VI, in 1798.
- number of teeth on a kitten.
- number of the French Département of Drôme.
- number of times the British Open golf tournament was held at St. Andrews course in Scotland, the most frequently used site.
- number of vertebrae in the human spine.
- the only number placed between a square number (25) and a cube number (27).
- record for kills by a Marine Corps ace fighter pilot in World War II, Captain Joseph Foss of the VMF-121 Squadron, flying a F4F.
- record number of enemy aircraft downed by an American fighter pilot in World War I, Captain Eddie Rickenbacker.
- retired baseball uniform number of:
    - ⇨ owner Gene Autry, retired by the Anaheim Angels.
    - ⇨ manager Johnny Oates, retired by the Texas Rangers.
    - ⇨ outfielder Billy Williams, retired by the Chicago Cubs.

- retired hockey shirt number of Peter Stastny, retired by the Quebec Nordiques.
- round in which Jess Willard KO'd Jack Johnson to become the heavyweight boxing champ, April 5, 1915, in Havana, Cuba.
- Title of the United States Code dealing with the Internal Revenue Code.
- twenty-six years:
  - ➪ age of the youngest delegate to the Constitutional Convention in May, 1787: Jonathon Dayton of New Jersey.
  - ➪ length of time Nelson Mandela spent in South African prison.

## Beethoven's *Sonata No. 26*

- "Das Lebewohl" sonata (in E-flat major).

## B-26

- the Marauder, a bomber that saw service against Japanese warships during the Battle of Midway in June 1942.

## Haydn's *Symphony No. 26*

- "Lamentatione" symphony (in D minor).

## longest winning streak in major league baseball— twenty-six games

- by the New York Giants, from their win over the Brooklyn Dodgers, July 7, to their loss to the Boston Braves, September 30, 1916.

## M-26

- U.S. military short-range artillery rocket armed with clusters of small grenade-like bombs that scatter and explode over a large area, used as an antipersonnel weapon.

## PDD-26

- Presidential Decision Directive twenty-six deals with U.S. Antarctic Policy, issued by President Bill Clinton on March 9, 1996.

## Proposition 26

- of Book I of Euclid's *Elements*; contains his proof of the AAS pattern of congruence of triangles.

**twenty-six**

- a dice game that was popular in stores and taverns in the Mid-western states in the 1950s. The customer paid a fee to play and to win tokens. The game consisted of throwing ten dice thirteen times and counting the number of times a pre-selected number, from 1 to 6, came up. The object is to make the point number at more than twenty-five or less than ten times. More extreme counts would win more tokens.

**26 August**

- Women's Equality Day, commemorating the passing of the nineteenth Amendment to the Constitution on this day in 1920, extending suffrage to women. President Ford proclaimed this day in 1974. Also known as Susan B. Anthony Day.

**26 Baku Commissars**

- (in Russian, *im 26 Bakinskikh Komissarov*): the name of a town in southeastern Azerbaijan and one in southwestern Kazakhstan.

**26 December**

- Boxing Day, also known as the Feast of St. Stephen, a holiday in the UK, Canada, and other Commonwealth nations. Originally a day on which merchants gave boxed gifts to their servants and employees, probably a precursor of Christmas bonuses. Traditionally also the day on which the clergy distributed the contents of alms boxes to the poor.

**Twenty-Six Dialogues with a Persian**

- transcript of conversations recorded by Manuel II, next-to-last emperor of the Byzantine Empire, in 1391, supposedly between him and the Sultan's Magistrate for the Ottoman Turks. The text is seen as offensive to and prejudiced against Islam, the religion of the Turkish Empire.

**twenty-six innings**

- longest major league baseball game, by innings; played by the Brooklyn Robins against the Boston Braves, May 1, 1920. Tied 1 to 1, the game was called because of darkness.

## 26 January

- Australia Day, national holiday commemorating the landing of the First Fleet at Sydney Cove on January 26, 1788.

## 26 miles, 385 yards

- length of a marathon. The actual distance the Greek soldier ran from Marathon to Athens, to tell Athenians about the great victory over the Persians, has been estimated as under 22 miles. When the Olympics were resumed in 1896 (they had been abolished in 393 by Roman emperor Theodosius I), a marathon race was included as an event. In that and succeeding years, Olympics marathons varied in length. In 1908, the games were held in London, and the British Olympic Committee decided to have the race run from Windsor Castle to the royal box in the sports stadium, a distance of 26 miles and 385 yards. This distance was standardized at the 1924 Olympics and has remained ever since, both in and out of Olympic races.

## Twenty-Sixth Amendment to the Constitution

- lowers the voting age to eighteen, ratified June 30, 1971.

## 26th of July Movement

- Fidel Castro's guerrilla movement that overthrew Fulgencio Batista and established a communist government in Cuba. The name honors several revolutionaries who were killed or captured in an unsuccessful attack on the Moncada Barracks in Santiago de Cuba on July 26, 1953, in an attempt to seize weapons and to capture the base radio station to announce the onset of the guerrilla movement.

## twenty-sixth president

- Theodore Roosevelt, September 14, 1901 to March 3, 1909.

## twenty-sixth state of the Union

- Michigan, admitted January 26, 1837.

# 027

- Decimal ASCII code for [escape].

# 27

- atomic number of the chemical element cobalt, symbol Co.
- the cube of 3.
- cubic feet in a cubic yard.
- international telephone calling code for South Africa.
- most consecutive home games won in NFL regular season play (football record): by the Miami Dolphins, 1971–74.
- most parts played by the same actor in a single film: Rolf Leslie in *Sixty Years a Queen*, a 1913 British movie biography of Queen Victoria.
- most rebounds in an NBA All-Star game (basketball record): by Bob Pettit, 1962.
- most strokes under par in an LPGA Tournament: by Annika Sorenstam, winning the Standard Register Ping at Moon Valley CC in Phoenix, Arizona, March 16, 2001.
- most World Series won (baseball record): by the New York Yankees, 1923–2009, vs. thirteen losses, 1921–2003.
- number of books in the New Testament.
- number of countries in the European Union (as of end of 2009). Started with the European Coal and Steel Community, established in 1951 by the Treaty of Paris, with six members: Belgium, France, West Germany, Italy, Luxembourg, and the Netherlands. In 1957, Treaty of Rome created the successor European Economic Community, and in 1967, the European Community superseded the EEC. In 1973, the EEC took in Denmark, Ireland, and the United Kingdom. Greece joined in 1981, and Spain and Portugal in 1986. The Treaty of Maastricht in 1992 provided for expanded cooperation among the member nations, proposed a common currency, and created the European Union, which was expanded in 1995 by the inclusion of Austria, Finland, and Sweden. Ten new countries joined the EU in 2004: Cyprus, the Czech Republic, Estonia, Hungary, Latvia, Lithuania, Malta, Poland, Slovakia, and Slovenia. On January 1, 2007, Romania and Bulgaria were added. As

of late 2010, Croatia, Iceland, Macedonia, and Turkey are candidates for admission.

- number of divisions (counties) in the state of Alaska.
- number of letters in the Spanish alphabet, the twenty-six of the English alphabet + Ñ.
- number of piano concertos composed by Wolfgang Amadeus Mozart.
- number of presidential electoral votes apportioned to the state of Florida. (May be reapportioned for the 2012 election.)
- number of satellites of the planet Uranus; five relatively large, eleven small dark inner ones discovered by *Voyager 2* in January 1986, and eleven other distant ones from later observations.
- number of the French Département of Eure.
- number of 0s in an octillion.
- retired baseball uniform number of:
  - ⇨ catcher Carlton Fisk, retired by the Boston Red Sox. Fisk's number 72 has also been retired by the Chicago White Sox.
  - ⇨ pitcher Catfish Hunter, retired by the Oakland Athletics.
  - ⇨ pitcher Juan Marichal, retired by the San Francisco Giants.

- retired basketball jersey number of Jack Twyman, retired by the Sacramento Kings.
- the smallest number requiring four syllables to verbalize.
- Title of the United States Code dealing with intoxicating liquors.
- 27 feet: width of a regulation singles tennis court.
- 27 hertz (cycles per second): lower limit of a piano's frequency range.
- twenty-seven years: length of the Peloponnesian War starting in 431 BC. An uneasy truce was signed in 421 BC, lasting until 415 BC, when Athens attacked Sicily and hostilities began anew. Peace was signed in 404 BC after the Spartan navy under Lysander defeated the Athenians at the Battle of Aegospotamos in 405 BC.

## 27 August

- legal holiday in Texas commemorating the 1908 birthday of President Lyndon B. Johnson, a native of Texas.

## 27A, Wimpole Street

- London residence of Professor Henry Higgins in the 1964 Warner Brothers film *My Fair Lady*, which starred Rex Harrison, Audrey Hepburn, and Stanley Holloway.

## 27 December 1947

- debut of *Howdy Doody Time*, 5:00 to 6:00 PM, on NBC-TV. The show ran until Septefmber 24, 1960 (a total of 2,343 episodes).

## "twenty-seven different wigs"

- in the nursery rhyme:

> Gregory Griggs, Gregory Griggs
> Had twenty-seven different wigs.
> He wore them up, he wore them down,
> He wore them everywhere in town;
> He wore them east, he wore them west,
> But he never could tell which he loved best.

## 27 January

- Vietnam Day, commemorating the signing of the peace agreement on this day in 1973.

## Twenty-Seventh Amendment to the Constitution

- deals with congressional compensation; ratified May 7, 1992. This Amendment was included with the first group of twelve proposed to the First Congress on September 25, 1789, ten of which became the Bill of Rights, the other two not ratified at the time by the necessary number of states.

## twenty-seventh president

- William H. Taft, March 4, 1909 to March 3, 1913.

**twenty-seventh state of the Union**

- Florida, admitted March 3, 1845.

### 27 Wagonloads of Cotton

- one-act play by Tennessee Williams, which he adapted to create the 1956 film, *Baby Doll,* directed by Elia Kazan and starring Carroll Baker, Eli Wallach, Karl Malden, and Mildred Dunnock. A sexual potboiler, it was excoriated by *Time* magazine as "possibly the dirtiest American film ever legally exhibited," was banned in many countries, including Sweden, and was denounced "for its carnal suggestiveness" by the Catholic Legion of Decency. It also received Academy Award nominations for Best Actress (Baker), Best Supporting Actress (Dunnock), Best Cinematography, and Best Adapted Screenplay.

## 27-1/3

- number of days in the moon's orbit around the Earth.

## 028

- Decimal ASCII code for [file separator].

## 28

- atomic number of the chemical element nickel, symbol Ni.
- average number of annual vacation days in the United Kingdom, compared to thirteen in the United States.
- days in the month of February, the shortest month of the Gregorian calendar, except for twenty-nine in leap years.
- days until your money will exceed one million dollars if you start with one cent and double the amount every day. On the twenty-eighth day your money will equal $1,342,177.28.
- most free throws scored in an NBA game (basketball record): twice, by Wilt Chamberlain for the Philadelphia Warriors against the New York Knicks, March 2, 1962; and by Adrian Dantley for the Utah Jazz against the Houston Rockets, January 5, 1984.
- most touchdowns scored rushing in a single NFL season (football record): by LaDainian Tomlinson for the San Diego Chargers, 2006.

- most U.S. Open tennis tournaments played: by Vic Seixas, 1940–42, 1944, 1946–69.
- number of African-American servicemen who earned a Medal of Honor in the Civil War.
- number of Giuseppe Verdi operas, including *Il Trovatore, Rigoletto, La Traviata, Falstaff,* and *Aida.*
- number of hymns of the Saiva saints; the Saiva are a sect of Indian Hindus that worships Shiva.
- number of letters in the Arabic and Esperanto alphabets.
- number of station stops on the first New York City subway. First line opened October 27, 1904, running 9.1 miles from City Hall to West 145th Street.
- number of teeth on a puppy.
- number of the French Département of Eure-et-Loir.
- number of years Jehu reigned over Israel in Samaria (2 Kings.10:36).
- record number of pinch hits in a baseball season: by John Vander Wal, Colorado Rockies, 1995.
- retired football jersey number of:
  - ⇨ Marshall Faulk, retired by the St. Louis Rams.
  - ⇨ Willie Galimore, retired by the Chicago Bears.
  - ⇨ Abner Hayes, retired by the Kansas City Chiefs.
  - ⇨ Curtis Martin, retired by the New York Jets
- the second perfect number (its divisors being 14, 7, 4, 2, and 1).
- sum of the first five prime numbers $(2 + 3 + 5 + 7 + 11)$.
- sum of the first seven integers $(1 + 2 + 3 + 4 + 5 + 6 + 7)$.
- Title of the United States Code dealing with the judiciary and judicial procedure.

## calendar cycle—twenty-eight years

- the same date will occur on the same day of the week in a twenty-eight-year cycle.

## 28 August 1963

- Martin Luther King, Jr. delivered his historic "I Have a Dream" speech at the March on Washington.

## 28, Dean Street

- London residence of Karl Marx.

### twenty-eight-hour Law (of 1877)

- Federal statute that governs humane handling of animals being transported across state lines. Provides that all animals, including those being raised for food, cannot be moved by surface transportation (excluding passage by air or water) "for more than 28 consecutive hours without being unloaded for five hours rest, water and food."

### twenty-eighth president

- Woodrow Wilson, March 4, 1913 to March 3, 1921.

### twenty-eighth state of the Union

- Texas, admitted December 29, 1845.

### 28 November 1925

- *The Grand Ole Opry* debuts on WSM radio, Nashville, Tennessee.

### twenty-eight parrot

- a subspecies (*Bernardius zonarius semitorquatus*) of the Australian ringneck parrot, found in southwest West Australia, distinguished by its black head and pale green belly. Its call is a noisy whistle that sounds like "twenty-eight, twenty-eight, twenty-eight," accompanied by loud chattering.

### 28-point type

- typographer's measure, known as "double English."

### 28 Up

- the fourth installment (1984) of the Michael Apted documentary on British TV following the lives of a group of British children in seven-year intervals.

## *029*

- Decimal ASCII code for [group separator].

# *29*

- age at which Gautama Buddha renounced his family and his material possessions.
- atomic number of the chemical element copper, symbol Cu.
- highest possible count in a hand of cribbage.
- most games lost by a pitcher in a season in the (baseball record): by Vic Willis for the Boston Beaneaters, 1905.
- most opponents' fumbles recovered in an NFL career (football record): by Jim Marshall for the Minnesota Vikings, 1961–79.
- number of counties in the state of Utah.
- number of days in February in a leap year.
- number of letters in the German alphabet.
- number of seats in the House of Representatives allotted to the state of New York (as of the 2000 Census).
- number of the French Département of Finistère.
- number of the track from which the "Chattanooga Choo-Choo" leaves the station.
- percentage of the Earth's surface that is land.
- retired baseball uniform number of infielder Rod Carew, retired by both the Anaheim Angels and the Minnesota Twins.
- retired football jersey number of Eric Dickerson, retired by the St. Louis Rams.
- retired hockey shirt number of Ken Dryden, retired by the Montreal Canadiens.
- Title of the United States Code dealing with labor.
- value of Roman numeral XXIX.

## B-29

- the Superfortress, a long-range, high-altitude U.S. bomber, the largest in service in World War II. Produced by Boeing, the B-29 had a crew of eleven. *Enola Gay,* the plane that dropped the first atomic bomb on Hiroshima, was a B-29.

## Beethoven's *Sonata No. 29*

- "Hammerklavier" (in B-flat major)

## PDD-29

- Presidential Decision Directive 29 deals with Security Policy Coordination, issued by President Bill Clinton on September 27, 1994.

## Proposition 29

- of Book I of Euclid's *Elements*; contains his controversial parallel postulate.

## Shakespearean Sonnet XXIX

- begins, "When in disgrace with fortune and men's eyes, I all alone beweep my outcast state . . ."

## twenty-nine

- a card game for four players. Cards are worth their face value, aces and face cards count one each. Each player in turn adds a card to those already played and announces the points accumulated until 29 is reached. Whoever makes the total 29 takes the trick. If a player is unable to add a card without exceeding 29, the play ends and the winner of the deal is the player who has taken the greatest number of tricks.

## 29, Fitzroy Square

- London residence of both George Bernard Shaw at the end of the nineteenth century, and Virginia Woolf early in the twentieth.

## Twentynine Palms

- a city in southern California, site of a Marine Corps base, about 70 miles east of San Bernardino. It is the home of Joshua Tree National Park.

## twenty-ninth president

- Warren G. Harding, March 4, 1921 to March 3, 1923.

## twenty-ninth state of the Union

- Iowa, admitted December 28, 1846.

# 29.5

- number of Earth years in Saturn's circuit of the sun.
- 29-½ feet: width of a regulation volleyball court.

# 030

- Dewey Decimal System designation for general encyclopedic works.
- medical classification code for leprosy.

# 30

- atomic number of the chemical element zinc, symbol Zn.
- the first sphenic number (in mathematics); the first positive integer that is the product of three distinct prime factors.
- in Hebrew, the numerical value of the letter *lamed*.
- international telephone calling code for Greece.
- most appearances in the NBA finals (basketball record): by the Los Angeles Lakers, with a 15-15 record through the 2008–09 season.
- most free throws scored in an NBA playoff game (basketball record): by Bob Cousy, for the Boston Celtics against the Syracuse Nationals, March 21, 1953.
- most runs scored by one team in a major league baseball game in the "modern era": by the Texas Rangers against the Baltimore Orioles, August 22, 2007, final score 30–3. In the pre-modern era, the Chicago Colts beat the Louisville Colonels, 36–7, on June 28, 1897.
- number of days in the months of April, June, September, and November.
- number of hurdles horses must jump in the Grand National Sweepstakes at Aintree Race Track in Liverpool, England. The horses make two circuits of the track, the first time jumping sixteen fences, the second time, fourteen.
- number of teeth on an adult cat.
- number of the French Département of Gard.
- number of tracks on *The Beatles*, the singing group's eponymous album, familiarly known as the *White Album*.
- number of 0s in a nonillion (American) or a quintillion (British).

- retired baseball uniform number of:
  - ⇨ first baseman Orlando Cepeda, retired by the San Francisco Giants.
  - ⇨ outfielder Tim Raines, retired by the Montreal Expos.
  - ⇨ pitcher Nolan Ryan, retired by the Anaheim Angels.

- retired basketball jersey number of:
  - ⇨ Bob Gross, retired by the Portland Trail Blazers.
  - ⇨ George McGinnis, retired by the Indiana Pacers.
  - ⇨ Terry Porter, retired by the Portland Trail Blazers.

- retired hockey shirt number of:
  - ⇨ Rogie Vachon, retired by the Los Angeles Kings.
  - ⇨ Mike Vernon, retired by the Calgary Flames.

- seahs in a homer (ancient Hebrew measure of dry capacity).
- sum of the first four square numbers $(1 + 4 + 9 + 16)$.
- 30 cubits: the height of Noah's ark (Genesis 6:15).
- thirty days: average incubation period of a duck or a goose.
- 30 degrees: size of each exterior angle of an equilateral dodecagon.
- 30 feet: depth of each end zone on a regulation football field.
- 30 inches: maximum circumference of a regulation basketball (minimum is 29-½ inches).
- thirty seconds: the running time of Samual Becket's Play, *Breath*, which has no actors and no dialogue. The play, written in 1969, was first performed in England, April 1970.
- thirty years:
  - ⇨ David's age when he began his reign over the Israelites (2 Samuel 5:4).
  - ⇨ Minimum age for election to the U.S. Senate; in addition, the person must have been a citizen for at least nine years.

- Title of the United States Code dealing with mineral lands and mining.
- value of Roman numeral XXX.

## Combat of the Thirty

- during the Hundred Years' War, with garrisons on all sides suffering debilitating losses, two opposing chiefs arrranged for a tournament/battle between thirty knights from each side. On March 27, 1351, Jean de Montfort (friendly to the English) and Charles de Blois (to the French) each sent thirty fighters to a pre-arranged combat between the chateaus of Josselin (pro-Blois) and Ploermel (pro-Montfort), the former led by Robert de Beaumanoir, the latter by Robert Bemborough. The French took the day, Bamborough and eight knights were killed, the others taken as slaves. During the battle, the wounded Beaumanoir asked for a drink of water, to which Geoffrey of Bouays is reputed to have made the historic reply, "Drink your blood, Beaumanoir, your thirst will be quenched soon enough." The battle had no effect on the war or the political issues of the time.

### Haydn's *Symphony No. 30*
- "Alleluia" symphony (in C major).

### thirtieth president
- Calvin Coolidge, March 4, 1923 to March 3, 1929.

### thirtieth state of the Union
- Wisconsin, admitted May 29, 1848.

### thirtieth wedding anniversary gift
- pearls are appropriate.

### "30"
- signifies the end of a news article; originated in the days of the telegraph, when the end of a transmission was indicated by "XXX".

### 30 April 1945
- Adolf Hitler commits suicide in his bunker in Berlin.

### 30 August
- Huey P. Long Day (born 1893), past governor of and senator from Louisiana; a public holiday in that state.

## "Thirty days hath September"

- the counting rhyme on the number of days in each month:

  Thirty days hath September,
  April, June, and November.
  All the rest have thirty-one,
  Though February's underdone
  With twenty-eight days, that's the time,
  Except in leap years, twenty-nine.

### thirty-day wonder

- a second lieutenant who has received his commission after only thirty days of officer training.

## 30 January

- (a) Franklin D. Roosevelt's birthday in 1882, a public holiday in Kentucky.
- (b) Royalist Fast Day, holiday in Virginia commemorating the beheading of Charles I on this day in 1649.

## 30 Lafayette Street

- address of the Brooklyn (New York) Academy of Music—with a four-screen theater, an opera house, plus several smaller theaters its first production dating from 1861.

## *30 Minutes*

- daytime-TV counterpart of evening's *60 Minutes*, this documentary series for youth dealt with controversial issues in a straightforward, hard-hitting manner. Aired on CBS-TV on Saturdays from September 1978 to August 1982.

## 30 October

- birthday of John Adams, 1735, second president of the United States.

### thirty-penny nail

- a carpenter's common nail, 4-½ inches long (coded 30d).

**thirty pieces of silver**

- the amount Judas Iscariot was paid for betraying Jesus (Matthew 26:15).

*30 Rock*

- comedy series on NBC-TV, starting in October 2006, with Tina Fey as the head writer of *The Girlie Show*, a weekly pastiche fashioned on *Saturday Night Live*. Created by Miss Fey, drawing on her experience as head writer for SNL. In support are Tracy Morgan, Jane Krakowski, and especially Alec Baldwin as an arrogant but erratic network executive.

*Thirty Seconds Over Tokyo*

- 1944 war picture portraying the Doolittle bombing raid of the Japanese capital on April 18, 1942. General Doolittle was played by Spencer Tracy.

**thirty sheets and thirty changes of garments**

- the prize for solving Samson's riddle (Judges 14:12).

*Thirtysomething*

- an ensemble series that ran on NBC-TV for eighty-five episodes, from September 1987 to May 1991, about a successful middle-class couple and their self-absorbed yuppie friends, all living in Philadelphia, trying to cope with everyday adult angst. Cast regulars included Mel Harris, Ken Olin, Timothy Busfield, Polly Draper, Peter Horton, Melanie Mayron, and Patricia Wettig. Too much of a soap for some viewers.

**The Thirty Tyrants**

- thirty aristocrats who, at the close of the Peloponnesian War in 404 BC, were appointed ,by Sparta to administer Athens. Their despotic rule led to an uprising, and they were overthrown by Thrasybulus in 403 BC.

*30 Years of Fun*

- 1963 Robert Youngson compilation of film sequences by great comedians of silent films Charlie Chaplin, Buster

Keaton, Laurel and Hardy, Harry Langdon, Charley Chase, etc.

**Thirty Years' Truce**
- a thirty-year treaty signed in 445 BC between the Athenian Empire and the Spartan Alliance. Within a decade the truce was breaking down, and in 431 BC, the Peloponnesian War broke out between Sparta and Athens.

***Thirty Years' View***
- alternatively titled *History of the Working of the American Government from 1820 to 1850,* this is the political autobiography of Thomas Hart Benton, known as "Old Bullion" for his strict financial policies.

**Thirty Years' War**
- collectively a series of religious wars in central Europe (1618–48), starting as a conflict between the Protestants of Northern Germany and the Roman Catholics of Southern Germany, then developing into a wider struggle involving most of the countries of Europe, and ending with the Treaty of Westphalia in 1648.

## *30.1*

- highest career average points scored per NBA game (basketball record): by Michael Jordan for the Chicago Bulls, 1984–98, and the Washington Wizards, 2001–03.

## *031*

- Decimal ASCII code for [unit separator].

## *31*

- Article of the New York State Vehicle and Traffic Law dealing with alcohol- and drug-related offenses.
- atomic number of the chemical element gallium, symbol Ga.
- highest score in an international soccer match, 31–0, by Australia over American Samoa, in World Cup Oceania

Zone Group 1 qualifiers, April 11, 2001. Australia was ranked seventy-fifth by FIFA, American Samoa 203rd, lowest team in the rankings.

- international telephone calling code for Netherlands.
- lunch-counter code for a glass of lemonade or orangeade.
- most assists in an NHL playoff year (hockey record): by Wayne Gretzky for the Edmonton Oilers, 1987–88.
- most points scored by the losing team in a Super Bowl game (football record): by the Dallas Cowboys, against the Pittsburgh Steelers (35), SB-XIII, 1979.
- most strokes under par in a 72-hole golf match on the PGA Tour: by Ernie Els, winning the Mercedes Championship in 2003.
- most three-point field goals made by both teams in an NBA game (basketball record): March 13, 2005, the Toronto Raptors (21) against the Philadelphia 76ers (10).
- most touchdowns scored in an NFL season (football record): by LaDainian Tomlinson for the San Diego Chargers, 2006.
- number of days in the months of January, March, May, July, August, October, and December.
- number of dynasties of the ancient Egyptian civilization.
- number of presidential electoral votes apportioned to the state of New York. (May be reapportioned for the 2012 election.)
- number of the French Département of Haute-Garonne.
- record number of goals allowed by a goaltender in All-Star hockey games: Patrick Roy in eleven games played between 1988 and 2003.
- retired baseball uniform number of:
  ⇨ pitcher Ferguson Jenkins, retired by the Chicago Cubs.
  ⇨ pitcher Greg Maddux, retired by both the Atlanta Braves and the Chicago Cubs.
  ⇨ outfielder Dave Winfield, retired by the San Diego Padres.

- retired basketball jersey number of:
  ⇨ Cedric Maxwell, retired by the Boston Celtics.
  ⇨ Reggie Miller, retired by the Indiana Pacers.
- retired football jersey number of:
  ⇨ William Andrews, retired by the Atlanta Falcons.
  ⇨ Jim Taylor, retired by the New Orleans Saints.

- retired hockey shirt number of:
  - ⇨ Grant Fuhr, retired by the Edmonton Oilers.
  - ⇨ Billy Smith, retired by the New York Islanders.

- sum of the first five powers of 2 $(2^0 + 2^1 + 2^2 + 8 + 16)$.
- thirty-one days: average gestation period of a domestic rabbit or a chipmunk.
- 31 miles: the length of the English Channel Tunnel (the "Chunnel"), 23 miles of which are under water, from Folkestone, England, to Sangatte, France, connecting London and Paris.
- Title of the United States Code dealing with money and finance.
- the value of a "Go" in the card game cribbage.
- value of Roman numeral XXXI.

## Haydn's *Symphony No. 31*

- "Horn Signal" symphony (in D major).

## longest pontificate of the Catholic Church

- Pope Pius IX served for thirty-one years, seven months, and three weeks, from June 16, 1846 to February 7, 1878.

## Mozart's *Symphony No. 31*

- "Paris" symphony (in D major).

## Section 31

- the secret police of the Federation on *Star Trek*.

## tanka

- the premier poetic form of the Japanese imperial court, consisting of five lines and a total of thirty-one syllables, in the pattern of five, seven, five, seven, seven.

## thirty-first president

- Herbert C. Hoover, March 4, 1929 to March 3, 1933.

## thirty-first state of the Union

- California, admitted September 9, 1850.

**thirty-one-day presidency**

- tenure of office for President William Henry Harrison before he died in office on April 4, 1841.

**31 de Janeiro**

- a town in the Vige province of Angola.

**31 Flavors**

- advertised by Baskin-Robbins Ice Cream, one for each day of the month. Actually, they have hundreds of additional flavors.

**31 July 1933**

- Jack Armstrong, the All-American Boy, debuts on CBS radio.

**31 October**

- Halloween.

**31 to 5**

- the odds against tossing a 6 or tossing an 8 in dice.

# 032

- Decimal ASCII code for [space].
- medical classification for diphtheria.

# 32

- Article of the Uniform Code of Military Justice (UCMJ) dealing with pre-trial investigations.
- atomic number of the chemical element germanium, symbol Ge.
- the fifth power of 2.
- fluid ounces in a quart.
- international telephone calling code for Belgium.
- most birdies in a 72-hole PGA tournament (gold record): twice, by Mark Calcavecchia in the 2001 Phoenix Open, and by Paul Gow in the 2001 B.C. Open.

- most losses at home in one NHL season (hockey record): by the San Jose Sharks, 1992–93.
- most pass completions in a Super Bowl game (football record): twice, by Tom Brady for the New England Patriots against the Carolina Panthers, SB-XXXVIII, 2004; and by Drew Brees for the New Orleans Saints against the Indianapolis Colts, SB-XLIV, 2010.
- number of black squares and number of white squares on a chess or checker board.
- number of boroughs (plus the City of London) that make up Greater London.
- number of bridges spanning the River Seine in Paris.
- number of cards in the deck for the European game, Poch.
- number of counties in Ireland—twenty-six in the republic of Ireland, six in Northern Ireland.
- number of Grand Cru vineyards in Burgundy—nine white and twenty-three red.
- number of letters in the Russian alphabet.
- number of piano sonatas composed by Beethoven.
- number of seats in the House of Representatives allotted to the state of Texas (as of the 2000 Census).
- number of teeth on a human adult: eight incisors, four cuspids, eight bicuspids, and twelve molars.
- number of the French Département of Gers.
- quarts in a bushel (dry measure).
- retired baseball uniform number of:
  ⇨ pitcher Steve Carlton, retired by the Philadelphia Phillies.
  ⇨ catcher Elston Howard, retired by the New York Yankees.
  ⇨ pitcher Sandy Koufax, retired by the Los Angeles Dodgers.
  ⇨ pitcher Jin Umbricht, retired by the Houston Astros.

- retired basketball jersey number of:
  ⇨ Fred Brown, retired by the Seattle SuperSonics.
  ⇨ Billy Cunningham, retired by the Philadelphia 76ers.
  ⇨ Sean Elliot, retired by the San Antonio Spurs.
  ⇨ Julius Erving, retired by the New Jersey Nets. Erving's number 6 with Philadelphia 76ers has also been retired.
  ⇨ Magic Johnson, retired by the Los Angeles Lakers.
  ⇨ Karl Malone, retired by the Utah Jazz.
  ⇨ Kevin McHale, retired by the Boston Celtics.

⇨ Bill Walton, retired by the Portland Trail Blazers.

⇨ Brian Winters, retired by the Milwaukee Bucks.

- retired football jersey number of:
  ⇨ Al Blozis, retired by the New York Giants.
  ⇨ Jim Brown, retired by the Cleveland Browns.

- retired hockey shirt number of Dale Hunter, retired by the Washington Capitals.
- shots in a quart (spirits measure).
- slang for a .32 caliber pistol.
- 32 feet: length of a squash court.
- Title of the United States Code dealing with the National Guard.

## Davis Cup wins

- originally the International Lawn Tennis Challenge, changed to the Davis Cup (after one of the tournament's founders) in 1945, the competition started in 1900 between men of the U.S. and Great Britain and has since grown to 125 nations competing as of 2009. Since its inception, the U.S. has won thirty-two times, more than any other country, the first in 1900, and most recently, in 2007. At this writing, Spain has won the last two years, 2008 and 2009. The 2010 finals are scheduled for Belgrade in December, Serbia against France.

## demisemiquaver

- thirty-second note, in musical notation.

## first woman elected to both the House of Representatives and the U.S. Senate

- Maine Republican Margaret Chase Smith served thirty-two years in Congress—in the House of Representatives, from June 1940 through the end of 1948, and in the U.S. Senate, from 1949 through 1972.

## in 32s

- describes a printing process whereby each printed sheet contains thirty-two book pages.

**most Grammys won in lifetime**
- by conductor Georg Solti, between 1958 and 1997.

**thirty-second president**
- Franklin Delano Roosevelt, March 4, 1933 to April 12, 1945.

**thirty-second state of the Union**
- Minnesota, admitted May 11, 1858.

**32 bars**
- the standard musical structure for popular music from the 1930s on, typically in four eight-bar units in the form AABA, the B called the bridge or the release.

**32 degrees Fahrenheit**
- temperature at which water freezes (=zero degrees Celsius).

**32mo**
- thirty-twomo or trigesimosecundo, the page size of a book from printer's sheets that are folded into thirty-two equal size leaves, approximately $3\frac{1}{4}$ x 5 to $5\frac{1}{2}$ inches.

**thirty-two points on a compass**
- each $11\frac{1}{4}$ degrees from the previous one.

## 033

- decimal ASCII code for [exclamation mark (!)].
- medical classification code for whooping cough.

## 33

- age of the youngest person ever to be awarded the Nobel Peace Prize: Rigoberta Menchú of Guatemala in 1992.
- atomic number of the chemical element arsenic, symbol As.
- international telephone calling code for France.
- most fair catches in an NFL season (football record): by Brian Mitchell for the Philadelphia Eagles, 2000.

- most passes caught in Super Bowl games, career (football record): by Jerry Rice for the San Francisco 49ers, SB-XXIII, 1989; SB-XXIV, 1990; and SB-XXIX, 1995, and for the Oakland Raiders, SB-XXXVI, 2002.
- number of beads on the Anglican form of rosary.
- number of cantos in each of the three parts of Dante's *Divine Comedy*.
- number of counties in Scotland: Aberdeen, Angus, Argyll, Ayr, Banff, Berwick, Bute, Caithness, Clackmannan, Dumfries, Dunbarton, East Lothian, Fife, Inverness, Kincardine, Kinross, Kirkcudbright, Lanark, Midlothian, Moray, Nairn, Orkney, Peebles, Perth, Renfrew, Ross and Comarty, Roxburgh, Selkirk, Stirling, Sutherford, West Lothian, Wigtown, and Zetland.
- number of counties in the state of New Mexico.
- number of days John Paul I served as pope, the shortest papal reign in the modern Catholic Church. Elected pope on August 26, 1978, John Paul was found dead on September 28, 1978.
- number of gods in Chinese Buddhism.
- number of medals and awards won by Audie Murphy, the most decorated soldier in U.S. military history, including the Medal of Honor. Portrayed himself in the 1956 film, *To Hell and Back*.
- number of peg holes in the board for the game of Solitaire as played in England. (The French version uses thirty-seven holes.)
- number of sanctuaries visited by Buddhist pilgrims in ancient Japan to venerate the Bodhisattva Avalokitesvara (Kannon-bosatsu) on the Pilgrimage to the Thirty-Three Holy Places of the Provinces of the West.
- number of the French Département of Gironde.
- number of years to construct the Panama Canal, which opened in 1914.
- number of 0s in a decillion.
- number that appears on bottles of Rolling Rock beer.
- record number of consecutive team wins in the NBA (basketball record): by the Los Angeles Lakers, November 5, 1971 to January 7, 1972.
- retired baseball uniform number of:

⇨ first baseman Eddie Murray, retired by the Baltimore Orioles.

⇨ pitcher Mike Scott, retired by the Houston Astros.

⇨ shortstop Honus Wagner, retired by the Pittsburgh Pirates.

- retired basketball jersey number of:
  ⇨ Kareem Abdul-Jabbar, retired by both the Los Angeles Lakers and the Milwaukee Bucks.
  ⇨ Alvan Adams, retired by the Phoenix Suns.
  ⇨ Larry Bird, retired by the Boston Celtics.
  ⇨ Patrick Ewing, retired by the New York Knicks.
  ⇨ Alonzo Mourning, retired by the Miami Heat.
  ⇨ Scottie Pippen, retired by the Chicago Bulls.
  ⇨ David Thompson, retired by the Denver Nuggets.

- retired football jersey number of:
  ⇨ Sammy Baugh, retired by the Washington Redskins.
  ⇨ Stone Johnson, retired by the Kansas City Chiefs.

- retired hockey shirt number of Patrick Roy, retired by the Colorado Avalanche.

- sum of the first eight Fibonacci numbers $(0 + 1 + 1 + 2 + 3 + 5 + 8 + 13)$.

- 33 inches:
  ⇨ Adult height of Charles Sherwood Stratton, who P. T. Barnum introduced to the public as Tom Thumb.
  ⇨ Height of the bar in women's 100-m hurdles, positioned 8.5-m apart.

- 33 miles: length of the narrow Khyber Pass between Pakistan and Afghanistan, which was the entrance route for several invaders of India.

- Title of the United States Code dealing with navigation and navigable waters.

**"Diga trente y tres"**

- said to the subject of a photograph, the Spanish equivalent of "Say cheese" in English.

## Heaven of the Thirty-Three Gods

- to which the Buddha ascended in order to convert his mother.

## longest winning streak in the National Basketball Association—thirty-three games

- by the Los Angeles Lakers, starting with their win over the Baltimore Bullets (110–106), November 5, 1971, and ending with their loss to the Milwaukee Bucks (120–104), January 7, 1972.

## thirty-third president

- Harry S. Truman, April 12, 1945 to January 20, 1953.

## thirty-third state of the Union

- Oregon, admitted February 14, 1859.

## "33"

- a phonograph record made to be played at thirty-three revolutions per minute. This new "LP" format—12-inch long-playing vinyl discs—was perfected by Columbia Records in 1948, meant to replace the 78 rpm shellac discs. Soon RCA Victor, Columbia's rival, began issuing seven-inch vinyl discs, played at 45 rpm ("EP"), which held as much sound as the older 78s. The 45s took over the market with the advent of rock and roll in the late '50s, but by the end of the 1960s, 33-rpm LPs were in the ascendancy.

## world's longest tunnel

- the Seikan Tunnel, a railroad tunnel under the Tsugaru Strait in Japan, is 33.5 miles long. Completed in 1988, it is also the world's deepest underwater tunnel. A tunnel called the Gotthard Base Tunnel is now in construction in the Swiss Alps, which, when finished in 2017–18, will extend 35.5 miles.

## *034*

- Decimal ASCII code for [double quote (")].

- medical classification code for streptococcal sore throat and scarlet fever.

## 34

- atomic number of the chemical element selenium, symbol Se.
- average number of annual vacation days in Brazil, compared to thirteen in the U.S.
- European size women's sweater equivalent to American size 10.
- international telephone calling code for Spain.
- magic constant in a 4 x 4 magic square:

| 1 | 12 | 8 | 13 |
| 14 | 7 | 11 | 2 |
| 15 | 6 | 10 | 3 |
| 4 | 9 | 5 | 16 |

- number of kills by the U.S. Navy's number-one ace fighter pilot in World War II, Captain David McCampbell of the VF-15 Squadron, flying an F6F.
- number of presidential electoral votes apportioned to the state of Texas. (May be reapportioned for the 2012 election.)
- number of the boat commanded by John Wayne in the 1945 film *They Were Expendable.*
- number of the French Département of Hérault.
- percentage of lead content in Waterford crystal (as opposed to the usual 24 percent).
- percentage of all dinner entrees in the U.S. that are completely homemade; 66 percent are not. Of the 34 percent, 7 percent are sandwiches. This is per a *Newsweek* survey in 2004.
- retired baseball uniform number of:
  ⇨ pitcher Rollie Fingers, retired by both the Milwaukee Brewers and the Oakland Athletics.
  ⇨ pitcher Nolan Ryan, retired by both the Houston Astros and the Texas Rangers.
  ⇨ outfielder Kirby Puckett, retired by the Minnesota Twins.

- retired basketball jersey number of:

⇨  Charles Barkley, retired by both the Philadelphia 76ers and the Phoenix Suns.
⇨  Austin Carr, retired by the Cleveland Cavaliers.
⇨  Mel Daniels, retired by the Indiana Pacers.
⇨  Hakeem Olajuwon, retired by the Houston Rockets.

- retired football jersey number of:
  ⇨  Earl Campbell, retired by the Tennessee Titans.
  ⇨  Walter Payton, retired by the Chicago Bears.
  ⇨  Joe Perry, retired by the San Francisco 49ers.
- 34 square miles: size of Manhattan Island in New York City.
- Title of the United States Code dealing with the navy.

## "A squat grey building of only thirty-four stories."

- the opening sentence of chapter one of Aldous Huxley's *Brave New World*.

## longest non-technical word in the English language

- with thirty-four letters, "supercalifragilisticexpialidocious," according to the Oxford English Dictionary, a nonsense word used especially by children, now chiefly expressing excited approbation; fantastic, fabulous. The word was made popular by the song of that name in Disney's 1964 film, *Mary Poppins*. The oft-quoted "antidisestablishmentarianism" has a mere twenty-nine letters.

## L34A1

- the Sterling, a submachine gun made by the Sterling Engineering Co. for the British navy in World War II.

## *Miracle on 34th Street*

- delightful 1947 Christmas story; film directed by George Seaton, with Maureen O'Hara, John Payne, and Natalie Wood, supporting Edmund Gwenn's Oscar-winning performance in the role of is-he-or-isn't-he Santa Claus. Remade in 1973 and 1994, but the original is still the best.

## MP-34

- 9mm submachine pistol widely used by German police and the German army in World War II. Originally listed as "9mm

(Mauser) Steyr Solothurm MOD 1934," the army logged it as "MP 34(O), Machinen Pistole 34 Osterreich."

## PDD-34

- Presidential Decision Directive 34 deals with Conventional Arms Transfer Policy, issued by President Bill Clinton on February 17, 1995.

## R-34

- designation for the British dirigible that made the first ever double air crossing of the Atlantic Ocean; westbound from East Fortune airbase in Scotland to Hazelhurst Field in Mineola, New York, July 2 to July 6, 1919; eastbound from Mineola to Pulham, England, July 10 to July 13. A historic voyage that became known as the "Forgotten Flight."

## thirty-fourth president

- Dwight D. Eisenhower, January 20, 1953 to January 20, 1961.

## thirty-fourth state of the Union

- Kansas, admitted January 29, 1861.

## 035

- Decimal ASCII code for [number sign (#)].
- medical classification code for erysipelas.

## 35

- atomic number of the chemical element bromine, symbol Br.
- average number of annual vacation days in Germany, compared to thirteen in the U.S.
- cubic feet of sea water in a displacement ton (measure of the displacement of ships) equal to a long ton of 2,240 pounds.
- lunch-counter code for a cherry Coke.

- most strikeouts pitched in a World Series (baseball record): by Bob Gibson, St. Louis Cardinals, against the Detroit Tigers in the seven-game 1968 Series.
- most times hit by a pitch in a "modern-era" season (baseball record): by Don Baylor, Boston Red Sox, 1986.
- number of bones broken by daredevil motorcyclist Evel Knievel, until his 1981 retirement.
- number of categories of magical powers bestowed by the spirits, according to the *Legemeton*, the Book of Spirits.
- number of English Pilgrims who sailed from their home in the Netherlands on the *Speedwell* on July 22, 1620. The ship proved unseaworthy and they switched to the *Mayflower* at Plymouth, England, sailing for the New World on September 16, with sixty-six additional passengers from Southampton and London. They arrived at Plymouth in the New World on December 21, 1620.
- number of the French Département of Ille-et-Vilaine.
- number of violin sonatas written by Mozart.
- retired baseball uniform number of:
  - ⇨ pitcher Randy Jones, retired by the San Diego Padres.
  - ⇨ pitcher Phil Niekro, retired by the Atlanta Braves.
  - ⇨ designated hitter/first baseman Frank Thomas, retired by the Chicago White Sox.

- retired basketball jersey number of:
  - ⇨ Roger Brown, retired by the Indiana Pacers.
  - ⇨ Darrell Griffith, retired by the Utah Jazz.
  - ⇨ Reggie Lewis, retired by the Boston Celtics.

- retired hockey shirt number of:
  - ⇨ Tony Esposito, retired by the Chicago Blackhawks.
  - ⇨ Mike Richter, retired by the New York Rangers.

- thirty-five years: minimum age of President of the United States, who (now) must be a natural-born citizen. The Constitution allows for the office to be held by a person who had been a citizen at the time of its adoption and had also been a resident for fourteen years, but those qualifications obviously no longer apply.
- Title of the United States Code dealing with patents.

## I-35

- major north-south interstate highway from Laredo to San Antonio, Fort Worth, Oklahoma City, Wichita, Kansas City, Des Moines, Minneapolis/St. Paul, and Duluth.

## longest-serving U.S. congresswoman—thirty-five years

- Edith Nourse Rogers, Republican from Massachusetts, served in the U.S. House of Representatives from June 30, 1925, to her death on September 10, 1960.

## longest undefeated streak in the National Hockey League—thirty-five games

- by the Philadelphia Flyers, starting with their October 14, 1979, win over the Toronto Maple Leafs, ending with their January 6, 1980 loss to the Minnesota North Stars. Total of twenty-five wins and ten ties.

## Mozart's *Symphony No. 35*

- "Haffner" symphony (in D major).

## thirty-fifth president

- John F. Kennedy, January 2, 1961 to November 22, 1963.

## thirty-fifth state of the Union

- West Virginia, admitted June 20, 1863.

## thirty-fifth wedding anniversary gift

- coral or jade is customary.

## thirty-five

- a card game in which the object is to hold a card count of at least thirty-five points in one suit in a hand of nine cards.

## thirty-five-plus

- age of the lead character on the CBS radio serial *The Romance of Helen Trent*—the story of a woman who sets out to prove that romance can live on at thirty-five . . . and even beyond.

**35 to 1**

- the odds against either throwing a 2 or throwing a 12 in dice.

*35 Up*

- the fifth installment (1991) of the Michael Apted documentary on British TV following the lives of a group of British children in seven-year intervals.

# *036*

- Decimal ASCII code for [dollar sign ($)].

# *36*

- atomic number of the chemical element krypton, symbol Kr.
- gallons in a beer barrel.
- highest number on a roulette wheel.
- inches in a yard.
- international telephone calling code for Hungary.
- lives lost when the German airship *Hindenburg* crashed and burned at Lakehurst, New Jersey, on May 6, 1937. There were ninety-seven aboard.
- most field goals scored in an NBA game (basketball record): by Wilt Chamberlain for the Philadelphia Warriors against the New York Knicks, March 2, 1962 (of sixty-three attempts).
- most triples hit in a baseball season: by Chief Wilson, Pittsburgh Pirates, 1912.
- number of black keys on a piano.
- number of colleges at Oxford University.
- number of counters for each player in the card game, Bell and Hammer (called Schimmel in Germany, where the game originated).
- number of counties in the state of Oregon.
- number of Doric columns around the Lincoln Memorial in Washington, D.C., representing the number of states in the Union at the time of Lincoln's death. Designed in the form

of a modified Greek temple, the Memorial was dedicated on May 30, 1922.
- number of the French Département of Indre.
- number of 0s in an undecillion (American) or a sextillion (British).
- record number of runs scored by one team in a baseball game, pre-1900: by the Chicago Colts against the Louisville Colonels (7), June 29, 1897.
- retired baseball uniform number of:
  ⇨ pitcher Gaylord Perry, retired by the San Francisco Giants.
  ⇨ pitcher Robin Roberts, retired by the Philadelphia Phillies.

- retired basketball jersey number of Lloyd Neal, retired by the Portland Trail Blazers.
- retired football jersey number of Mack Lee Hill, retired by the Kansas City Chiefs.
- the square of 6.
- sum of the first eight integers $(1 + 2 + 3 + 4 + 5 + 6+ 7 + 8)$.
- sum of the first three cubic numbers $(1 + 8 + 27)$.
- 36 degrees: size of each exterior angle of an equilateral decagon.
- 36 feet: width of a regulation tennis court for doubles play.
- 36 inches:
  ⇨ diameter of each of the four cables supporting the George Washington Bridge across the Hudson River connecting New York City with Fort Lee, New Jersey. Each cable contains 61 strands, each composed of 434 wires.
  ⇨ height of the bar in men's 400-m hurdles, positioned 35-m apart.

- 36 million miles: average distance of the planet Mercury from the sun.
- Title of the United States Code dealing with patriotic societies and observances

**longest tenure on the U.S. Supreme Court**
- thirty-six years (and 209 days), by Justice William O. Douglas, April 17, 1939 to November 12, 1975.

## MAS 36

- French rifle in wide production before the country collapsed in World War II. This was the last bolt-action rifle to be introduced by a major army.

## Mozart's *Symphony No. 36*

- "Linz" symphony (in C major).

## PDD-36

- Presidential Decision Directive 36 deals with U.S. Policy on the Ocean Environment, issued by President Bill Clinton on April 5, 1995.

## *The Thirty-Six Dramatic Situations*

- a descriptive list of plot ideas intended as a catalog of all possible dramatic themes. Written by Georges Polti, intended as a source for writers, dramatists, etc.

## *36 Hours*

- suspenseful 1965 film, written and directed by George Seaton, in which James Garner is a captured American intelligence officer in World War II, from whom his German captors want to learn about the upcoming Allied invasion of the continent. Rod Taylor plays the German examiner and Eva Marie Saint fills the distaff role.

## Thirty-Six Immortal Poets

- compilation of the works of thirty-six notable Japanese poets, thirty-one men and women, writing in the seventh to eleventh centuries, collected by the poet Fujiwara no Kinto in the eleventh century. Later the subject of a well-known two-panel screen painting by Sakai Hoitsu, done in ink, color, and gold on paper, probably at the beginning of the nineteenth century.

## Thirty-Six Immortal Women Poets

- group that flourished in Imperial Japan in the ninth through thirteenth centuries, capturing women's attitudes toward court life.

## 36-point type

- typographer's measure, known as "double great primer."

## *Thirty-Six Strategies*

- a Chinese collection of thirty-six proverbs intended to illustrate military tactics and strategies, structured (based on the hexagrams of the *I-Ching*) in six chapters of six proverbs each, each proverb followed by a short explanation of its military application.

## thirty-sixth president

- Lyndon B. Johnson, November 22, 1963 to January 20, 1969.

## thirty-sixth state of the Union

- Nevada, admitted October 31, 1864.

## Thirty-Six Views of Mount Fuji

- a series of woodcut prints by the master Japanese artist, Katsushika Hokusai, done about 1826–33.

## *037*

- Decimal ASCII code for [percent sign (%)].
- medical classification code for tetanus.

## *37*

- age at which Lou Gehrig died from amyotrophic lateral sclerosis.
- atomic number of the chemical element rubidium, symbol Rb.
- average number of annual vacation days in France, compared to thirteen in the U.S.
- European size women's shoes equivalent to American size 7.
- in Buddhism, the number of divisions of *Arhatship*, the state of being worthy.
- number of elephants Hannibal led across the Alps (with 35,000 men) to invade Italy in 218 BC.

- number of KO's scored by Muhammad Ali (Cassius Clay) in his professional boxing career (out of sixty-one fights, fifty-six wins, and five losses).
- number of peg holes in the board for the game of Solitaire as played in France. (In England the board traditionally has thirty-three peg holes.)
- number of plays written by William Shakespeare.
- number of pockets on a foreign roulette wheel, 1 through 36 plus 0. American roulette wheels have thirty-eight, an 00 pocket added to the above.
- number of propositions in Book III of Euclid's *Elements*, on circles.
- number of the French Département of Indre-et-Loire.
- retired baseball uniform number of manager Casey Stengel, retired by both the New York Yankees and the New York Mets.
- retired football jersey number of:
  ⇨ Jimmy Johnson, retired by the San Francisco 49ers.
  ⇨ Doak Walker, retired by the Detroit Lions.

- Title of the United States Code dealing with pay and allowances of the uniformed services.

**cloth yard**
- a former measure of 37 inches.

**37 degrees Celsius**
- commonly accepted normal body temperature (=98.6 degrees F).

**thirty-seventh president**
- Richard M. Nixon, January 20, 1969 to August 9, 1974. Nixon resigned his presidency after the so-called Watergate scandal.

**thirty-seventh state of the Union**
- Nebraska, admitted March 1, 1867.

# 038

- Decimal ASCII code for [ampersand (&)].
- medical classification code for septicemia.

# 38

- atomic number of the chemical element strontium, symbol Sr.
- liters in a bath (ancient Hebrew liquid measure).
- most consecutive games scoring field goals (football record): by Matt Stover, for the Baltimore Ravens, 1999–2001.
- most power-play goals in NHL playoffs, career (hockey record): by Brett Hull for the St. Louis Blues (27), Dallas Stars (7), and Detroit Red Wings (4), between 1988 and 2004.
- number of days the Israelites wept for Moses in the plains of Moab (Deuteronomy 34:8).
- number of letters in the Armenian alphabet.
- number of pockets on an American roulette wheel, numbers 1 through 36, plus 0 and 00. Foreign roulette wheels frequently have only thirty-seven pockets, no 00.
- number of the French Département of Isère.
- slang for a .38 caliber pistol.
- Title of the United States Code dealing with veterans' benefits.

## C-38

- designation of the Civil Marriage Act, the bill that legalized same-sex marriage in Canada. Introduced in February 2005, it became law on July 20, 2005.

## MAS 38

- a 7.65mm (Long) Model 1938 SMG submachine gun used by France in World War II. One of the first such weapons to have a folding trigger.

## Mozart's *Symphony No. 38*

- "Prague" symphony (in D major).

## MP-38

- 9mm Parabellum submachine gun used by the German Army in World War II. Called a Schmeiser, although Otto Schmeiser did not design it. It was produced by the Haenel firm, of which a man named Schmeiser was general manager.

## P-38

- the Lockheed Lightning, a World War II long-range fighter aircraft, the "Fork-Tailed Devil," with its distinctive pair of fuselages and its center pod cockpit. Later redesignated F-38.

## shortest war in history—thirty-eight minutes

- between Britain and Zanzibar, on August 27, 1896. After the Anglo-friendly Sultan of Zanzibar died, his brother-in-law Khalid bin Bargash seized the throne and would not give it up to Hamoud bin Mohammed, the choice of the Colonial British. When Bargash ignored the British 9:00 AM ultimatum to abdicate, they attacked his palace at 9:02 AM and sunk his only warship. By 9:40 AM Bargash had surrendered and fled. The war thus lasted for thirty-eight minutes.

## 38th parallel

- 38 degrees north latitude, the dividing line between North and South Korea established at the Potsdam Conference in 1945.

## thirty-eighth president

- Gerald R. Ford, August 9, 1974 to January 20, 1977.

## thirty-eighth state of the Union

- Colorado, admitted August 1, 1876.

## 039

- Decimal ASCII code for [single quote (')].

# 39

- atomic number of the chemical element yttrium, symbol Y.
- international telephone calling code for Italy and Vatican City.
- number of books in the Old Testament.
- number of counties in the state of Washington.
- number of the French Département of Jure.
- number of zeroes in a duodecillion.
- retired baseball uniform number of Roy Campanella, retired by the Los Angeles Dodgers.
- retired football jersey number of:
  ⇨ Larry Csonka, retired by the Miami Dolphins.
  ⇨ Hugh McElhenny, retired by the San Francisco 49ers.

- sum of the first three powers of 3: $3^1 + 3^2 + 3^3$.
- sum of the squares of the first four primes ($1^2 + 2^2 + 3^2 + 5^2$).
- Title of the United States Code dealing with postal service.

## BB39

- number of the battleship *Arizona*, sunk by Japanese forces at Pearl Harbor in Honolulu, Hawaii, on December 7, 1941, precipitating World War II.

## Jack Benny's purported age

- which remained unchanged for his entire career. Actually, he was born in Waukegan, Illlinois, on February 14, 1894.

## PDD-39

- Presidential Decision Directive 39 deals with U.S. Policy on Counterterrorism, issued by President Bill Clinton on June 21, 1995.

## Pier 39

- San Francisco's forty-five-acre playground, on the Bay, for both tourists and residents. The west marina had been home to the world-famous California sea lions, with as many as 900 in residence. The Pier is described as "a festival marketplace with more than 110 shops, thirteen full-service restaurants with Bay views, and numerous fun-filled attractions," including an oversize arcade and a 3-D adventure ride.

### The Thirty-Nine Articles

- the basic statement of faith for Anglican and Episcopal churches, "all churches of his majesty's dominions," set forth in 1563. The Articles enunciate some of the major differences between Anglican and Roman Catholic doctrine.

### 39 Crenshaw, Tuckahoe, New York

- home of the Findlay family on the CBS-TV show *Maude*.

### *The Thirty-Nine Steps*

- an exciting spy story by John Buchan (1915); in 1935, as *The 39 Steps*, turned into a classic movie directed by Alfred Hitchcock, starring Robert Donat and Madeleine Carroll.

### thirty-ninth president

- James (Jimmy) Carter, January 20, 1977 to January 20, 1981.

### thirty-ninth state of the Union

- North Dakota, admitted November 2, 1889.

## 040

- Decimal ASCII code for open parenthesis [ ( ].

## 40

- age of Atlanta Diamondback pitcher Randy Johnson when he pitched a perfect game against the Braves at Atlanta's Turner Field on May 18, 2004. Johnson, a five-time winner of the Cy Young Award as the best pitcher in the league, is the oldest to throw a perfect game, the seventeenth in baseball history.
- the age at which life begins, according to a well-known cliché.

- atomic number of the chemical element zirconium, symbol Zr.
- buqshas in a riyal, currency of the Yemen Arab Republic.
- cubic feet in a freight ton.
- forty days: average incubation period of a kangaroo.
- 40 degrees: size of each exterior angle of an equilateral nonagon.
- 40 feet:
  ⇨ distance between the stakes in horseshoes.
  ⇨ preferred width of a high school basketball court.

- 40 inches: minimum height of the wooden wall surrounding a field hockey rink.
- forty years: age of Muhammad in on AD 610, when he emerged as a prophet and founder of Islam.
- in Hebrew, the numerical value of the letter *mem.*
- international telephone calling code for Romania.
- in the Bible the number forty appears frequently, in different contexts:
  ⇨ rain fell upon the earth for forty days and forty nights (Genesis 7:12).
  ⇨ Isaac was forty years old when he took Rebekah as his wife (Genesis 25:20).
  ⇨ the children of Israel ate manna for forty years until they came to the land of Canaan (Exodus 16:35).
  ⇨ Moses spent forty days and forty nights with the Lord on Mount Sinai (Exodus 34:28).
  ⇨ forty years that the Israelites wandered through the wilderness (Numbers 14:33).
  ⇨ forty years that the Lord delivered the Israelites into the hands of the Philistines because they did evil in His eyes (Judges 13:1).
  ⇨ forty years that David reigned over Israel (2 Samuel 5:4).
  ⇨ length in cubits of the temple Solomon built in Jerusalem (1 Kings 6:17).
  ⇨ forty years that Solomon reigned in Jerusalem over all Israel (1 Kings 11:42).
  ⇨ forty days and forty nights that Elijah journeyed to Mount Horeb (1 Kings 19:8).
  ⇨ forty years the Lord prophesies desolation for Egypt (Ezekiel 29:12).

⇨ forty days Jonah gives Nineveh to repent its evil ways (Jonah 3:4).

⇨ forty days and forty nights that Jesus fasted in the wilderness (Matthew 4:2).

⇨ forty days Christ spent on earth after the Resurrection (Acts 1:3).

- most field goals made in an NFL season (football record): by Neil Rackers for the Arizona Cardinals, 2005.

- most free throws scored by a team in an NBA All-Star game (basketball record): by the East All-Stars, 1959.

- most points scored by one player in an NFL game (football record): by Ernie Nevers for the Chicago Cardinals against the Chicago Bears' six points, November 28, 1929. Nevers scored all the Cardinals points with six rushing touchdowns and four points after.

- most runs batted in World Series games (baseball record): by Mickey Mantle for the New York Yankees, 1951–64.

- number of cat breeds recognized as purebreds by the Cat Fanciers Association.

- number of enemy planes shot down by the most successful American fighter pilot of World War II, Major Richard I. Bong, flying a P-38 in the Pacific Theater of Operations.

- number of members required for a quorum in the British House of Commons.

- number of propositions in Book VII of Euclid's *Elements,* on fundamentals of number theory, and in Book XI on solid geometry.

- number of teeth on an adult male horse (mares have thirty-six).

- number of the French Département of Landes.

- the only number whose letters, when written out, are in alphabetical order.

- rap slang for a 40-ounce bottle of malt liquor.

- retired baseball uniform number of:
    ⇨ manager Dale Murtaugh, retired by the Pittsburgh Pirates.
    ⇨ pitcher Don Wilson, retired by the Houston Astros.

- retired basketball jersey number of:
    ⇨ Byron Beck, retired by the Denver Nuggets.
    ⇨ Jason Collier, retired by the Atlanta Hawks.

⇨ Bill Laimbeer, retired by the Detroit Pistons.

- retired football jersey number of:
  - ⇨ Tom Brookshier, retired by the Philadelphia Eagles.
  - ⇨ Mike Haynes, retired by the New England Patriots.
  - ⇨ Joe Morrison, retired by the New York Giants.
  - ⇨ Gale Sayers, retired by the Chicago Bears.
  - ⇨ Pat Tillman, retired by the Arizona Cardinals.

- rods in a furlong.
- Title of the United States Code dealing with public buildings, property, and works.
- value in the game of pinochle for a meld of king and queen in trumps, called royal marriage or common in trumps.
- value in the game of pinochle for a meld of the queen of spades and the jack of diamonds, called a pinochle.
- value of medieval Roman numeral F.
- value of Roman numeral XL.
- yards in a bolt of cloth.

## A40

- a motorway in the British transport network that connects England and Wales.

## "Ali Baba and the Forty Thieves"

- one of the most memorable stories in *Arabian Nights*, in which the woodcutter Ali Baba overhears the magic password, "Open Sesame," providing him entrance into a cave of treasures, and makes off with untold riches. Converted to a film in 1944, with Turhan Bey, Maria Montez, and Jon Hall.

## Field of the Forty Footsteps

- located at the back of the British Museum in London, at the northeast extremity of Upper Montague Street, once called Southampton Fields. Here, it is said, two brothers fought one another, having taken opposite sides in the Duke of Monmouth's rebellion. Both were killed, but their feet left forty impressions in the ground, where no grass would grow for many years.

**the Forties**

- a well-known gang of thieves in 1870s England.

**fortieth president**

- Ronald Reagan, January 20, 1981 to January 20, 1989.

**fortieth state of the Union**

- South Dakota, admitted November 2, 1889.

**fortieth wedding anniversary gift**

- rubies are appropriate.

**40 acres and a mule**

- phrase describing the intended reparations to ex-slaves following the Civil War. Union general William Tecumseh Sherman, following his march through Georgia to the sea, proposed to the War Department that each freed slave be given a tract of confiscated land and a surplus animal from the Union forces, the intent being "to enable them to establish a peaceable agricultural settlement." Congress overrode the plan and all such land was returned to its original white plantation owners.

**forty-cents-per-hour minimum wage**

- for workers engaged in interstate commerce, raised from twenty-five cents by the Wage and Hours Act, signed by President Roosevelt on June 25, 1938.

**forty days of Lent**

- length of the Christian season of Lent (originally called Quadragesima), starting on Ash Wednesday and ending at noon on the day before Easter, Holy Saturday; actually six and a half weeks, but Sundays, being the Lord's Day, are not counted; thus, forty days.

### Forty Deuce

- 1982 film directed by Paul Morrissey, starring Orson Bean and Kevin Bacon, about drug-addict street hustlers in New York City. Set in the infamous area around 42nd Street in the '80s, before the sleaze in that part of the city was cleaned up. The film is notable for its split-screen final sequence showing concurrent activities in a hotel room.

### forty-four work week

- became effective on October 24, 1940. The shorter work week was voted by Congress as part of the Fair Labor Standards Act of 1938.

### The Forty Immortals

- the forty members of the French Academy.

### forty jacks

- in the card game pinochle, a meld of four jacks, one from each suit, is worth forty points.

### Forty Martyrs

- a group of Christian soldiers who, in the fourth century, were condemned by Roman emperor Licinius to be frozen to death for their faith, near Sebaste in Lesser Armenia.

### Forty Mile

- a town in western Yukon Territory, Canada, about 60 miles northwest of Dawson.

### Fortymile River

- a waterway in eastern Alaska.

### forty-penny nail

- a carpenter's common nail, 5 inches long (coded 40d).

### 40-point type

- typographer's measure, known as "double paragon."

**40-second play clock**

- refers to the rule in football that a team must snap the ball to begin a play within forty seconds after the previous play has ended. After a time-out or other stoppage of play, a 25-second play clock starts when the ball is spotted and signaled ready for play.

*Forty Shades of Blue*

- 2005 film directed by Ira Sachs about a well-to-do music producer (Rip Torn) and his much younger Russian live-in girlfriend (Dina Korzun), whose relationship is upset by the arrival of his estranged son from an earlier marriage (Darren E. Burrows).

**forty whacks**

- the number of blows Lizzie Borden dealt her mother, said the doggerel of the period:

  > Lizzie Borden took an ax
  > And gave her mother forty whacks.
  > When she saw what she had done
  > She gave her father forty-one.

**forty winks**

- slang for a short nap.

**I-40**

- major east-west interstate highway from Barstow to Flagstaff, Alburquerque, Amarillo, Oklahoma City, Little Rock, Memphis, Nashville, Winston-Salem, Greensboro, and Wilmington (North Carolina).

*Life Begins at Forty*

- Will Rogers as a small-town newspaperman in a 1935 film with a simple story line that allows him to observe and comment on American foibles.

## M40

- a motorway in the British transport network that runs from London to Birmingham.

## *North Dallas Forty*

- a realistic look at what it means to be a professional football player, complete with the aches and pains of the life of an athlete both on and off the field. Directed by Frank Yablans, who also co-wrote the screenplay. Nick Nolte, Mac Davis, and Bo Svenson are the players in this 1979 satire of the game;, G. D. Spradlin and Charles Durning are the coaches.

## P-40

- the Curtiss Warhawk (or Tomahawk), a U.S. fighter-bomber, the shark-toothed plane that was the mainstay of the Flying Tigers in China.

## quadragenarian

- a person of between forty and forty-nine years of age.

## Red 40

- common name of the food additive FD & C Red Dye # 40.

## Top 40

- music industry term for the most popular songs of the moment, sometimes limited to a particular type of music, sometimes nonspecific for the most frequently played or best-selling pop records. Instituted by radio exec Todd Storz on station WTIZ in New Orleans in 1953, in what has come to be called Top 40 Radio. This has become the format of several radio stations around the country in different music genres.

## *A Traveler at Forty*

- one of Theodore Dreiser's autobiographical works (1913).

## WD-40

- a spray lubricant.

## 041

- Decimal ASCII code for close parenthesis [ ) ].

## 41

- Article of the New York State Tax Law covering Taxpayers' Bill of Rights.
- atomic number of the chemical element niobium, symbol Nb.
- charlton Heston's ID number as a galley slave in the film, *Ben-Hur.*
- European size men's shirt equivalent to American size 16.
- international telephone calling code for Switzerland.
- lunch-counter code for a lemonade.
- most rebounds in an NBA playoff game (basketball record): by Wilt Chamberlain for the Philadelphia 76ers against the Boston Celtics, April 5, 1967.
- a number believed to be hidden in much of the music of J. S. Bach, supposedly encoded as a type of signature. Using simple number substitution for Latin letters (+ K), "41" could be determined as: J=I (9) + S (18) + B (2) + A (1) + C (3) + H (8) = 41. Bach alone would equal 2 + 1 + 3 + 8 = 14, another number prominent in his works.
- number of signers of the Mayflower Compact, the first attempt at government for the New England colony, on November 11, 1620.
- number of symphonies composed by Mozart.
- number of the PT boat commanded by Robert Montgomery in the 1945 film *They Were Expendable.*
- number of the French Département of Loir-et-Cher.
- number of years that Asa reigned in Jerusalem (1 Kings 15:10).
- record number of games won by a pitcher in a season in the "modern" baseball era: by Jack Chesbro, New York Yankees, 1904. (Pre-1901, Old Hoss Radbourn won 59 in 1884.)
- retired baseball uniform number of:
  - ⇨ third baseman Eddie Mathews, retired by the Atlanta Braves.
  - ⇨ pitcher Tom Seaver, retired by the New York Mets.

- retired basketball jersey number of Wes Unseld, retired by the Washington Wizards.
- retired football jersey number of Brian Piccolo, retired by the Chicago Bears.
- sum of the first six prime numbers $(2 + 3 + 5 + 7 + 11 + 13)$.
- Title of the United States Code dealing with public contracts.
- uniform number assigned to forty-one-year-old pitcher Randy Johnson by the New York Yankees when he signed a two-year contract for $32 million in January 2005.
- value of Roman numeral XLI.

**forty-first president**

- George H. W. Bush, January 20, 1989 to January 20, 1993.

**forty-first state of the Union**

- Montana, admitted November 8, 1889.

**Mozart's *Symphony No. 41***

- "Jupiter" symphony (in C major).

**No. 41, Albert Square**

- residence of Pat Evans (née Harris), on the BBC-TV series *EastEnders*.

**Rule 41**

- National Hockey League Rulebook item that deals with unsportsmanlike conduct.

## *042*

- Decimal ASCII code for [asterisk (*)].
- medical classification code for HIV disease.

## *42*

- age of Elvis Presley when he died.

- apartment number of Fox Mulder on the TV series *The X-Files.*
- atomic number of the chemical element molybdenum, symbol Mo.
- average number of annual vacation days in Italy, compared to thirteen in the U.S.
- 42 feet: length of the right arm of the Statue of Liberty.
- forty-two hours: the length of time Juliet is put to sleep by the potion supplied by Friar Lawrence, in Act IV, Scene 1, of *Romeo and Juliet.*
- 42 inches:
  - ⇨ maximum length of the bat in professional baseball.
  - ⇨ height of the basket backboard on a regulation basketball court.
  - ⇨ height of the bar in men's 110-m high hurdles, positioned 10 yards apart.

- forty-two months: length of the 1929–33 Great Depression.
- 42 square miles: area of the borough of the Bronx, New York City.
- gallons in a petroleum barrel.
- largest margin of victory in an NBA finals (basketball record): Chicago Bulls over the Utah Jazz (96–54), Game 3, June 7, 1998.
- maximum total number of any one gift received in the song "The Twelve Days of Christmas" (either geese a-laying or swans a-swimming).
- minimum age to hold office as a consul in the Roman Republic.
- most career runs scored in World Series games (baseball record): by Mickey Mantle for the New York Yankees, 1951–64.
- most passes intercepted in an NFL season (football record): by George Blanda for the Houston Oilers, 1962.
- most penalty minutes served in an NHL playoff game (hockey record): by Dave "The Hammer" Schultz for the Philadelphia Flyers against the Toronto Maple Leafs, April 22, 1976.
- most points scored in an NBA All-Star game (basketball record): by Wilt Chamberlain, 1962.
- number of generations from Abraham to Jesus (Matthew 1:17).

- number of holes in a Connect Four playing structure.
- number of months the beast will exercise power over the Earth (Revelation 13:5).
- number of teeth on an adult dog.
- number of the French Département of Loire.
- number of 0s in a tredecillion (American) or a septillion (British).
- retired baseball uniform number of pitcher Bruce Sutter, retired by the St. Louis Cardinals (and of Jackie Robinson; see below).
- retired basketball jersey number of:
  ⇨ Connie Hawkins, retired by the Phoenix Suns.
  ⇨ Nate Thurmond, retired by both the Cleveland Cavaliers and the Golden State Warriors.
  ⇨ James Worthy, retired by the Los Angeles Lakers.

- retired football jersey number of:
  ⇨ Charlie Conerly, retired by the New York Giants.
  ⇨ Ronnie Lott, retired by the San Francisco 49ers.
  ⇨ Sid Luckman, retired by the Chicago Bears.

- Title of the United States Code dealing with the public health and welfare.
- total number of pips, or spots, on a pair of dice (1 + 2 + 3 + 4 + 5 + 6, twice).

## baseball uniform number 42

- uniform number of second baseman Jackie Robinson, retired by the Los Angeles Dodgers in 1972. On April 15, 1997, at a ceremony at Shea Stadium, the number 42 was retired from all major league baseball. The only players allowed to continue wearing number 42 were those who already wore it, who were "grandfathered in," namely: Butch Huskey of the Mets, Mike Jackson of the Indians, Scott Karl of the Brewers, Jose Lima of the Astros, Mariano Rivera of the Yankees, Mo Vaughn of the Red Sox, Lenny Webster of the Orioles, Tom Goodwin of the Royals, and Marc Sagmoen of the Texas Rangers. As of the 2010 season, the only player still wearing number 42 was Mariano Rivera.

## forty-second president

- William (Bill) Clinton, January 20, 1993 to January 20, 2001.

**forty-second state of the Union**

- Washington, admitted November 11, 1889.

*42nd Street*

- the granddaddy of all backstage musicals, this 1933 film has Dick Powell singing and Ruby Keeler tapdancing while Busby Berkeley moves hundreds of chorines through intricate kaleidoscopic maneuvers photographed by overhead cameras. When the leading lady breaks her ankle, impresario Warner Baxter delivers the now-famous line to Keeler: "You're going out a youngster, but you've got to come back a star."

**Forty-Two**

- also known as Texas Forty-Two, this is a trick-taking game played with a set of double-six dominoes. Two teams of two players each bid to predict the number of points they will accumulate through play. The first team to reach 250 points wins.

**42 Below**

- a vodka produced in New Zealand.

*42 Up*

- the sixth installment (1998) of the Michael Apted documentary on British TV following the lives of a group of British children in seven-year intervals.

**043**

- Decimal ASCII code for [plus (+)].

**43**

- age at which President John F. Kennedy was inaugurated—the youngest man to win the American presidency—in 1960, beating Republican Richard M. Nixon. He was also the first Roman Catholic, and the youngest to die in office.

- an American football team may have forty-three players, but only eleven are allowed on the playing field at any given time.
- atomic number of the chemical element technetium, symbol Tc.
- forty-three cars: maximum number to start in a NASCAR auto race, per a 2001 rule.
- international telephone calling code for Austria.
- most bases on balls taken in World Series games (baseball record): by Mickey Mantle, New York Yankees, 1951–64.
- number of cities played by The Rolling Stones in their 1994 concert tour, which grossed a record-breaking $121.2 million.
- number of the French Département of Haute-Loire.
- retired baseball uniform number of pitcher Dennis Eckersley, retired by the Oakland Athletics.
- retired basketball jersey number of:
  - ➪ Brad Daugherty, retired by the Cleveland Cavaliers.
  - ➪ Jack Sikma, retired by the Seattle SuperSonics.

- retired football jersey number of Jim Norton, retired by the Tennessee Titans.
- Title of the United States Code dealing with public lands.

**forty-third president**
- George W. Bush, January 20, 2001 to January 20, 2009.

**forty-third state of the Union**
- Idaho, admitted July 3, 1890.

**Haydn's *Symphony No. 43***
- "Mercury" symphony (in E-flat major).

**record twenty-four-hour rainfall in the U.S.**
- 43 inches of rain fell in Alvin, Texas, July 25–26, 1979.

**Rule 43**
- National Hockey League Rulebook item that deals with "Attempt to Injure."

## Sonnet 43

- from Elizabeth Barrett Browning's "Sonnets from the Portuguese," starts "How do I love thee? Let me count the ways."

## X-43A

- NASA's scramjet that set a new world record for a jet aircraft— Mach 9.6, almost 7,000 miles per hour—on November 16, 2004. The flight was aided by a Pegasus booster rocket.

## 044

- Decimal ASCII code for [comma (,)].

## 44

- atomic number of the chemical element ruthenium, symbol Ru.
- European size men's shoes equivalent to American size 11.
- forty-four days: average incubation period of a gray squirrel.
- 44 feet: length of a regulation badminton court.
- 44 kilograms: maximum body weight for women weightlifters in category 1, Olympic competition.
- international telephone calling code for the United Kingdom.
- lunch-counter code for a cup of coffee.
- memorable number in Syracuse University football history. Worn by running backs Jim Brown, Ernie Davis, and Floyd Little, all of whom made All-American while at Syracuse.
- number of platforms in Grand Central Terminal in New York City; designed with two levels, there are thirty platforms and forty-one tracks on the upper level, and fourteen platforms and twenty-six tracks on the lower.
- number of "Sonnets from the Portuguese," by Elizabeth Barrett Browning.
- number of the French Département of Loire-Atlantique.
- retired baseball uniform number of:
  - ⇨ outfielder Hank Aaron, retired by both the Milwaukee Brewers and the Atlanta Braves.

⇨ outfielder Reggie Jackson, retired by the New York Yankees. Jackson's number 9 has also been retired by the Oakland Athletics.
⇨ first baseman Willie McCovey, retired by the San Francisco Giants.

- retired basketball jersey number of:
  ⇨ George Gervin, retired by the San Antonio Spurs.
  ⇨ Dan Issel, retired by the Denver Nuggets.
  ⇨ Sam Lacey, retired by the Sacramento Kings.
  ⇨ Jerry West, retired by the Los Angeles Lakers.
  ⇨ Paul Westphal, retired by the Phoenix Suns.

- retired football jersey number of:
  ⇨ Floyd Little, retired by the Denver Broncos.
  ⇨ Pete Retzlaff, retired by the Philadelphia Eagles.

- slang for a .44 caliber pistol, such as the .44 Magnum. This is the caliber of the pistol Frankie uses to shoot Johnny, in the song "Frankie and Johnny."
- slang for a prostitute (from rhyming slang for a whore).
- Title of the United States Code dealing with public printing and documents.

**forty-four-hour work week**
- maximum work hours (for workers engaged in interstate commerce) prescribed by the Wage and Hours Act, signed by President Roosevelt on June 25, 1938.

**44-point type**
- typographer's measure, known as "meridian" or "4-line small pica."

**44s**
- children (CB slang).

**The Forty Fours**
- islands in the Chatham Island Group, southeast of New Zealand.

## forty-fourth president
- Barack H. Obama, January 20, 2009 to —.

## forty-fourth state of the Union
- Wyoming, admitted July 10, 1890.

## Haydn's *Symphony No. 44*
- "Trauer" symphony (in E minor).

## I-44
- east-west interstate highway from Wichita Falls (Texas) to Oklahoma City, Tulsa, Springfield, and St. Louis.

## PDD-44
- Presidential Decision Directive 44 deals with Heroin Control Policy, issued by President Bill Clinton in November 1995.

## *045*
- Decimal ASCII code for [minus or dash (–)].
- medical classification code for acute poliomyelitis.

## *45*
- atomic number of the chemical element rhodium, symbol Rh.
- 45 degrees: size of each exterior angle of an equilateral octagon.
- 45 inches in an ell: an obsolete English unit of length used mainly for measuring cloth.
- 45 meters: minimum length of runway for Olympic long-jump competition.
- 45 yards: longest punt return in a Super Bowl game (football record): by John Taylor for the San Francisco 49ers against the Cincinnati Bengals, SB-XXIII, 1989.
- international telephone calling code for Denmark.

- kilograms in 100 pounds.
- largest winning margin in a Super Bowl game (football record): San Francisco 49ers, 55, Denver Broncos, 10, in SB-XXIV, 1990, at the Superdome in New Orleans, Louisiana. However, in the 1940 NFL Championship, pre-Super Bowl, the spread was greater: The Chicago Bears shut out the Washington Redskins (73–0) at Griffith Stadium in D.C. on December 8, 1940.
- most career wins on the Champions golf tour (formerly the Senior PGA): by Hale Irwin, 1995–2007.
- most consecutive batters retired by a pitcher (baseball record): by Mark Buehrle for the Chicago White Sox, July 18 to July 28, 2009.
- most goals by one team in an NHL playoff series (hockey record): by the Edmonton Oilers against the Chicago Blackhawks, in a six-game series, 1985.
- most passes completed in an NFL game (football record): by Drew Bledsoe for the New England Patriots against the Minnesota Vikings, November 13, 1994 (OT).
- most points scored in NFL Pro Bowl games, career (football record): by Morten Andersen, for the NFC, 1985–88, 1990, 1992, and 1996.
- most turnovers in an NBA game (basketball record): by the San Francisco 49ers against the Boston Patriots, March 19, 1971.
- number of enemy standards captured by Napoleon's forces at the Battle of Austerlitz in December 1805.
- number of stones having magical powers as described in the second book of *The Secrets of Albertus Magnus*.
- number of the French Département of Loiret.
- number of zeroes in a quattuordecillion.
- retired baseball uniform number of pitcher Bob Gibson, retired by the St. Louis Cardinals.
- retired basketball jersey number of:
  - ⇨ Geoff Petrie, retired by the Portland Trail Blazers.
  - ⇨ Rudy Tomjanovich, retired by the Houston Rockets.
- retired football jersey number of Ernie Davis, retired by the Cleveland Browns.
- slang for a .45 caliber pistol.
- sum of the first nine integers $(1 + 2 + 3 + 4 + 5 + 6 + 7 + 8 + 9)$.
- Title of the United States Code dealing with railroads.

**auction forty-five**

- a form of the card game Spoil Five, played by four or six players in two teams of partners.

**forty-fifth state of the Union**

- Utah, admitted January 4, 1896.

**forty-fifth wedding anniversary gift**

- sapphire is customary.

**"45"**

- "extended play" (EP) phonograph records made to be played at forty-five revolutions per minute, introduced by RCA Victor in 1949 to compete with Columbia Records' new 33-rpm "LP." The 45-rpm records found their greatest success with the advent of rock and roll in the late 1950s, but were in decline by the end of the 1960s, as the 12-inch LP grew in popularity.

**Forty-Five**

- a trick-taking card game played with a standard deck of fifty-two cards, popular in Ireland and in the Merrimack Valley in northern Massachusetts and southern New Hampshire. Game is scored by points, determined by the number of tricks taken.

**The Forty-Five**

- the second rebellion of the Jacobites, beginning in July 1745, with the Young Pretender, Charles Edward (Bonnie Prince Charlie), proclaiming his father as James VIII of Scotland and James III of England, and ending the Jacobite threat with their decisive defeat at Culloden Moor, April 16, 1746.

**_Forty-Five Minutes from Broadway_**

- Broadway musical of 1906, with music and lyrics by George M. Cohan; the story of a small-town who unexpectedly inherits a fortune. Introduced the song "Mary's a Grand Old Name."

### *Forty-Five Minutes from Hollywood*

- a popular radio show of the mid-thirties that offered condensed versions of soon-to-be-released movies, with the films' stars' parts played by various radio actors and actresses.

### Haydn's *Symphony No. 45*

- "Farewell" symphony (in F-sharp minor).

### longest word in the English language

- generally thought to be a forty-five-letter word which designates a form of lung disease, "pneumonoultramicroscopicsilicocolvanoconiosis." The word appears in both the Oxford English Dictionary and Webster's Third New International Dictionary.

### 45, Albert Square

- residence of Pauline Fowler, on the BBC-TV series, *EastEnders.*

### world-record twelve-hour rainfall

- 45 inches fell at Foc-Foc on the Indian Ocean island of La Réunion, January 8, 1956, during tropical cyclone Denise.

## *046*

- Decimal ASCII code for [dot (.)].

## *46*

- atomic number of the chemical element palladium, symbol Pd.
- international telephone calling code for Sweden.
- longest winning streak in tennis matches: by Guillermo Vilas, January 11–October 1, 1977.
- number of chromosomes in the human nucleus.
- number of columns surrounding the Parthenon in Athens, Greece.
- number of counties in the state of South Carolina.

- a number of interest to numerologists, who point out that Shakespeare was forty-six when the King James Bible was published. Psalm 46 has as its forty-sixth word "shake," and the forty-sixth word from the back end is "spear."
- number of peaks of the Adirondack mountain range. Climbers who have conquered all of them are called "forty-sixers."
- number of the French Département of Lot.
- number of years that a state of war existed between Israel and Jordan until the peace of 1994.
- percentage of the Earth's surface that is the Pacific Ocean.
- retired football jersey number of Don Fleming, retired by the Cleveland Browns.
- Title of the United States Code dealing with shipping.

## Code 46

- 2003 film that tells a futuristic love story in which government investigator Tim Robbins, inquiring into forged travel permits (a violation of Code 46), has a steamy one-night affair with Samantha Morton, the subject of his investigation. Michael Winterbottom directed.

## 46 defense

- a football defensive formation in which the scrimmage line is packed with six defenders (the four linemen and two linebackers) and backed by the middle linebacker and strong safety, prepared to attack whomever is carrying the ball. Devised by the Chicago Bears defensive coordinator, Buddy Ryan, the "46" is named for the jersey number of Doug Plank, who moved from safety to linebacker position alongside Mike Singletary when the Bears used this formation.

## forty-sixth state of the Union

- Oklahoma, admitted November 16, 1907.

## 46 to 1

- the odds against being dealt three of a kind in five-card poker.

## 047

- Decimal ASCII code for [forward slash (/)].

## 47

- atomic number of the chemical element silver, symbol Ag.
- 47 inches: height of Johnny (John Roventini), the page boy in Philip Morris ads, 1933–74.
- forty-seven years: tenure of J. Edgar Hoover as director of the Federal Bureau of Investigation, from 1924 to 1972, after serving as assistant director from 1921 to 1924.
- international telephone calling code for Norway.
- lives lost on the American battleship Iowa through an explosion in a gun turret on April 19, 1989.
- most points scored in one hockey Stanley Cup series: by Wayne Gretzky for the Edmonton Oilers in 1985. Gretzky scored seventeen goals and thirty assists in eighteen games, against the Los Angeles Kings and the Winnipeg Jets in the first two playoff rounds, against the Chicago Blackhawks in the semis, and then in the finals against the Philadelphia Flyers, which the Oilers won four games to one, the last on May 10 in Edmonton.
- number of miracles performed by Jesus listed in the New Testament.
- number of strings on a harp.
- number of the French Département of Lot-et-Garonne.
- percentage of suppers eaten at home in the U.S. that do not involve turning on the stove, as per a *Newsweek* survey in 2004.
- retired baseball uniform number of:
  ⇨ pitcher Tom Glavine, retired by the Atlanta Braves.
- Title of the United States Code dealing with telegraphs, telephones, and radio-telegraphs.

### AK-47

- the standard weapon for the military of Soviet and Soviet-bloc countries, its designation is short for "automatic Kalashnikov (its inventor) 1947." Available with both automatic and semiautomatic capabilties, using 7.62mm ammunition, it was used by all communist armies until replaced in the 1960s by the AKM.

## B-47

- the Stratojet plane, once the backbone of the American nuclear deterrent until it was retired from service in 1966.

## C-47

- the Skytrain, the military version of the Douglas DC-3 aircraft, extensively used in World War II to carry supplies and personnel.

## Forty-Seven Ronin

- a group of Japanese samurai who, in 1703, assassinated an official named Kira Yoshinaka to avenge his having disgraced their master, Lord Asano, and causing his suicide. After much national debate, the ronin (masterless samurai) committed seppuku (suicide), as they had been ordered to do. Their action has since become a national legend, and they have attained heroic stature as models of loyalty. They are celebrated in national stories, plays, works of art, and a two-part film of 1941–42, directed by Kenji Mizoguchi.

## 47 rue des Ecoles

- address of the Sorbonne in Paris, originally founded as a college of theology in the 1200s.

## 47 Society

- a group at Pomoma College, California, that advances the (factitious) conviction that the number 47 occurs in nature significantly more frequently than any other random number.

## forty-seventh state of the Union

- New Mexico, admitted January 6, 1912.

## '47 Workshop

- George Pierce Baker's enterprise at Harvard (1905–25) that encouraged development of theater craftsmen—including stage designers, dramatists, directors—and that provided a nurturing venue for staging plays by such students as Eugene O'Neill, S. N. Behrman, Sidney Howard, Philip Barry, and Tom Wolfe.

## Group 47

- (Gruppe 47): a literary association in Germany after World War II; founded in 1947, it disbanded in 1977 because of political differences within the group.

## Haydn's *Symphony No. 47*

- The "Palindrome" symphony (in G major).

## longest winning streak in NCAA football—forty-seven games

- by the University of Oklahoma Sooners under coach Bud Wilkinson, starting with a 19–14 win over the Texas Longhorns on October 10, 1953, and ending with a 0–7 loss to Notre Dame on November 16, 1957. The Sooners were named National Champions in 1955 and 1956.

## P-47

- U.S. warplane in World War II, a fighter-bomber called the Thunderbolt, produced by Republic Aircraft.

## Proposition 47

- of Book I of Euclid's *Elements*; contains his proof of the Pythagorean Theorem.

## *048*

- Decimal ASCII code for [zero (0)].

## *48*

- atomic number of the chemical element cadmium, symbol Cd.
- 48 acres: area covered by Grand Central Terminal in New York City.

- 48 kilograms: maximum body weight for women weightlifters in category 2, Olympic competition.
- forty-eight years: age of Julio Franco on May 4, 2007, when he hit a home run for the New York Mets, the oldest major league baseball player to do so.
- international telephone calling code for Poland.
- in the military, a weekend liberty; a 48-hour pass.
- liters in a saa, unit of dry capacity in Algeria.
- most complete games pitched in a "modern-era" season (baseball record): by Jack Chesbro, New York Yankees, 1904. Pre-1901 the record was seventy-five, by Will White with the Cincinnati Reds in 1879.
- most games lost by a pitcher in a pre-modern-era season (baseball record): by John Coleman, Philadelphia Phillies, in 1883.
- most individual points scored in Super Bowl games, career (football record): by Jerry Rice, for the San Francisco 49ers in SB-XXIII, 1989; SB-XXIV, 1990; and SB-XXIX, 1995; and for the Oakland Raiders, SB-XXXVII, 2003.
- most points scored by a team in a Pro Football Hall of Fame game: by the Baltimore Colts over the Pittsburgh Steelers' 17 points, September 6, 1964.
- most wins by a goal tender in regular-season games in one NHL season (hockey record): by Martin Brodeur with the New Jersey Devils, 2006–07.
- number of cards in a pinochle deck.
- number of contiguous United States.
- number of freckles on Howdy Doody's face, representing the forty-eight states in the 1950s.
- number of the French Département of Lozère.
- number of preludes and fugues written for the keyboard by J. S. Bach.
- number of propositions in Book I of Euclid's *Elements*, on the fundamentals of geometry.
- number of 0s in a quindecillion (American) or an octillion (British).
- record number of touchdown passes in a Canadian Football League season, by Doug Flutie for the Calgary Stampeders, 1994.

- Title of the United States Code dealing with territories and insular possessions.

### The First 48

- a TV reality series, debuting on A&E in 2004, that deals with real cases of homicide and featuring the detectives that work to solve them. Done on the streets of various cities, these fast-paced shows get behind the scenes of actual investigations. The theme is spelled out by a voice-over accompanying the opening scenes: "For homicide detectives, the clock starts ticking the moment they are called. Their chance of solving a case is cut in half if they don't get a lead within the first forty-eight hours."

### 48

- the popular name for the 48 preludes and fugues of Bach's *Well-Tempered Clavier I and II* (24 in each). The name derives from Bach's design of a prelude and a fugue in each of the major and minor keys, totaling 48.

### 48, Doughty Street

- Charles Dickens's London residence in the late 1830s, where he wrote *Oliver Twist.*

### Forty-Eighters

- Germans who traveled to other parts of the world, especially the United States and Australia, after the revolutions of 1848 failed to bring reform to the German government. Several were hunted for their revolutionary activity and had to leave to start new lives elsewhere. In the U.S. many fought in the Civil War, most on the Union side because they disapproved of slavery.

### 48HRS

- an entertaining 1982 film amalgam of action and comedy starring Eddie Murphy and Nick Nolte, directed by Walter Hill.

### 48 Hours Mystery
- CBS-TV program that presents "true-crime" documentaries. First aired on January 19, 1988.

### forty-eighth state of the Union
- Arizona, admitted February 14, 1912.

### 48mo
- forty-eightmo, the page size of a book from printer's sheets that are folded into forty-eight equal sections, each about $2\text{-}\frac{1}{2}$ x 4 inches.

### 48-point type
- typographer's measure, known as "canon" or "4-line pica."

### Haydn's Symphony No. 48
- "Maria Theresia" symphony (in C Major).

## 049
- Decimal ASCII code for [the number one (1)].

## 49
- atomic number of the chemical element indium, symbol In.
- fewest team points scored in an NBA game (basketball record): by the Chicago Bulls against the Miami Heat, April 10, 1999.
- international telephone calling code for Germany.
- number of days Gautama sat beneath the Bodhi tree until he attained enlightenment and became a Buddha.
- number of the French Département of Maine-et-Loire.
- record number of runs scored by both teams in a baseball game: by the Chicago Cubs (twenty-six) against the Philadelphia Phillies (twenty-three), at Cubs Park (later Wrigley Field), August 25, 1922.
- retired baseball uniform number of:
  - ⇨ pitcher Larry Dierker, retired by the Houston Astros.
  - ⇨ pitcher Ron Guidry, retired by the New York Yankees.

- span between Jubilee Years as ordained by the Bible in Leviticus (25:8–54).
- the square of 7.
- Title of the United States Code dealing with transportation.

### *The Crying of Lot 49*

- 1965 obscurant novel by Thomas Pynchon in which Lot 49 refers to an auction lot of postage stamps, and "crying" means bidding. The story follows Oedipa Maas, a woman who attempts to act as executor of her ex-lover's estate, through a maze of odd experiences involving truly bizarre characters that may or may not point to a world-wide conspiracy.

### FN-49

- a semi-automatic rifle designed for use by police, developed by the Belgian company Fabrique Nationale (FN).

### forty-nine degrees latitude

- dividing line between the United States and Canada.

### 49er

- a person who took part in the California gold rush of 1849.

### 49ers

- name of the San Francisco professional football team.

### forty-ninth state of the Union

- Alaska, admitted January 3, 1959.

### *49th Parallel*

- exciting 1941 British film about German sailors whose U-boat is sunk off the coast of Canada. All-star cast includes Eric Portman, Anton Walbrook, Leslie Howard, Raymond Massey, Laurence Olivier, and Findlay Currie.

### *49 Up*

- seventh in an award-winning series of documentaries filmed by British director Michael Apted, who followed a group of seven-year-old British children from different social strata and diverse ethnic backgrounds as they advanced from childhood to adulthood. This unique series started with the 1964 *Seven Up*, followed in seven-year intervals by new episodes looking at the changes that had occurred in their lives during the previous seven-year period. In this, the 2005 edition, the subjects at forty-nine years of age, the major finding is that although education is an important determinant of employment success, the British class system still has a strong influence in pre-ordaining life's course. In this episode, twelve of the original fourteen subjects are still participating, forty-two years after the series began.

## Haydn's *Symphony No. 49*

- "La Passione" symphony (in F minor).

## *Ladder 49*

- 2004 film directed by Jay Russell, starring John Travolta as veteran fire-fighter, Captain Mike Russell, who takes probationary fireman Jack Morrison (Joaquin Phoenix) under his wing. Has some spectacular rescue scenes.

## PDD-49

- Presidential Decision Directive 49 deals with National Space Policy, issued by President Bill Clinton on September 19, 1996.

## percentage of the U.S. population that is male

- actually 49.1 percent, according to the 2000 census.

## Rule 49

- National Hockey League Rulebook item that deals with clipping.

## *49-1/2*

- 49-½ feet, length (by 16-½ feet width) of a petanque court in the U.S.

## *050*

- Decimal ASCII code for [the number two (2)].
- medical classification code for smallpox.

- record number of times a batter was hit by pitchers in a modern-era baseball season: Ron Hunt, Montreal Expos, 1971.

## 50

- atomic number of the chemical element tin, symbol Sn.
- bonus points collected in Scrabble by using all seven tiles in one turn.
- 50 cubits: the width of the ark the Lord tells Noah to build (Genesis 6:15).
- fifty days: average gestation period of a mink.
- 50 degrees Celsius = 122 degrees Fahrenheit.
- 50 degrees Fahrenheit = 10 degrees Celsius.
- 50 feet:
  ⇨ Length of a horseshoes playing surface;
  ⇨ Preferred width of a regulation basketball court.

- in Hebrew, the numerical value of the letter *nun*.
- most consecutive years playing in the golf Masters Tournament: by Arnold Palmer, 1955–2004.
- most touchdown passes in an NFL season (football record): by Tom Brady for the New England Patriots, 2007 (breaking Peyton Manning's record of forty-nine in 2004).
- the number of Argonauts who sailed on Jason's ship, the *Argo*.
- the number of Nereids (sea-deity granddaughters of Oceanus) in Greek mythology. They lived at the bottom of the sea and were proficient in spinning and weaving.
- number of stars on the American flag, each representing a state of the Union.
- number of states in the United States.
- number of the French Département of Manche.
- record number of hockey games scoring three or more goals (hat tricks): by Wayne Gretzky, for four different teams, 1979–99 (thirty-seven three-goal games, nine four-goal games, and four five-goal games).
- record number of home runs pitched in a baseball season: by Bert Blyleven, Minnesota Twins, 1986.

- record number of times a batter was hit by pitchers in a modern-era baseball season: Ron Hunt, Montreal Expos, 1971.
- retired baseball uniform number of coach Jimmie Reese, retired by the Anaheim Angels.
- retired basketball Jersey number of David Robinson, retired by the San Antonio Spurs.
- retired football jersey number of:
  - ⇨ Mike Singletary, retired by the Chicago Bears.
  - ⇨ Ken Strong, retired by the New York Giants.
- score of a dartboard bull's-eye.
- shekels in a mina (ancient Hebrew measure of weight).
- the smallest number that can be written as the sum of two squares in two different ways: $50 = 7^2 + 1^2 = 5^2 + 5^2$.
- table number in the Club Room at The Stork Club where columnist Walter Winchell, the New York gossip columnist, held court.
- Title of the United States Code dealing with war and national defense.
- value of Roman numeral L.
- the winning score in the first televised basketball game. University of Pittsburgh beat Fordham, 50–37, on February 28, 1940, at Madison Square Garden, in the game televised by station W2XBS, New York City.

## Fiftieth Ordinance
- British government's 1928 ruling permitting Hottentots in the Cape Colony to be landowners and abolishing previous restrictions on free movement.

## fiftieth state of the Union
- Hawaii, admitted August 21, 1959.

## fiftieth wedding anniversary gift
- gold is customary.

## fifty black ships
- according to *The Iliad*, "fifty black ships" crossed from Athens to fight the Trojans.

## .50 caliber

- heavy machine gun favored by U.S. armed forces.

## 50 Cent

- stage name of gangsta rapper Curtis James Jackson III, with two albums: *Get Rich or Die Tryin'* and *The Massacre*—that went multi-platinum almost immediately on their release. A product of Queens, New York, 50 Cent spent his early years in the drug trade, during which he survived being shot several times.

## fifty-cent piece

- half-dollar; U.S. coin in circulation since 1794.

## fifty Danaids

- in Greek mythology, the fifty daughters of Danaus who were forced to marry the fifty sons of their uncle Aegyptus, with whom Danaus had clashed. On their wedding night, at the behest of their father, forty-nine of the Danaids slayed heir bridegrooms; only one, Hypermnestra, spared her husband.

## Fifty Decisions

- document drawn up under Emperor Justinian of Rome early in the sixth century which attempted to develop a consistent body of jurisprudence from the decisions of the great jurists of the previous several centuries. Intended to settle points of disagreement among judges and to eliminate contradictions and omit repetitions. Published with the authority of a statute. Led the way to the publication, in late 533, of the authoritative *Pandects*.

## fifty-dollar lane

- passing lane on a highway (CB slang).

## fifty dollars

- (a) denomination of U.S. paper money that bears the portrait of Ulysses S. Grant.

- (b) denomination of U.S. savings bond that bears the picture of George Washington.
- (c) denomination of U.S. treasury bond that bears the picture of Thomas Jefferson.

**50-50**
- divided equally; partitioned into equal shares (each 50 percent).

*50 First Dates*
- 2004 film starring Adam Sandler and Drew Barrymore, directed by Peter Segal. A romantic comedy with a twist: He has to romance her anew each day because she has no short-term memory, resulting from a brain condition, and she forgets each day's events as she sleeps.

**50mm**
- focal length of the standard lens in 35mm photography.

**fifty-penny nail**
- a carpenter's common nail, $5\frac{1}{2}$ inches long (coded 50d).

**50-star flag**
- became the official flag of the United States on July 4, 1960, accommodating the admission of Hawaii to the Union.

**"50 Ways to Leave Your Lover"**
- hit recording by Paul Simon; broke nationally in December 1975, spent three weeks as number one on the charts in February 1976, and was certified gold in March.

**50, Wimpole Street**
- London residence of poet Elizabeth Barrett, before marrying Robert Browning.

## Jubilee Year

- according to the Old Testament (Leviticus 25:8–54), the year that follows seven successive Sabbatical years (the Sabbatical year being the seventh year of a seven-year cycle) shall be treated as a Jubilee Year, in which all debts between Jews are to be canceled, slaves are to be emancipated, land sold in poverty shall be reverted to its original owner, and cultivation of the land shall be withheld. The intent of the Jubilee Year was thus to forgive the debts and restore lost property or personal liberty to those who had suffered such loss through poverty or other adverse circumstances.

## longest tenure as a baseball manager

- fifty years for Connie Mack, with the Philadelphia Athletics, retiring on October 8, 1950.

## M-50

- four-pound incendiary bomb using magnesium, developed by the U.S. military during World War II.

## Studio 50

- CBS television studio, originally the Hammerstein Theater built in 1927, leased by CBS in 1950, and converted for TV use for *The Ed Sullivan Show*. Site of The Beatles' only live telecast in the U.S., on February 9, 1964. In 1971 the *Sullivan Show* went off the air and the theater was unused until 1993, when CBS bought it, did a multimillion-dollar renovation, and turned it over to David Letterman's *The Late Show*, which debuted on August 10, 1993.

## 50.4

- highest scoring average in an NBA season (basketball record): by Wilt Chamberlain, for the Philadelphia Warriors, 1961–62.

## 051

- Decimal ASCII code for [the number three (3)].

# 51

- atomic number of the chemical element antimony, symbol Sb.
- fifty-one years: the span of John Wayne's film career, from his first—as a stand-in for Francis X. Bushman in the 1925 silent *Brown of Harvard*—to his last feature, *The Shootist*, 1976. His first credited film was *Hangman's House* in 1928.
- international telephone calling code for Peru.
- lives lost when the Ialian liner *Andrea Doria* collided with the Swedish liner *Stockholm* off Nantucket Island, Massachusetts, on July 26, 1956.
- lunch-counter code for a cup of hot chocolate (fifty-two = two cups, etc.)
- most aces hit in a match (tennis record): twice, by Joachim Johansson, losing to Andre Agassi in the fourth round of the 2005 Australian Open; and by Ivo Karlovic, losing to Daniele Bracciali in the first round of the 2005 Wimbledon.
- most gold medals won at the 2008 Summer Olympics in Beijing: by China. U.S. was second, with thirty-five.
- most times hit by a pitch in a season (baseball record): Hughie Jennings, Baltimore Orioles, 1896.
- number of London churches designed by Sir Christopher Wren following the Great Fire of London in 1666. (And four others outside the city.)
- number of the French Département of Marne.
- number of 0s in a sexdecillion.
- number on the locomotive used by the escaping soldiers in the 1965 Sinatra movie, *Von Ryan's Express*.
- record number of grand prix races won by Formula One driver, Alain Prost, 1980–93—over 25 percent of all the races he entered.
- retired football jersey number of:
  - ⇨ Dick Butkus, retired by the Chicago Bears.
  - ⇨ Sam Mills, retired by the Carolina Panthers.

## Area 51

- a supersecret military area (part of NTS, Nevada Test Site) in the vicinity of Groom Lake, Nevada, where the government has tested advanced technology for several decades. Also reputed to be the repository of captured extraterrestrial technology.

## fifty-first state

- in the U.S., an epithet for Puerto Rico, since Hawaii achieved statehood in 1959.

## *Fifty-One Tales*

- a collection of short stories written by Lord Dunsany.

## 51 Fifth Avenue, Apartment 11D

- New York City address of Paul and Jamie Buchman on the NBC-TV sitcom *Mad About You.*

## fifty-one minutes

- shortest nine-inning major league baseball game on record; the New York Giants over the Philadelphia Phillies, 6 to 1, on September 28, 1919.

## "Highway 51 Blues"

- 1962 folk song recording by Bob Dylan, a track in his eponymous Columbia Records album, *Bob Dylan.*

## I-51

- north-south interstate running from Louisiana to Wisconsin, touching Jackson, Mississippi, Memphis, Decatur, Madison, and Wausau, Wisconsin.

## longest consecutive scoring streak in professional hockey—fifty-one games

- by Wayne Gretzky, for the Edmonton Oilers, October 5, 1983, to January 8, 1984 (61 goals and 92 assists, for a total of 153 points).

## percentage of the U.S. population that is female

- Actually 50.9 percent, according to the 2000 census.

## P-51

- U.S. warplane in World War II, a fighter called the Mustang, made by North American Aviation, used for high-altitude escort to bombers B-17 and B-24. By the end of World War II, P-51s had destroyed more enemy aircraft than any other fighter in Europe.

## 51-½

**longest-serving U.S. senator—51-½ years**

- Robert C. Byrd, Democrat of West Virginia, became the longest-serving U.S. senator on Monday, June 12, 2006. On that day, Senator Byrd put in his 17,327th day in that office, surpassing Strom Thurmond of South Carolina. After winning an unprecedented ninth elected term, Senator Byrd died on June 28, 2010 at age ninety-two, having served in the Senate for fifty-one years, five months, twenty-six days.

## 052

- (a) Decimal ASCII code for [the number four (4)].
- (b) medical classification code for chicken pox.

## 52

- age of the oldest player in NHL history (hockey record): Gordie Howe when he retired in 1997.
- atomic number of the chemical element tellurium, symbol Te.
- the Aztec culture functioned in fifty-two-year cycles, each new cycle marking symbolic rebirth of the world and celebrated by the joyous New Fir Ceremony.
- international telephone calling code for Mexico.
- 52 kilograms: maximum body weight for men weightlifters in category 1 or women weight-lifters in category 3, Olympic competition.
- number of cards in a poker deck.
- number of counties in England and Wales.
- number of the French Département of Haute-Marne.

- number of people killed in riots between Irish Catholics and Irish Protestants in New York City, July 12, 1871.
- number of white keys on a piano.
- percentage of the labor force in agriculture in 1870, compared with 4 percent in 1970.
- retired basketball jersey number of Buck Williams, retired by the New Jersey Nets.
- weeks in a year.

## B-52

- the Stratofortress, U.S. long-range heavy bomber with eight jet engines, developed after World War II. Originally intended as a vehicle to deliver an atomic bomb as far away as the Soviet Union, the B-52 proved to be extremely versatile as a conventional bomber and maritime reconnaissance aircraft.

## 52nd Street

- center of jazz clubs in New York City in the 1930s and '40s.

## *52nd Street*

- Billy Joel's 1979 Grammy-winning Album of the Year.

## 52 pick-up

- a practical-joke card game in which the uninitiated has to pick up the cards of a deck that have been scattered by the person perpetrating the joke.

## 52-20 Club

- informal name given to unemployed GIs after World War II, who were entitled to receive twenty dollars a week for fifty-two weeks.

## Rule 52

- National Hockey League Rulebook item that deals with "Deliberate Injury of Opponents."

## 52-½

- 52- ½ feet: length of a beach volleyball court

# 053

- Decimal ASCII code for [the number five (5)].

# 53

- atomic number of the chemical element iodine, symbol I.
- 53 miles: width of the Bering Strait, separating the U.S. from Russia.
- 53 yards: longest field goal scored in an NFL Pro Bowl game (football record): by David Akers for the NFC, 2003.
- international telephone calling code for Cuba.
- most three-point field goals scored in NBA finals games (basketball record): by Robert Horry for five different teams, 1993–2008.
- number of counties in the state of North Dakota.
- number of maps in the first modern atlas, the 1570 *Theatrum Orbis Terrarum*, by Abraham Ortelius of Antwerp.
- number of seats in the House of Representatives allotted to the state of California (as of the 2000 Census).
- number of the French Département of Mayenne.
- number of vessels in the Spanish fleet opposing Sir Richard Grenville's one ship in the 1591 naval battle off the Azores, celebrated in Lord Tennyson's stirring poem, "The Revenge."
- number painted on the Volkswagen Beetle called The Love Bug in the Disney movies.
- retired baseball uniform number of pitcher Don Drysdale, retired by the Los Angeles Dodgers.
- retired basketball jersey number of Mark Eaton, retired by the Utah Jazz.
- retired football jersey number of Mick Tingelhoff, retired by the Minnesota Vikings.

**Fifty-Three Stations of the Tokaido**

- (*Tokaido Gojusan-tsugi*): a series of woodblock prints by Ando Hiroshige representing the stops on the Tokaido, the coastal road of ancient Japan between Kyoto, the old imperial capital and residence of the emperor, and Edo (now Tokyo), the eastern capital and headquarters of the period's military leadership.

## Haydn's *Symphony No. 53*

- "Imperial" symphony (in D major).

## 53 $\frac{1}{3}$

- 53 1/3 yards (160 feet): width of regulation football field.

## 054

- Decimal ASCII code for [the number six (6)].
- medical classification code for herpes simplex.

## 54

- according to the folks at *People* magazine, Princess Diana was pictured on the cover an incredible fifty-four times—more than double that of any other celebrity.
- atomic number of the chemical element xenon, symbol Xe.
- 54 feet: the width of the proscenium arch at the Metropolitan Opera House in New York City. The height of the proscenium is also 54 feet.
- 54 yards: longest field goal in a Super Bowl game (football record), by Steve Christie for the Buffalo Bills against the Dallas Cowboys, SB-XXVIII, 1994.
- fifty-four years: life expectancy in 1920 as determined by the Bureau of Public Health.
- international telephone calling code for Argentina.
- most batters hit by a pitcher in a pre-modern-era season (baseball record): by Phil Knell, Columbus Solons, in 1891.
- most strikeouts taken in World Series games (baseball record): by Mickey Mantle for New York Yankees, 1951–64.
- most times stealing home in career (baseball record): by Ty Cobb, 1905–28, fifty for the Detroit Tigers, and four with the Philadelphia Phillies.
- number of colored squares visible on a Rubik's Cube.
- number of letters in the Sanskrit alphabet, each having a masculine and a feminine form, shiva and shakti (together, 108 letters).
- number of nations represented at the 1944 Chicago Conference convened to arrange "for the immediate establishment of provisional world air routes and services."

- number of nations that sent teams to Saudi Arabia to compete in the first Islamic Solidarity Games (aka, the Muslim Olympics) in April 2005 in Mecca. The Games comprised eighteen individual and team sports with 6,500 athletes, all male, competing.
- number of the French Département of Meurthe-et-Moselle.
- number of 0s in a septendecillion (American) or a nonillion (British).
- retired football jersey number of Bob Johnson, retired by the Cincinnati Bengals.
- sum of the first nine Fibonacci numbers (0 + 1 + 1 + 2 + 3 + 5 + 8 + 13 + 21).

### *Car 54, Where Are You?*

- situation comedy about the misadventures of New York City police officers Gunther Toody (Joe E. Ross) and Francis Muldoon (Fred Gwynne). Ran on NBC-TV from September 17, 1961, to September 8, 1963.

### C-54

- the Skymaster, a World War II airplane from Douglas Aircraft, adapted for military service from the DC-4. The DC-4, carrying forty-four passengers, was in wide use into the 1960s.

### 54 BC

- date of the Roman invasion of Britain under the leadership of Julius Caesar.

### "Fifty-Four Forty or Fight"

- slogan used in the 1840s by expansionist advocates who asserted American ownership of the entire Oregon Territory, including the area claimed by Great Britain between latitudes 49° N and 54° 40" N, the border with Russian Alaska. The phrase became a campaign rallying cry of the 1844 presidential campaign of James Polk.

### 54 Pearl Street

- address of Fraunces Tavern in New York City, site of Washington's farewell address to the officers of the Continental Army, December 4, 1783.

## longest-serving U.S. representative—fifty-five years

- John D. Dingell, Democrat from Michigan, became the longest-serving member of the U.S. House of Representatives on February 11, 2009, when he began his 19,420th day in that job, surpassing James L. Whitten of Mississippi. Taking office on December 13, 1955, Representative Dingell succeeded his father, who had represented Michigan's 15th district since 1933.

## *055*

- Decimal ASCII code for [the number seven (7)].
- medical classification code for measles.

## *55*

- atomic number of the chemical element cesium, symbol Cs.
- 55 feet square: base of the Washington Monument on the Mall in D.C.
- international telephone calling code for Brazil.
- lunch-counter code for a glass of root beer.
- most in-the-park home runs, career (baseball record): by Jesse Burkett; with five teams, 1890–1905.
- most personal fouls committed by a team in an NBA playoff game (football record): by the Syracuse Nationals against the Boston Celtics, March 21, 1953.
- most points scored by the winning team in an NFL Pro Bowl game (football record): fifty-five by the NFC (vs. fifty-two for the AFC), 2004.
- most points scored by the winning team in a Super Bowl game (football record): by the San Francisco 49ers against the Denver Broncos' 10 points, SB-XXIV, 1990.
- most rebounds in an NBA game (basketball record): by Wilt Chamberlain, Philadelphia Warriors against the Boston Celtics, November 24, 1960.
- number of counties in the state of West Virginia.
- number of delegates who attended the Constitutional Convention, which convened on May 25, 1787, at Philadelphia's State House (Independence Hall).

- number of presidential electoral votes apportioned to the state of California. (May be reapportioned for the 2012 election.)
- number of the French Département of Meuse.
- number of wives of the Mormon leader, Brigham Young; he married the first, eighteen-year-old Miriam Works, on October 8, 1824, and the last, sixty-five-year-old Hannah Tapfield, on December 8, 1872.
- record number of times Richard Nixon appeared on the cover of *Time* magazine.
- sum of the first five square numbers $(1 + 4 + 9 + 16 + 25)$.
- sum of the first ten integers $(1 + 2 + 3 + 4 + 5 + 6 + 7 + 8 + 9 + 10)$.

## fifty-fifth wedding anniversary gift

- emerald is customary.

## 55 Cancri

- a star, similar to our sun, that is the center of a solar system with two orbiting planets, 55 Cancri B and 55 Cancri C. About forty-five light-years from Earth, but bright enough to be seen from it, the star is in the constellation Cancer.

## *55 Days at Peking*

- director Nicholas Ray's 1963 film version of the 1900 Boxer Rebellion in China, starring Charlton Heston, Ava Gardner, and David Niven, with Flora Robson playing the Dowager Empress Tzu Hsi. Notable for the Peking reconstruction sets by production designers Veniero Colasanti and John Moore.

## Haydn's *Symphony No. 55*

- "Schoolmaster" symphony (in E-flat major).

## I-55

- major north-south interstate highway from New Orleans to Jackson, Memphis, St. Louis, Springfield, and Chicago.

## *Music 55*

- a 1955 summer replacement show on CBS-TV featuring Stan Kenton with his big band, and weekly guest stars, playing jazz, standards, and contemporary hits.

## *056*

- (a) Decimal ASCII code for [the number eight (8)].
- (b) medical classification code for rubella.

## *56*

- Article of the New York State Education Law covering charter schools.
- atomic number of the chemical element barium, symbol Ba.
- 56 kilograms: maximum body weight for men weightlifters in category 2 or women weight-lifters in category 4, Olympic competition.
- international telephone calling code for Chile.
- most fumbles recovered in an NFL career (football record): by Warren Moon for the Houston Oilers, 1984–93; then three other teams, 1994–2000.
- number of professional fights won by Muhammad Ali (of sixty-one fought), thirty-seven by knockouts.
- number of Medals of Honor awarded for meritorious service in the 1914 Mexican campaign.
- number of Aubrey holes at Stonehenge, a concentric ring of pits about 285 feet in diameter surrounding the stone structures. The subject of many theories, their purpose is still unknown.
- number of counties in the state of Montana.
- number of curls Shirley Temple wore in her hair as a child star. They were set every night by her mother, Gertrude, thus maintaining the exact number.
- number of Sherlock Holmes stories written by Sir Arthur Conan Doyle.
- number of signers of the Declaration of Independence.
- number of the French Département of Morbihan.

- police slang for the time off that takes the place of a weekend. (Used by police who ordinarily work on Saturdays and Sundays.)
- retired football jersey number of:
  - ⇨ Bill Hewitt, retired by the Chicago Bears.
  - ⇨ Joe Schmidt, retired by the Detroit Lions.
  - ⇨ Lawrence Taylor, retired by the New York Giants.

## longest hitting streak in major league baseball–fifty-six consecutive games

- by Joe DiMaggio (New York Yankees), from May 15 to July 16, 1941. In 223 times at bat, he hit sixteen doubles, four triples, and fifteen homers.

## Fifty Six

- a city in Stone County, Arkansas.

## 057

- Decimal ASCII code for [the number nine (9)].

## 57

- atomic number of the chemical element lanthanum, symbol La.
- international telephone calling code for Colombia.
- number of the French Département of Moselle.
- number of 0s in a novemdecillion.
- pounds in a cubic foot of ice.
- retired football jersey number of:
  - ⇨ Steve Nelson, retired by the New England Patriots.
  - ⇨ Jeff Van Note, retired by the Atlanta Falcons.

## Heinz 57

- brand designation for a line of foods produced by the H. J. Heinz Company of Pittsburgh, Pennsylvania. Although the company advertises a nominal fifty-seven products, it actually produces several thousand.)

**oldest american woman to give birth to twins**

- on November 9, 2004, Aketa St. James delivered a boy, Gian, and a girl, Francesca, at Mt. Sinai Hospital in New York City, just three days shy of her fifty-seventh birthday.

*Passenger 57*

- Wesley Snipes as an anti-terrorist specialist who confronts an airplane highjacker (Bruce Payne) in midflight. A 1992 film directed by Kevin Hooks.

## 57-1/2

**longest-serving member of U.S. Congress—fifty-seven and a half years**

- Robert C. Byrd, Democrat of West Virginia, served in the U.S. House of Representatives for six years (January 1953 to January 1959), and in the U.S. Senate for over fifty-one years (January 1959 to his death on June 28, 2010). This record of tenure surpasses that of Arizonan Carl Hayden, who served fifteen years in the House and forty-two years in the Senate.

## 058

- Decimal ASCII code for [colon (:)].

## 58

- atomic number of the chemical element cerium, symbol Ce.
- fewest pitches thrown in a nine-inning shutout (baseball record): by Red Barrett, Boston Braves against the Cincinnati Reds, August 10, 1944.
- 58 hertz (cycles per second): lower limit of a bassoon's frequency range.
- 58 inches: maximum height of a horse ($14\frac{1}{2}$ hands at the shoulder) to be classified as a pony.
- international telephone calling code for Venezuela.
- liters in a saa, unit of dry capacity in Algeria.

- most points scored by a team in an NBA quarter (basketball record): by the Buffalo Braves against the Boston Celtics, October 20, 1972 (fourth quarter).
- number of counties in the state of California.
- number of the French Département of Nièvre.
- retired football jersey number of Derrick Thomas, retired by the Kansas City Chiefs.
- sum of the first seven prime numbers (2 + 3 + 5 + 7 + 11 + 13 + 17).

**longest-running stage play**
- fifty-eight years of Agatha Christie's *The Mousetrap*, as of the end of 2010; it opened at London's Ambassadors Theatre on November 25, 1952, and moved to St. Martin's Theatre in March 1974, where it still resides.

## *58.7*

- number of Earth days in a day on the planet Mercury.

## *059*

- Decimal ASCII code for [semi-colon (;)].

## *59*

- age of the oldest champion steer roper, Ike Rude, in 1953 competition.
- atomic number of the chemical element praseodymium, symbol Pr.
- 59 feet: length of a regulation volleyball court.
- fifty-nine years that France had not won the Davis Cup of tennis, until 1991; France was twice runner-up in that period (1933 and 1982).
- highest number included in the New York State Lottery.
- lowest 18-hole golf score by a woman in an LPGA tournament: by Sweden's Annika Sorenstam in the Standard Register Ping Tournament at the Moon Valley Country Club in Phoenix, Arizona, March 16, 2001. Sorenstam also set a record for thirteen birdies in a round.

- lowest golf score for an 18-hole round in a PGA tournament: three times, by Al Geiberger in the second round of the Memphis Classic, on the Colonial Golf Course in Memphis, Tennessee, June 10, 1977; by Chip Beck in the third round of the Las Vegas Invitational, on the Sunrise Golf Course, Las Vegas, Nevada, October 11, 1991; and by David Duval in the final round of the Bob Hope Chrysler Classic, La Quinta, California, on January 24, 1999.
- most consecutive scoreless innings pitched in a season (baseball record): by Orel Hershiser, Los Angeles Dodgers, August 30 to September 28, 1988.
- most points scored by an NBA player in a half (basketball record): by Wilt Chamberlain for the Philadelphia Warriors, second half against the New York Knicks, March 2, 1962.
- most wins by a pitcher in one season (baseball record): by Charley "Old Hoss" Radbourn for Providence Grays in 1884. (See *41* for modern-era.)
- number of beads in a Catholic rosary.
- number of Medals of Honor awarded for meritorious service in the Boxer Rebellion.
- number of members in the first House of Representatives in 1789.
- number of the French Département of Nord.
- percentage of Americans who say religion is very important to them—compared with 33 percent of English, 30 percent of Canadians, and 27 percent of Italians (according to Pew Global Attitudes Project survey report of December 2002).
- years in the reign of George III of Great Britain (1760–1820), second longest only to Queen Victoria.

## Haydn's *Symphony No. 59*

- "Feuer" symphony (in A major).

## I-59

- north-south interstate highway from New Orleans to Birmingham and Chattanooga.

# 060

- Decimal ASCII code for [less than (<)].
- medical classification code for yellow fever.

# 60

- atomic number of the chemical element neodymium, symbol Nd.
- base of the number system used by the ancient Babylonians.
- calories in 6 fluid ounces of café au lait.
- grains in a dram (troy or apothecary).
- in Hebrew, the numerical value of the letter *samekh*.
- international telephone calling code for Malaysia.
- minas in a talent (ancient measure of weight).
- minims in a fluid dram.
- minutes in a degree (angular measurement).
- minutes in an hour.
- number of home runs hit by Babe Ruth in his best (154-game) season, in 540 times at bat, the last on September 30, 1927. Of the 60, 28 were hit at Yankee Stadium. Roger Maris then hit 61 in the 1961 season, but did it in a 162-game schedule, hitting only his 59th by the 154th game. The current record of 73 is held by Barry Bonds, San Francisco Giants, in 2001, surpassing Mark McGwire's 70 with the St. Louis Cardinals in 1998.
- number of the French Département of Oise.
- number of zeroes in a novemdecillion (American) or a decillion (British).
- retired football jersey number of:
  - ⇨ Chuck Bednarik, retired by the Philadelphia Eagles.
  - ⇨ Tommy Nobis, retired by the Atlanta Falcons.

- seconds in a minute (time and angular measurement).
- 60 cubits: length of the temple Solomon built (1 Kings 6:2).
- 60 degrees: size of each interior angle of an equilateral triangle.
- sixty dollars: the cost of Mediterranean Avenue, the cheapest property on the Monopoly game board.
- 60 feet:

  ⇨ width of regulation tennis playing field.

  ⇨ Length of a regulation bowling alley, from foul line to center of the head pin.

- 60 inches: maximum length of a regulation hockey stick, measured from the heel to the end of the shaft.
- 60 kilograms: maximum body weight for men weightlifters in category 3 or women weight-lifters in category 5, Olympic competition.
- sixty minutes:
    - ⇨ length of a regulation football game, divided into four quarters of fifteen minutes each.
    - ⇨ length of a regulation hockey game, divided into three periods of twenty minutes each.
- 60 yards: width of a regulation field hockey or lacrosse playing field.
- value of Roman numeral LX.

## CVA-60

- number of the U.S. Navy large aircraft carrier USS *Saratoga*.

## E60

- a major European highway running from Brest, France, to Constanta, Romania.

## Haydn's *Symphony No. 60*

- "Il Distratto" symphony (in C major).

## like sixty

- (adv., informal) with great speed, ease, or energy.

## PDD-60

- Presidential Decision Directive 60 deals with Nuclear Weapons Employment Policy, issued by President Bill Clinton in November 1997.

## sexagenarian

- a person between sixty and sixty-nine years of age.

**sixtieth wedding anniversary gift**

- diamonds are appropriate.

**60 degrees**

- measurement of each exterior angle of an equilateral hexagon, or each interior angle of an equilateral triangle.

*Sixty Glorious Years*

- 1938 sequel to the biopic *Victoria the Great*, starring Anna Neagle as the Queen and Anton Walbrook as Prince Albert, following the events of Victoria's reign—the longest in British history—until her death in 1901. Directed by Herbert Wilcox.

**Sixty Mile**

- a town in western Yukon Territory.

*60 Minutes*

- long-running CBS-TV news magazine, featuring several CBS news people. Debuted September 24, 1968, and is still on the air at this writing.

**sixty queens**

- in the card game pinochle, a meld of four queens, one from each suit, is worth sixty points.

**sixty-penny nail**

- a carpenter's common nail, 6 inches long (coded 60d).

**60-point type**

- typographer's measure, known as "5-line pica."

*Sixty Years a Queen*

- 1913 movie biography of Queen Victoria in which British actor Rolf Leslie played twenty-seven different roles.

### *Studio 60 on the Sunset Strip*

- debuting on NBC-TV on August 5, 2006, this ensemble series offered a backstage look at the workings of a fictional weekly comedy show, much like SNL. Created by Aaron Sorkin and featuring Matthew Perry, Bradley Whitford, Amanda Peet, and Steven Webber. Final show aired on June 28, 2007.

## *60-1/2*

- 60-½ feet: pitching distance from home plate on a regulation baseball field.

## *061*

- Decimal ASCII code for [equal sign (=)].

## *61*

- atomic number of the chemical element promethium, symbol Pm.
- international telephone calling code for Australia.
- most free throws scored by one team in an NBA game (basketball record): by the Phoenix Suns against the Utah Jazz, April 9, 1990 (OT).
- most points scored in an NBA finals game (basketball record): by Elgin Baylor for the Los Angeles Lakers against the Portland Trail Blazers, April 14, 1962.
- number of Roger Maris's home runs in the 1961 season, breaking Babe Ruth's record of 60. Maris played in a season of 162 games (590 at-bats), Ruth in a season of 154 games. He hit his 61st off Boston's Tracy Stallard on October 1, 1961, having already passed his 155th game. (The Yankees beat Boston, 1–0.) Maris's record in turn was broken by Sammy Sosa's 64 home runs in 1999.
- number of the French Département of Orne.
- retired football jersey number of Bill George, retired by the Chicago Bears.
- sixty-one days: average gestation period of a domestic dog.
- 61 hertz (cycles per second): lower limit of a French horn's frequency range.

- 61 square miles: size of the principality of Liechtenstein.

## P-61

- U.S. warplane in World War II, called the Black Widow, produced by Northrup. The first aircraft designed specifically as a night-fighter.

## *61\**

- 2001 baseball movie directed by Billy Crystal. Set in the summer of 1961, when Roger Maris (Barry Pepper) and Mickey Mantle (Thomas Jane) compete to break Babe Ruth's record of sixty single-season home runs.

## 61, Charington Gardens

- London residence of Margot and Tony Wendice (Grace Kelly and Ray Milland) in the 1954 Hitchcock movie *Dial M for Murder*, based on a 1952 teleplay by Frederick Knott.

## 062

- Decimal ASCII code for [greater than (>)].

## 62

- atomic number of the chemical element samarium, symbol Sm.
- international telephone calling code for Indonesia.
- most pitching saves in a season (baseball record): by Francisco Rodriguez, Los Angeles Angels, 2008.
- most wins in one NHL season (hockey record): by the Detroit Red Wings, 1995–96 season.
- number of counties in the state of New York.
- number of goddesses who accompany Demchok, the Tibetan god known as Supreme Bliss.
- number of moons of the planet Saturn, nine known before 1900, augmented by others discovered by unmanned space probes, notably by *Voyager 1* and *Voyager 2* flybys, starting in 1980, and by the Cassini mission starting in 2004, and by advanced telescopy beginning in 2000.

- number of nations that eventually signed the Kellogg-Briand Peace Pact of 1928, which outlawed wars and agreed to settle international disputes by arbitration.
- number of self-portraits painted by Rembrandt van Rijn.
- number of the French Département of Pas-de-Calais.
- record number of team wins in an (eighty-two-game) NHL season (hockey record): by the Detroit Red Wings, 1995–96.
- record number of women's tennis Grand Slam titles, won by Margaret Smith Court between 1960 and 1975: twenty-four singles (eleven Australian Opens, five French Opens, five U.S. Opens, and three Wimbledons), nineteen women's doubles, and nineteen mixed doubles. A close second was Martina Navratilova with fifty-nine: eighteen singles, thirty-one women's doubles, and ten mixed doubles.
- sixty-two days: average gestation period of a skunk.
- 62 feet: average depth of Lake Erie, the shallowest of the Great Lakes.
- 62 miles:
  - ⇨ = 100 kilometers.
  - ⇨ length of Kiel Ship Canal , built in northern Germany to provide safe passage for the German navy between the Baltic and the North Seas.

**longest reign in Japan's history—sixty-two years**
- that of Emperor Hirohito, who died in 1989.

# 063

- Decimal ASCII code for [question mark (?)].

# 63

- Article of the New York State Education Law covering salaries of teachers and supervisors.
- atomic number of the chemical element europium, symbol Eu.
- best score for one 18-hole round in a Masters Tournament: by Nick Price in the third round, 1986, and by Greg Norman in the first round, 1996.
- best score for one 18-hole round in a U.S. Open, four times: Johnny Miller, in the fourth round, 1973 at the Oakmost

Country Club, Pennsylvania; Jack Nicklaus and Tom
Weiskopf, both in the first round, 1980 at the Baltusrol
Golf Club in Springfield, New Jersey; Vijay Singh, in the
second round at the Olympia Fields in Olympia Fields,
Illinois.

- centimeters in an alin: unit of length in Iceland.
- international telephone calling code for Philippines.
- most points scored in an NBA playoff game (basketball
  record): by Michael Jordan for the Chicago Bulls against the
  Boston Celtics (2 OT), April 20, 1986.
- most wild pitches thrown in a season (baseball record): by
  Bill Stemmeyer, Boston Beaneaters, 1886.
- number of known satellites of the planet Jupiter. Four of
  the nearest—Io, Europa, Ganymede, and Callisto—were
  discovered by Galileo in 1610; twenty-one of the others were
  first observed in 2003.
- number of the French Département of Puy-de-Dôme.
- retired football jersey number of:
  - ⇨ Willie Lanier, retired by the Kansas City Chiefs.
  - ⇨ Mike Munchak, retired by the Tennessee Titans.
  - ⇨ Lee Roy Selmon, retired by the Tampa Bay Bucs.

- sixty-three days: average gestation period of a domestic cat.
- 63 yards:
  - ⇨ longest field goal in NFL history (football record): twice,
    by Tom Dempsey for the New Orleans Saints against the
    Detroit Lions, November 8, 1970, and by Jason Elam
    for the Denver Broncos against the Jacksonville Jaguars,
    October 25, 1998. The Dempsey kick was the last play of
    the game and won it for the Saints, 19—17; Dempsey had
    an artificial right foot, his kicking foot.
  - ⇨ longest punt in a Super Bowl game (football record): by
    Lee Johnson for the Cincinnati Bengals, SB-XXIII, 1989.

- U.S. gallons in a hogshead.

## BB63

- registry of the battleship *Missouri*, on which the Japanese
  signed their surrender ending World War II; signed in
  Tokyo Bay, September 2, 1945.

**CVA-63**

- registry of the U.S. Navy aircraft carrier, USS *Kitty Hawk*.

**Haydn's *Symphony No. 63***

- "La Roxelane" symphony (in C major).

**longest-reigning monarch of Great Britain**

- Queen Victoria ruled for sixty-three years, from 1837 to 1901.

**P-63**

- U.S. warplane, the Bell Kingcobra, used on the Russian front in World War II.

**sixty-three shalakapurushas**

- a series of spiritual and temporal leaders, prominent figures in Jain universal history.

## *064*

- Decimal ASCII code for ["at" symbol (@)].

## *64*

- atomic number of the chemical element gadolinium, symbol Gd.
- dry pints in a bushel.
- fluid drams in a cup.
- international telephone calling code for New Zealand.
- members of the House of Representatives in the first Congress (1789–91), thirty-eight Federalist and twenty-six Democrat-Republican.
- number of counties in the state of Colorado and of parishes in the state of Louisiana.
- number of pice in a rupee (former Indian currency).
- number of squares on a checkerboard (eight columns of eight rows each)—32 black and 32 white. (The same for a chess board.)
- number of the French Département of Pyrénées-Atlantiques.

- 64 miles per hour: minimal wind force to be catagorized a severe storm, as per the Beaufort Wind Scale.
- rap slang for a 64-ounce bottle of malt liquor.
- the sixth power of 2.
- the smallest sqube (i.e., both a perfect square [$8^2$] and a perfect cube [$4^3$]).
- the square of 8.
- the third power of 4.

## Commodore 64

- an early 8-bit home computer, commonly known as the C64, introduced in August 1982.

## CVA-64

- number of the U.S. Navy aircraft carrier, USS *Constellation*.

## Haydn's *Symphony No. 64*

- "Tempora mutantur" symphony (A major).

## hemidemisemiquaver

- sixty-fourth note, in musical notation.

## in 64s

- describes a printing process whereby each printed sheet contains sixty-four book pages.

## I-64

- east-west interstate highway from St. Louis to Louisville, Charleston (West Virginia), Charlottesville, Richmond, and Norfolk.

## PDD-64

- Presidential Decision Directive 64 deals with Humanitarian Demining, issued by President Bill Clinton in May 1998.

## Rule 64

- National Hockey League Rulebook item that deals with hooking.

## Six Four Incident

- in China, refers to the Tiananmen Square protests of June 4, 1989. In early May approximately 100,000 students and workers peacefully assembled in Beijing, protesting government corruption and demanding democratic reforms. The government declared martial law, and on June 4, the Chinese army confronted the demonstrators with live ammunition, killing several hundred civilians and injuring several thousand.

## sixty-four-dollar question

- (informal) the most difficult or significant question, usually in a sequence.

## *$64 Question*

- radio quiz show, originally called *Take It or Leave It* when first broadcast on CBS in 1940.

## 64mo

- sixty-fourmo, the page size of a book from printer's sheets that are folded into sixty-four equal sections, about 2 x 3 inches.

## *$64,000 Question*

- the first big-money TV quiz show, an inflated descendant of radio's *$64 Question*, on CBS-TV, hosted by Hal March from June 7, 1955 to November 2, 1958.

## "When I'm Sixty-Four"

- a Beatles song, written by Paul McCartney and John Lennon, recorded on the Beatles' album, *Sgt. Pepper's Lonely Hearts Club Band*.

## *065*

- Decimal ASCII code for [capital letter A].

## *65*

- atomic number of the chemical element terbium, symbol Tb.

- international telephone calling code for Singapore.
- largest number of costume changes by one performer in a film: Elizabeth Taylor in the 1963 Twentieth Century Fox epic, *Cleopatra*.
- magic constant of a 5 x 5 magic square:

| 9  | 3  | 22 | 16 | 15 |
|----|----|----|----|----|
| 2  | 21 | 20 | 14 | 8  |
| 25 | 19 | 13 | 7  | 1  |
| 18 | 12 | 6  | 5  | 24 |
| 11 | 10 | 4  | 23 | 17 |

- number of college basketball teams in the NCAA Men's Division I Basketball Championship that compete in the annual three-week single elimination tournament known as March Madness.
- number of the French Département of Hautes-Pyrénées.
- retired football jersey number of Elvin Bethea, retired by the Tennessee Titans.
- 65 miles per hour: speed limit for highways in several states.
- 65 yards: width of a Canadian football playing field.
- usual age of retirement in the U.S., UK, Germany, and several other countries.

**CVN-65**

- number of the U.S. Navy nuclear aircraft carrier USS *Enterprise*.

**I-65**

- major north-south interstate highway from Mobile to Birmingham, Nashville, Louisville, Indianapolis, and Gary (Indiana).

**Rule 65**

- National Hockey League Rulebook item that deals with icing the puck.

*Sixtyfive Roses*

- a memoir of her terminally ill sister, Pam, written by Heather Summerhayes Cariou. Title derives from a child's mispronunciation of "cystic fibrosis."

# 066

- Decimal ASCII code for [capital letter B].

# 66

- atomic number of the chemical element dysprosium, symbol Dy.
- feet in a surveyor's chain (100 feet in an engineer's chain).
- international telephone calling code for Thailand.
- most penalties, both teams, in an NHL playoff game (hockey record): the Detroit Red Wings (thirty-three) against the St. Louis Blues (thirty-three), April 12, 1991. St. Louis won, 6–1.
- number of books in the Protestant Bible (thirty-nine Old Testament and twenty-seven New Testament).
- number of counties in the state of South Dakota.
- number of the French Département of Pyrénées-Orientales.
- retired football jersey number of:
  - ⇨ Ray Nitschke, retired by the Green Bay Packers.
  - ⇨ Bulldog Turner, retired by the Chicago Bears.

- retired hockey shirt number of Mario Lemieux, retired by the Pittsburgh Penguins.
- 66 feet: height of the Sphinx at Giza, Egypt.
- 66 yards: longest kickoff return in an NFL Pro Bowl game (football record): by Michael Bates for the NFC, 2000.
- sum of the first eleven integers.
- team number of George Gipp, the Gipper, who played football for Notre Dame, 1917–20.

**oldest woman to give birth**

- sixty-six-year-old Romanian Adriana Iliescu delivered a three-pound, three-ounce baby girl, Eliza Maria, by cesarean section, at the Giulesti Maternity Hospital in Bucharest on January 16, 2005. Iliescu was artificially inseminated with sperm and egg from anonymous donors, and was initially carrying triplets. Two of the fetuses died and a cesarean was done at the thirty-third week.

## Route 66

- "Main Street of America" formerly the main road between Chicago and the West Coast, its 2,448 miles pass through St. Louis, Tulsa, Oklahoma City, Amarillo, Albuquerque, Flagstaff, San Bernadino, Los Angeles, ending in Santa Monica. Commissioned in 1926, when U.S. numbered highways first came into existence, it was deactivated in 1984 with the advent of the national interstate highway system. Also nicknamed "The Mother Road" and "Will Rogers Highway," it was popularized by the song "(Get Your Kicks on) Route 66," written by Bobby Troup.

## *Route 66*

- adventure series on CBS-TV from October 1960 to September 1964, in the mode of a road picture. Martin Milner starred, with help from George Maharis and Glenn Corbett.

## sechsundsechzig

- a German card game.

## Sixty-six

- two-handed variation of the card game Bezique, in which the object is to score sixty-six points before the opponent. Points are accumulated by the rank of the cards taken in tricks. Four-handed Sixty-six is a partnership form of the game.

## *067*

- Decimal ASCII code for [capital letter C].

## *67*

- age of George Washington at his death, December 14, 1799.
- atomic number of the chemical element holmium, symbol Ho.
- most doubles hit in a season (baseball record): by Earl Webb with the Boston Red Sox, 1931.

- most field goals by a team in an NBA playoff game (basketball record): three times, by the Milwaukee Bucks against the Philadelphia 76ers, March 30, 1970; the San Antonio Spurs against the Denver Nuggets, May 4, 1983; the Los Angeles Lakers against the Denver Nuggets, May 22, 1985.
- number of counties in each of the states of Alabama, Florida, and Pennsylvania.
- number of the French Département of Bas-Rhin.
- number worn by partners Jane Fonda and Michael Sarrazin in the dance marathon in the 1969 movie *They Shoot Horses, Don't They?*
- record number of penalty minutes in one NHL game (hockey record): by Randy Holt for the Los Angeles Kings against Philadelphia Flyers, March 11, 1979.
- 67 million miles: average distance of the planet Venus from the sun.

## CV-67

- number of the U.S. Navy aircraft carrier USS *John F. Kennedy*.

## GAM-67

- a U.S. decoy missile, originally the Radioplane B-67 Crossbow.

### longest engagement in history (arguably)

- Stephen Pile, in *The Book of Heroic Failures*, relates that in 1900, Octavio Guillen met the girl who would one day be his wife. Two years later he announced his engagement to Adriana Martinez. In 1969, both eighty-two years of age, they finally married in Mexico City after an engagement of sixty-seven years.

## *67.5*

- 67.5 kilograms: maximum body weight for men weightlifters in category 4 or women weightlifters in category 6, Olympic competition
- Section of the California Penal Code dealing with bribery of public officer or employee.

## 068

- Decimal ASCII code for [capital letter D].

## 68

- atomic number of the chemical element erbium, symbol Er.
- greatest winning margin in an NBA game (basketball record): the Cleveland Cavaliers over the Miami Heat (148–80), December 17, 1991.
- meters in a zirah, unit of textile length in Lebanon.
- number of the French Département of Haut-Rhin.
- sixty-eight days: average gestation period of a guinea pig.
- 68 feet: length of the B-17D Flying Fortress bomber of World War II.
- 68 percent: area of normal curve lying between −1 and +1 standard deviations (more precisely, 68.2 percent).

### CVN-68
- number of the U.S. Navy nuclear aircraft carrier USS *Nimitz*, named after Admiral Chester Nimitz.

### Rule 68
- National Hockey League Rulebook item that deals with interference by or with spectators.

## 069

- Decimal ASCII code for [capital letter E].
- Dewey Decimal System designation for writings on museum science.

## 69

- age at which President Ronald Reagan was inaugurated, the oldest man to become president.
- atomic number of the chemical element thulium, symbol Tm.

- highest numbered channel in the UHF analog TV broadcast spectrum; the next band, channels 70 to 83, are reserved for use by cellular phones.
- most combined goals in an NHL playoff series (hockey record): the Edmonton Oilers, forty-four, the Chicago Blackhawks, twenty-five, in a six-game series, 1985.
- most minutes played in an NBA game (basketball record): by Dale Ellis for the Seattle SuperSonics against the Milwaukee Bucks (5 OT), November 9, 1989.
- number of electors—all of those voting—choosing George Washington to be the first President of the United States. Of the seventy-three eligible, four were absent.
- number of the French Département of Rhône.
- section of the California Penal Code dealing with resisting arrest.
- 69 feet: height of Roy Lichtenstein's *Mural with Blue Brushstroke,* mounted in the five-story atrium of the Equitable Center on Seventh Avenue in New York City.
- sixty-nine: average age in the British House of Lords, as of September 2009.

**CVN-69**

- number of the U.S. Navy nuclear aircraft carrier USS *Dwight D. Eisenhower.*

**The Fighting 69th**

- the 69th Infantry Division, made up mostly of Irish New Yorkers. Its history dates back to the 69th New York State Volunteer Regiment, which fought in the Revolution and the War of 1812. It served in the Civil War as the 69th New York State Militia (NYSM), which split into two Volunteer Regiments, the 69th New York State Volunteers (NYSV), the nucleus of the "Irish Brigade," and the 69th New York National Guard (NYNG), the beginnings of the "Irish Legion." In World War I, the 69th was redesignated the 165th Infantry, and it was incorporated into the 42nd (Rainbow) Division. In World War II, it served with distinction in the South Pacific at Makin, Saipan, and Okinawa.

## *The Fighting 69th*

- 1940 patriotic film set in the World War I French trenches, with James Cagney starring as a swaggering recruit from Brooklyn, and real-life characters Father Duffy, portrayed by Pat O'Brien, "Wild Bill" Donovan (later head of the OSS), played by George Brent, and Jeffrey Lynn as poet Joyce Kilmer.

## Haydn's *Symphony No. 69*

- "Laudon" symphony (in C major).

## I-69

- north-south interstate highway from Indianapolis to Fort Wayne, Lansing, and Port Huron (Michigan).

## the most prolific mother on record

- the wife of Feodor Vassilyev, an eighteenth-century Russian peasant from the town of Shuya, delivered sixty-nine children. In a total of twenty-seven confinements between 1725 and 1765, she gave birth to sixteen pairs of twins, seven sets of triplets, and four sets of quadruplets, but never a singleton. Of them all, sixty-seven survived infancy. Her husband's second wife bore him an additional eighteen children. (The case is authenticated by records in the Imperial Academy of St. Petersburg.)

## M-69

- a six-pound incendiary containing napalm, developed by the U.S. military during World War II.

## "69 Année érotique"

- ("69 Erotic Year"); song written by Serge Gainsbourg in 1968; recorded by Gainsbourg and Jane Birkin in 1969, quickly becoming a hit in France. The title refers to both the sexual liberation of the period and the 69 sex position.

## soixante-neuf

- sexual activity between two persons involving reciprocal simultaneous oral sexual stimulation.

## Vat 69

- a brand of blended Scotch whiskey produced by William Sanderson.

## *070*

- Decimal ASCII code for [capital letter F].
- Dewey Decimal System designation for books on journalism, publishing media.

## *70*

- atomic number of the chemical element ytterbium, symbol Yb.
- calories in one ounce of red or black caviar.
- in Hebrew, the numerical value of the letter *ayin*.
- medical classification code for viral hepatitis.
- most games played in NBA finals, career (basketball record): by Bill Russell for the Boston Celtics, 1957–70.
- most punt returns in an NFL season (football record): Danny Reese for the Tampa Bay Buccaneers, 1979.
- number of children, all male, of Papatuanuku (Papa), the Earth goddess, and her husband Ranginui (Rangi), the sky god, according to Maori legend. Their children became the gods of the Maori.
- number of disciples Jesus sent out to heal the sick and spread the Gospel (Luke 10:1).
- number of elders of Israel asked by the Lord to share his burden (Numbers 11:16–17).
- number of the French Département of Haute-Saône.
- retired football jersey number of:
  - ⇨ Art Donovan, retired by the Indianapolis Colts.
  - ⇨ Charlie Krueger, retired by the San Francisco 49ers.
  - ⇨ Jim Marshall, retired by the Minnesota Vikings.
  - ⇨ Ernie Stautner, retired by the Pittsburgh Steelers.
  - ⇨ Al Wistert, retired by the Philadelphia Eagles.

- 70 pounds: weight of the hump and body harness worn by Lon Chaney in the 1923 silent movie, *The Hunchback of Notre Dame*.

- seventy shekels: the offering made by each of the twelve princes of Israel to the dedication of the altar of the tabernacle (Numbers 7).
- 70 yards: minimum width of a soccer field for international matches.
- seventy years: normal life span according to the Bible: "The days of our years are three score and ten . . ." (Psalms 90:10).
- value of medieval Roman numeral S.
- value of Roman numeral LXX.

## *Boccaccio '70*

- 1962 Italian film with three unconnected stories directed by three different directors: Vittorio De Sica, Federico Fellini, and Luchino Visconti. Casts include Sophia Loren, Anita Ekberg, and Romy Schneider. Little more than a vehicle to expose the actresses' physical attributes.

## copyright protection term

- under the 1976 Copyright Act, copyright protection endures for seventy years after an author's death for books published after January 1, 1978.

## CVN-70

- number of the U.S. Navy nuclear aircraft carrier USS *Carl Vinson*, named after U.S. Representative Carl Vinson.

## Group 70

- a nonprofit international organization of people from all walks of life who seek to develop an appreciation of the beauty of astronomy and to popularize its study worldwide.

## heptagenarian, septuagenarian

- a person between seventy and seventy-nine years of age.

## I-70

- major east-west interstate highway from Cove Fort (Utah) to Grand Junction, Denver, Topeka, Kansas City, St. Louis, Indianapolis, Columbus, Wheeling, and Baltimore.

## Quorum of the 70

- in the Church of Jesus Christ of Latter-day Saints (Mormons), one of the General Authorities is the priesthood council of seventy members which assists the Quorum of the twelve and the First Presidency. The seventy elders are called upon to preach the gospel and to build up the Church under the direction of the twelve.

## Septuagint

- a Greek version of the Old Testament, traditionally said to be translated by seventy or seventy-two Jewish scholars in Alexandria two centuries before Christ.

## seventieth wedding anniversary gift

- platinum is customary.

## seventy-'leven

- archaic term for a fairly large indeterminate number.

## 70 Mile House

- a town on Cariboo Highway in south-central British Columbia, Canada.

## 071

- Decimal ASCII code for [capital letter G].

## 71

- atomic number of the chemical element lutetium, symbol Lu.
- most lifetime hits in World Series games (baseball record): by Yogi Berra for the New York Yankees, 1947–63.
- most losses in an NHL (eighty-four-game) season (hockey record): by the San Jose Sharks, 1992–93.
- most passes completed by both teams in an NFL game (football record): the New England Patriots (forty-five)

against the Minnesota Vikings (twenty-six), November 13, 1994.

- number of the French Département of Saône-et-Loire.
- percentage of the Earth's surface that is water.
- retired football jersey number of Walter Jones, retired by the Seattle Seahawks.
- seventy-one months: duration of the depression of 1815–21, encompassing the first major banking crisis in U.S. history.
- 71 square miles: land area of the Borough of Brooklyn in New York City.

## CVN-71

- number of the U.S. Navy nuclear aircraft carrier USS *Theodore Roosevelt.*

## I-71

- north-south interstate highway from Louisville to Cincinnati, Columbus, and Cleveland.

## Sanhedrin

- the highest court of the ancient Jews, composed of seventy-one Torah sages who had the authority both to adjudicate criminal cases, including the prescription of capital punishment, and to interpret the Torah and resolve questions of religious law. (Sometimes called the Great Sanhedrin, to distinguish it from the localized Lesser Sanhedrin [q.v. under twenty-three]).

## Shakespearean Sonnet LXXI

- begins, "No longer mourn for me when I am dead . . ."

## SR-71

- the Lockheed Blackbird, a high-altitude reconnaissance aircraft developed as a successor to the U-2 spy plane.

## *072*

- Decimal ASCII code for [capital letter H].

# 72

- atomic number of the chemical element hafnium, symbol Hf.
- half a gross.
- highest team score in an NFL game (football record): by the Washington Redskins' 72–41 win over the New York Giants, November 27, 1966.
- most free throws scored by both teams in an NBA All-Star game (basketball record): 1971. The West beat the East, 108–107.
- number of beads on the Franciscan rosary.
- a number central to Jain faith; the age at which Mahavira, Jainism's founder, achieved Nirvana.
- number of counties in the state of Wisconsin.
- number of disciples appointed by Jesus Christ (Luke 10:1).
- number of German aircraft claimed shot down by Captain William Avery "Billy" Bishop, Canada's number-one ace fighter pilot in World War I. Some military historians dispute the claim.
- number of points in a printer's inch.
- number of scenes on the Bayeux Tapestry.
- number of the French Département of Sarthe.
- par on most golf courses.
- record number of team wins in a professional basketball season: by the Chicago Bulls in the 1995–96 season (72–10).
- retired baseball uniform number of catcher Carlton Fisk, retired by the Chicago White Sox. Fisk's number 27 has also been retired by the Boston Red Sox.
- 72 inches: width of the basket backboard on a regulation basketball court.
- seventy-two months: duration of the protracted depression of 1837–43.
- 72 pounds: IGFA confirmed record weight for a lake trout, caught by Lloyd E. Bull in Great Bear Lake, Northwest Territories, Canada, on August 9, 1995.
- seventy-two years (and 280 days), age of the oldest winner of an Olympic medal: Swedish Oscar Swahn won silver in a shooting event at the 1920 Olympics in Belgium.

## CVN-72

- number of the U.S. Navy nuclear aircraft carrier USS *Abraham Lincoln.*

## longest-reigning monarch of France

- Louis XIV, the Sun King, ruled for seventy-two years, assuming the throne at five years old in 1643 and wearing the crown until 1715.

## rule of 72

- important concept in understanding compound bank interest. Serves two purposes:
    - ⇨ to estimate number of years to double an investment through annual compounding at a specified interest rate, divide 72 by the annual rate. E.g., @ 7%, 72 ÷ 7 = 10.28 years.
    - ⇨ to estimate the interest rate needed to double an investment in a prescribed number of years, divide 72 by the number of years. E.g., for 12 years, 72 ÷ 12 = 6%.

## 72 degrees

- size of each exterior angle of an equilateral pentagon.

## 72-point type

- typographer's measure, known as "6-line pica."

## The Seventy-Two Spirits of Solomon

- from Agares to Zepar, the demons that can be summoned by a sorceror, as cataloged in the *Lemegeton,* the Book of the Spirits.

## T-72

- a main battle tank of Russian forces. Over 50,000 have been built, and several upgrades incorporated, since its first appearance in 1972.

## world-record twenty-four-hour rainfall

- 72 inches fell at La Réunion Island, March 15–16, 1952.

# 073

- Decimal ASCII code for [capital letter I].

# 73

- atomic number of the chemical element tantalum, symbol Ta.
- largest margin of victory in an NFL championship game (football record): 73 points. In 1940, pre-Super Bowl, Chicago Bears shut out Washington Redskins (73–0) at Griffith Stadium in D.C. on December 8, 1940.
- most games pitched post-season, career (baseball record): by Mariano Rivera, New York Yankees, 1995 through 2007.
- number of books in the Catholic Bible (forty-six Old Testament and twenty-seven New Testament).
- number of elevators in the Empire State Building, including six freight elevators.
- number of kills claimed by British World War I air ace Major Edward "Mick" Mannock of the Royal Flying Corps, more than any other British pilot. Some observers put the tally somewhat lower; best estimate is about sixty-one.
- number of the French Département of Savoie.
- record number of home runs hit in a baseball season: by Barry Bonds, San Francisco Giants, in 2001, surpassing Mark McGuire's seventy with the St. Louis Cardinals in 1998.
- record number of team losses in a professional basketball season: by the Philadelphia 76ers in the 1972–73 season (9-73).
- retired football jersey number of:
  - ⇨ John Hannah, retired by the New England Patriots.
  - ⇨ Joe Klecko, retired by the New York Jets.
  - ⇨ Leo Nomellini, retired by the San Francisco 49ers.

- 73 meters: length of the Sphinx.
- 73 miles: length of Hadrian's Wall across northern England.
- 73 miles per hour: minimal wind force to be categorized as a hurricane, as per the Beaufort Wind Scale.

- 73 yards: longest punt in an NFL Pro Bowl game (football record): by Shane Lechler for the AFC, 2002.
- telegrapher's ending to a message, meaning "best wishes," "regards," etc.

## CVN-73

- number of the U.S. Navy nuclear aircraft carrier USS *George Washington.*

## Haydn's *Symphony No. 73*

- "La Chasse" symphony (in D major).

## Rule 73

- National Hockey League Rulebook item that deals with "Obscene Language or Gestures."

## 73 BC

- year in which gladiator Spartacus led a revolt against the Romans.

## 73rds

- signoff, meaning "best wishes" (CB slang).

## SM-73

- the Fairchild Bull Goose, a surface-to-surface missile.

## Winchester 73

- "the rifle that won the West," a quick-fire weapon prized by those who used it, such as Buffalo Bill and Frank James. It held fifteen rounds of .44 caliber bullets, which could all be fired in sixty seconds.

## *Winchester 73*

- subtitled "The Story of a Rifle," a 1950 Western movie directed by Anthony Mann and starring James Stewart and Shelley Winters.

# *074*

- Decimal ASCII code for [capital letter J].

# *74*

- atomic number of the chemical element tungsten (wolfram), symbol W.
- most field goals scored by an NBA team in one game (basketball record): by the Detroit Pistons against the Denver Nuggets, December 13, 1983 (3 OT).
- most one-point conversions after touchdown in an NFL season (football record): by Stephen Gostkowski for the New England Patriots, 2007.
- number of letters in the Khmer alphabet of Cambodia, the longest in the world.
- number of the French Département of Haute-Savoie.
- retired football jersey number of:
  ⇨ Bruce Matthews, retired by the Tennessee Titans.
  ⇨ Merlin Olsen, retired by the St. Louis Rams.

- 74 miles per hour, minimal wind speed of a category 1 hurricane (up to 95 mph), as per the Saffir-Simpson Hurricane Scale.

## CVN-74

- number of the U.S. Navy nuclear aircraft carrier USS *John C. Stennis,* named after U.S. Senator John C. Stennis from Mississippi.

## I-74

- east-west interstate highway from Galesburg (Illinois) to Peoria, Indianapolis, and Cincinnati.

## longest winning streak in game show history—seventy-four telecasts

- most consecutive wins on a TV quiz show, by thirty-year-old Ken Jennings on *Jeopardy!,* from June 2 through November 29, 2004, for a total accretion of $2,525,700. Ended by Mr. Jennings's loss to Nancy Zerg when he couldn't identify the firm whose 70,000 seasonal white-collar employees work only four months a year. ("What is H&R Block?")

## longest winning streak in women's singles tennis matches—seventy-four matches

- by Martina Navratilova in 1984, begun with a February win over Nancy Yeargin in the U.S. Indoor Championship matches, ended by a December 6 loss to Helen Suková in the semifinals at the Australian Open.

## Seventy-four

- a two-deck ship-of-the-line with heavy fire power but also good sailing characteristics. Designed to carry seventy-four of the largest available guns (36-pounders), these ships were developed by the French early in the eighteenth century and became accepted craft in all European navies by about 1800.

## XC-74

- the Douglas Globemaster, military airplane designed to carry 125 troops and a crew of thirteen. First flown in 1945, it was not as widely accepted as the commercial aircraft DC-7.

## *075*

- Decimal ASCII code for [capital letter K].

## *75*

- atomic number of the chemical element rhenium, symbol Re.
- calories in a large egg or in 6 ounces of orange juice or in a 3-½ ounce glass of red wine.
- the highest number that can appear on a bingo card.
- largest composite score in a Super Bowl game (football record): the San Francisco 49ers' forty-nine to the San Diego Chargers' twenty-six, at SB-XXIX, 1995.
- most complete games pitched in a pre-modern-era season (baseball record): by Will White, Cincinnati Red Stockings, 1879.

- most touchdowns scored by a team in one season (football record): by the New England Patriots, 2007.
- most yards made on a rushing play in a Super Bowl game (football record): by Willie Parker for the Pittsburgh Steelers against the Seattle Seahawks, SB-XL, February 5, 2006.
- most World Series games played (baseball record): by Yogi Berra, with the New York Yankees, 1947–63.
- nowhere in the UK is more than 75 miles from the sea.
- number of counties in the state of Arkansas.
- number of post offices in the United States at the time of the first census (1790).
- number of the French Département of Paris.
- the number worn on Harpo Marx's jersey in the football game at the end of the 1932 Marx Brothers film, *Horse Feathers*.
- seventy-five dollars: penalty for landing on the Luxury Tax square in Monopoly.
- retired football jersey number of Deacon Jones, retired by the St. Louis Rams.
- 75 feet: average length of a blue whale, the Earth's largest mammal.
- 75 kilograms: maximum body weight for men weightlifters in category 5 or women weight-lifters in category 7, Olympic competition.
- 75 miles per hour: speed of the winning car, driven by Ray Harroun, in the first Indy 500 race at Indianapolis, Indiana, on Memorial Day, 1911.

**CVN-75**

- number of the U.S. Navy nuclear aircraft carrier USS *Harry S. Truman*.

**French 75**

- (a) a large artillery piece, the French 75mm cannon of World War I.
- (b) a cocktail made with champagne, dry gin, lemon juice, and powdered sugar.

**I-75**

- major north-south interstate highway from Miami Lakes (Florida) to Naples, Tampa, Chattanooga, Knoxville, Cincinnati, Toledo, Detroit, and Sault Ste. Marie (Michigan).

## *076*

- Decimal ASCII code for [capital letter L].

## *76*

- atomic number of the chemical element osmium, symbol Os.
- most goals scored by a player in a rookie season (hockey record): by Teemu Selanne for the Winnipeg Jets, 1992–93.
- most times sacked in an NFL season (football record): David Carr for the Houston Texans, 2002.
- number of the French Département of Seine-Maritime.
- retired football jersey number of Lou Groza, retired by the Cleveland Browns.
- 76 feet: length of an American bocce court (by 12 feet wide), as prescribed by the World Bocce Association. (In other countries, governed by the Confederation Boccistica Internationale, the standard is 90 feet long by 13 feet wide).
- 76 yards: greatest width of the Grand Canal in Venice.
- seventy-six years: average period of Halley's Comet. Next appearance due in 2061.
- "That's the spirit!"

**CVN-76**

- number of the U.S. Navy nuclear aircraft carrier USS *Ronald Reagan*.

**greatest recorded twenty-four-hour snowfall in North America**

- 76 inches at Silver Lake, Colorado, April 14–15, 1921. The accumulation reached 87 inches in 27-½ hours, and 95 inches in 32-½ hours.

## I-76

- east-west interstate highway from Denver to Route 80 in Nebraska; then again from Akron, to Harrisburg and Philadelphia.

## Rule 76

- National Hockey League Rulebook item that deals with "Physical Abuse of Officials."

## "76"

- a brand name of gas stations under the Conoco-Phillips umbrella.

## 76 drops

- capacity of a teaspoon (cooking measures).

## 76ers

- name of the Philadelphia professional basketball team.

## 76 trombones

- instruments that "led the big parade" in Meredith Willson's 1957 Broadway musical *The Music Man*, starring Robert Preston and Barbara Cook.

## *Spirit of 76*

- classic painting by Archibald McNeal Willard (from 1875 or 1876), showing two drummers and a fifer marching through a battlefield in the Revolutionary War. It now hangs in Abbot Hall in Marblehead, Massachusetts.

## *077*

- Decimal ASCII code for [capital letter M].

## *77*

- atomic number of the chemical element iridium, symbol Ir.

- football jersey number of Red ("The Galloping Ghost") Grange at University of Illinois, 1922–25, which he kept when he played professionally with the Chicago Bears.
- number of counties in the state of Oklahoma.
- number of female Earth Tengri, the second level of Mongol dieties, beneath the Heavenly Tengri.
- number of stories in New York City's Chrysler Building.
- number of swimmers saved by young Ronald Reagan in the seven summers he served as a lifeguard, according to his biography.
- number of the French Département of Seine-et-Marne.
- number of times Lamech shall be avenged (Genesis 4:24).
- retired basketball jersey number of Jack Ramsay, retired by the Portland Trail Blazers.
- retired football jersey number of:
  - ⇨ Red Grange, retired by the Chicago Bears.
  - ⇨ Stan Mauldin, retired by the Arizona Cardinals.
  - ⇨ Jim Parker, retired by the Indianapolis Colts.
  - ⇨ Korey Stringer, retired by the Minnesota Vikings.

- retired hockey shirt number of Ray Bourque, retired by both the Boston Bruins and the Colorado Avalanche.
- smallest number requiring five syllables to verbalize.
- sum of the first eight prime numbers ($2 + 3 + 5 + 7 + 11 + 13 + 17 + 19$).

## Charter 77

- Czech human rights manifesto, published in January 1977, signed by 230 prominent Czech intellectuals committed to human rights, including Vaclav Havel.

## CVN-77

- number of the U.S. Navy aircraft carrier USS George *H. W. Bush*.

## Group of 77

- a loose coalition of developing countries organized to promote its members' collective economic interests, to collaborate as a joint negotiating force in the United Nations, and to advance economic and technical cooperation among developing nations. Established in 1964, the association has since expanded to include 131 member nations as of the end of 2010. Eight other states are listed as former members, including New Zealand and Mexico, now members of the OECD, and Malta and Romania, now part of the EU. Sometimes known as G-77.

## I-77

- north-south interstate highway from Columbia to Charlotte, Charleston (West Virginia), Akron, and Cleveland.

## Operation 077

- the clandestine plan that serves as the agenda for the 1946 film, *13 Rue Madeleine*.

## 77 rue de Varenne

- address of the Rodin Museum in Paris.

## *77 Sunset Strip*

- TV detective series starring Efrem Zimbalist Jr., Roger Moore, and Edd "Kookie" Burns. Played on ABC-TV from October 1958 to September 1964.

## *Talking Heads: 77*

- name of the first album issued by the group Talking Heads.

## *078*

- Decimal ASCII code for [capital letter N].

## *78*

- atomic number of the chemical element platinum, symbol Pt.
- lives lost in crash of a TWA Constellation in Hinsdale, Illinois, four minutes after takeoff from Midway Airport in Chicago, September 1, 1961, the worst single plane crash in U.S. commercial aviation to that point.

- most aces served in a tennis match: by Ivo Karlovic against Radek Stepanek in a Davis Cup semifinal on September 18, 2009. However, Karlovic lost the match, 3–2.
- most free throws scored in NBA All-Star games, career (basketball record): by Elgin Baylor, 1959–70.
- number of cards in a tarot deck, four suits of fourteen cards each, plus twenty-two symbolic trumps.
- number of crime novels written by Dame Agatha Christie (1890–1976), the best-selling author of the twentieth century, from *The Mysterious Death of Styles* (1920), introducing detective Hercule Poirot, to *Sleeping Murder* (1976), the last of the Miss Marple series. Her crime books have sold over 2 billion copies in forty-four languages. She has also written sixteen plays, and, using the pseudonym Mary Westmacott, six romance novels.
- number of the French Département of Yvelines.
- retired football jersey number of:
  ⇨ Bobby Bell, retired by the Kansas City Chiefs.
  ⇨ Jackie Slater, retired by the St. Louis Rams.
  ⇨ Bruce Armstrong, retired by New England Patriots.

- seventy-eight dollars: the per capita national debt in 1866, following the Civil War.
- 78 feet: length of the playing area on a regulation tennis court.
- 78 percent of the air we breathe is nitrogen.
- sum of the first twelve integers.
- total number of gifts received on the last day, according to the carol "The Twelve Days of Christmas."

## CVN-78

- number of the U.S. Navy aircraft carrier USS Gerald R. Ford.

## Rule of 78

- method for determining the amount of monthly interest due at any point in the repayment period of a one-year installment loan. The number 78 represents the sum of the numbered months for a year, the first month being 1, the second 2, etc., until 12 for the last month. In a Rule of 78 loan, the first month's interest is computed as 12/78 of the total interest due for the year, the second month's is 11/78, etc. The advantage to the lender is in the accelerated receipt of interest, even though the total interest for the period of the loan is the same as in a simple interest loan. To the disadvantage of the borrower is the extra interest already paid if the loan is paid off early.

## I-78

- east-west interstate highway from Harrisburg to Allentown, and Elizabeth (New Jersey).

## "78"

- a phonograph record played at 78 revolutions per minute, an older standard for records. By the late 1940s, 78s started being replaced by the new LPs (33 rpm) and EPs (45 rpm). By the 1960s, production of 78-rpm records was virtually over.

## 079

- Decimal ASCII code for [capital letter O].

## 79

- atomic number of the chemical element gold, symbol Au.
- most complete games played in an NBA season (basketball record): by Wilt Chamberlain for the Philadelphia Warriors, 1961–62.
- number of the French Département of Deux-Sèvres.
- retired football jersey number of:
    ⇨ Jim Hunt, retired by the New England Patriots.
    ⇨ Bob St. Clair, retired by the San Francisco 49ers.

- AD 79: Mount Vesuvius erupted, burying the cities of Pompeii and Herculaneum.
- 79th floor of the Empire State Building in New York City was the site of a spectacular accident on July 28, 1945, when a B-25 Mitchell bomber flown by Lieutenant Colonel William

F. Smith crashed into the north side of the building. It happened on a Saturday, so, fortunately, few people were in the building. Nonetheless, at least fourteen people died in addition to the pilot and his crewman.

## I-79

- north-south interstate highway from Charleston (West Virginia) to Pittsburgh, and Erie (Pennsylvania).

## 79 Wistful Vista

- residence of Fibber McGee and Molly on their NBC radio show, 1935–52.

## 080

- Decimal ASCII code for [capital letter P].
- medical classification code for louse-borne (epidemic) typhus.

## 80

- atomic number of the chemical element mercury, symbol Hg.
- chains in a survey mile.
- 80 grains: weight of a U.S. nickel.
- 80 yards: maximum width of a soccer field for international matches.
- equal to fourscore.
- in Hebrew, the numerical value of the letter *pe*.
- ISBN group identifier for books published in the Czech Republic and Slovakia.
- maximum number of residents in a thorp, in the game Dungeons & Dragons.
- minims in a teaspoon (liquid measure).
- number of brothers of Okuni-nushi in the Japanese myth about courting the beautiful princess, Yagami-hime.
- number of kills claimed by German air ace Manfred von Richthofen (known as The Red Baron) in World War I.
- number of the French Département of Somme.

- photographic filter that corrects for excessive redness under tungsten lighting. Available in several degrees of intensity as 80A, 80B, and 80C.
- retired football jersey number of:
  - ⇨ Cris Carter, retired by the Minnesota Vikings.
  - ⇨ Steve Largent, retired by the Seattle Seahawks.
  - ⇨ Jerry Rice, retired by the San Francisco 49ers.

- upper age limit for cardinals voting in papal elections.
- value of medieval Roman numeral R.
- value of Roman numeral LXXX.

### Around the World in Eighty Days

- in Jules Verne's 1873 novel, Phileas Fogg and his loyal servant Passepartout win a bet by circling the globe in only eighty days, an incredible feat for its time. Filmed in 1956 with David Niven as Fogg, the marvelous Mexican mimic Cantinflas as his bumbling valet, Shirley MacLaine filling out the distaff role, and cameo appearances by several dozen of filmdom's elite, the movie captured the Oscars for Best Film, Screenplay, Cinematography, and Editing.

### eighty kings

- in the card game pinochle, a meld of four kings, each of a different suit, is worth eighty points.

### Eighty Mile Beach

- a long stretch of waterfront land on the Indian Ocean in the northwestern part of Western Australia.

### Eighty Years' War

- the Dutch Revolt against Spain, lasting intermittently from 1568 to 1648, in which the northern Protestant states of the Netherlands joined the Union of Utrecht and achieved independence from Spain in 1581 as the United Provinces of the Netherlands. Hostilities continued on and off until the Treaty of Muster, finalized on January 30, 1648, confirmed the integrity of the Dutch Republic.

## F-80

- (a) The Lockheed "Shooting Star," the first USAF aircraft to exceed 500 miles per hour in level flight. Designed as a high-altitude interceptor, the F-80 jet was used extensively as a fighter-bomber in the Korean War. On November 8, 1950, an F-80C, flown by Lieutenant Russell J. Brown, shot down a Russian-built MiG-15 in the world's first all-jet-fighter air battle.
- (b) high-rated Nikon 35mm film camera.

## I-80

- major east-west interstate highway from San Francisco to Sacramento, Reno, Salt Lake City, Cheyenne, Omaha, Des Moines, Toledo, Youngstown, Stroudsburg, and Fort Lee (New Jersey).

## MD-80

- Boeing twinjet aircraft, in service since 1980.

## octogenarian

- a person between eighty and eighty-nine years of age.

## T-80

- a main battle tank of Soviet forces, in production since 1984. The first Soviet tank powered by a gas-turbine engine and the first to incorporate a laser range-finder.

## 80.4

- percentage of the U.S. population that has a high school education or beyond, according to the 2000 Census.

## 081

- Decimal ASCII code for [capital letter Q].

## 81

- age at which George Cukor directed his fiftieth and final film, *Rich and Famous* (1981).

- age of the oldest person ever to receive a regular Academy Award: Jessica Tandy for Best Actress in *Driving Miss Daisy*, in 1990.
- atomic number of the chemical element thallium, symbol Ti.
- 81 milligrams: the strength of children's aspirin.
- 81 miles per hour: fastest recorded speed on a bicycle on a flat road, by Sam Whittingham at the World Human-Powered Speed Challenge in Nevada, October 5, 2002.
- the fourth power of 3.
- international telephone calling code for Japan.
- ISBN group identifier for books published in India (shared with 93).
- lunch-counter code for a glass of water (eighty-two = two glasses, etc.).
- minimum number of residents in a hamlet, in the game Dungeons & Dragons.
- most interceptions in an NFL career (football record): by Paul Krause for the Washington Redskins, 1964–67, and Minnesota Vikings, 1968–79.
- number of the French Département of Tarn.
- photographic filter that balances daylight films to electronic flash. Intensifies reds and oranges and produces a generally warming effect. Good for shooting fall foliage. Available in several degrees of intensity as 81A, 81B, and 81C.
- retired football jersey number of Doug Atkins, retired by the New Orleans Saints.
- the square of 9.

**age of the oldest delegate to the Constitutional Convention in May 1787**

- Benjamin Franklin, representing Pennsylvania.

**I-81**

- major north-south interstate highway from Knoxville to Roanoke, Harrisburg, Scranton, Binghamton, Syracuse, Watertown, and the Thousand Islands Bridge to Ivy Lea, Ontario.

## *082*

- Decimal ASCII code for [capital letter R].

## *82*

- atomic number of the chemical element lead, symbol Pb.
- international telephone calling code for Republic of Korea (South Korea).
- ISBN group identifier for books published in Norway.
- liters in a kilderkin (British unit of capacity).
- most consecutive games scoring hundred team points or more in a season (basketball record): by the Denver Nuggets, October 30, 1981 to April 17, 1982 (entire season).
- most kickoff returns in an NFL season (football record): by MarTay Jenkins for the Arizona Cardinals, 2000.
- number of counties in the state of Mississippi.
- number of games each team plays in a season of NBA basketball or NHL hockey.
- number of the French Département of Tarn-et-Garonne.
- record number of career golf matches won in PGA Tournaments: by Sam Snead, 1937–79. In 1980 Snead joined the Champions Tour.
- retired football jersey number of Raymond Berry, retired by the Indianapolis Colts.

**Haydn's *Symphony No. 82***
- the "Bear" symphony (in C major).

## *82.5*

- 82.5 kilograms, maximum body weight for men weightlifters in category 6 or women weightlifters in category 8, Olympic competition.

## *083*

- Decimal ASCII code for [capital letter S].

# 83

- atomic number of the chemical element bismuth, symbol Bi.
- Imperial gallons in 100 U.S. gallons.
- ISBN group identifier for books published in Poland.
- most at-bats by a pinch hitter in a season (baseball record): by Lenny Harris, New York Mets, 2001.
- number of counties in the state of Michigan.
- number of the French Département of Var.

**Haydn's *Symphony No. 83***

- The "Hen" symphony (in G minor).

# 084

- decimal ASCII code for [capital letter T].

# 84

- atomic number of the chemical element polonium, symbol Po.
- eighty-four days: the length of time the Cuban fisherman in Ernest Hemingway's *The Old Man and the Sea* (1952) goes without a catch. On the eighty-fifth day he hooks a giant marlin.
- 84 feet: preferred length of a high school basketball court.
- international telephone calling code for Republic of Vietnam.
- ISBN group identifier for books published in Spain.
- a naval prison (World War II navy usage).
- lives lost in a fire at the MGM Grand Hotel in Las Vegas on November 21, 1980.
- most career post-season wins by a baseball manager: Joe Torre, for the New York Mets, 1977–81; the Atlanta Braves, 1982–84; the St. Louis Cardinals, 1990–95; the New York Yankees, 1996–2007; and the Los Angeles Dodgers, 2008–09.
- most shutouts by a major league soccer goalie: by Kevin Hartman for Los Angeles and Kansas City, 2003–09.
- number of Earth years in Uranus's circuit of the sun.

- number of Mahasiddhas ("Great Magicians"), Indian yogis believed to have possessed magical powers.
- number of the French Département of Vaucluse.

## Eighty Four

- a town in Pennsylvania, in Washington County, about 25 miles southwest of Pittsburgh.

## 84 Avenue Foch, Paris

- location of the counterintelligence branch of the Gestapo, known as the *Sicherheitsdienst* (or SD), during the German occupation of Paris in World War II.

## *84 Charing Cross Road*

- sensitive film about the growing friendship between a New York City woman and a London book dealer who supplies her with rare volumes. Anne Bancroft, Anthony Hopkins, and Judi Dench star in this 1987 movie directed by David Jones.

## Haydn's *Symphony No. 84*

- "In Nomine Domini" (in E-flat major).

## I-84

- east-west interstate highway from Portland (Oregon) to Boise, Ogden (Utah), and to Eisenhower Highway (Route 80), east of Salt Lake City.

## P-84

- U.S. warplane, the Republic Thunderjet (or Thunderstreak), that saw action in the Korean War. The P-84 is also in the air forces of several NATO countries.

# *085*

- Decimal ASCII code for [capital letter U].

# 85

- Article of the Uniform Code of Military Justice (UCMJ) dealing with desertion.
- atomic number of the chemical element astatine, symbol At.
- 85 feet: width of a regulation ice hockey rink.
- 85 miles: the distance covered by the infamous Bataan Death March of American and Filipino prisoners taken by the Japanese at the fall of Bataan in the Philippines. More than 70,000 POWs left Mariveles on April 10, 1942; less than 54,000 reached Camp O'Donnell six days later. After the war, Japanese commander Masaharu Homma was convicted of atrocities and executed on May 3, 1946.
- 85 pounds: record IGFA confirmed size of a great barracuda, caught off San Juan, Puerto Rico, by John W. Helfrich, on April 11, 1992.
- 85 yards: longest pass completion in a Super Bowl game (football record): by Jake Delhomme to Muhsin Muhammad for the Carolina Panthers against the New England Patriots, at SB-XXXVIII, February 1, 2004. Also the Super Bowl's longest play from scrimmage. Though the play resulted in a touchdown, the Panthers nonetheless lost to New England, 32–29.
- ISBN group identifier for books published in Brazil.
- number of letters in the Cherokee alphabet developed by Sequoya (George Guess).
- number of the French Département of Vendée.
- photographic filter that corrects for the bluish cast of daylight when used with tungsten film.
- retired baseball uniform number of team executive August Busch Jr., retired by the St. Louis Cardinals.
- retired football jersey number of:
    - ⇨ Chuck Hughes, retired by the Detroit Lions.
    - ⇨ Jack Youngblood, retired by the St. Louis Rams.

## Haydn's *Symphony No. 85*
- "La Reine" symphony (in B-flat major).

## I-85

- major north-south interstate highway from Montgomery to Atlanta, Charlotte, Greensboro, and Petersburg (Virginia).

## Ochocinco

- Chad Javon Ochocinco, wide receiver for football's Cincinnati Bengals, was born Chad Javon Johnson, but just before the 2008 season, he changed his name to match the 85 number on his team jersey.

## Rule 85

- National Hockey League Rulebook item that deals with slashing.

## *086*

- Decimal ASCII code for [capital letter V].

## *86*

- article of the Uniform Code of Military Justice (UCMJ) dealing with absence without leave.
- atomic number of the chemical element radon, symbol Rn.
- bartender's code for an unwelcome customer, meaning "Get rid of him." (May be from Article 86 of the New York State Liquor Code, which defined the circumstances in which a bar patron need be refused service, or 86'ed.)
- code for "cancel the order" in restaurant slang, or refuse to serve (an unwanted customer).
- code for "out of stock."
- number of office floors in the Empire State Building. Add the mooring mast, equivalent to fourteen floors, and two basements, and the total height equals 102 stories.
- international telephone calling code for People's Republic of China.
- ISBN group identifier for books published in Serbia, Montenegro, and shared by Bosnia and Herzegovina (with

9958), Croatia (with 953), Macedonia (with 9989), and Slovenia (with 961).

- major Group 86 under the U.S. Department of Labor consists of membership organizations of several types.
- number of the French Département of Vienne.
- number of years Boston baseball fans waited until the Red Sox won the World Series in 2004, in four straight games against the St. Louis Cardinals, after a remarkable come-back from being down three games to none in the American League pennant race against the New York Yankees. Boston's last World Series win was in 1918 against the Chicago Cubs in six games. They won again in 2007, beating the Colorado Rockies in four games.
- retired football jersey number of Buck Buchanan, retired by the Kansas City Chiefs.

**Agent 86**

- Agent Maxwell Smart, played by Don Adams on the TV comedy series *Get Smart*. Ran on NBC from September 1965 to September 1969, and CBS, September 1969 to September 1970).

**Bill 86**

- introduced into the Quebec legislature in 2005 to permit the use of English on outdoor public signs in the Province of Quebec, as long as the French-language portion is more prominent.

**86'ed**

- slang for eliminated, dropped, discarded.

**86 proof**

- describes an alcoholic drink having a 43 percent alcohol content (i.e., most Scotch whiskey).

**P-86**

- the Sabre, the famous "MiG-killer" of the Korean War. Produced by North American Aircraft, this was the first swept-wing U.S. jet fighter.

# 087

- Decimal ASCII code for [capital letter W].

# 87

- atomic number of the chemical element francium, symbol Fr.
- eighty-seven hours is the running time of the longest film ever made: *Cure for Insomnia*, in which poet L. D. Graham reads in its entirety a poem of 4,080 lines. The film was done at the School of the Art Institute of Chicago, January 31 to February 3, 1987.
- 87 yards: longest return of an interception in an NFL Pro Bowl game (football record): by Deion Sanders for the NFC, 1999.
- ISBN group identifier for books published in Denmark.
- number of counties in the state of Minnesota.
- number of the French Département of Haute-Vienne.
- photographic filter that blocks visible light and captures only lower-frequency infrared light.
- retired football jersey number of Dwight Clark, retired by the San Francisco 49ers.

**Eighty-Seven Immortals**

- the largest mural in the world, at 3,200 square meters, this picture is carved on the surface of a dam of the Panshan Reservoir in China. Based on the masterpiece of Tang Dynasty artist Wu Daozi, which hangs in Xu Beihong Museum in Beijing.

**"four score and seven years"**

- period (counted from 1776) referred to in Lincoln's Gettysburg Address (November 19, 1863).

**G-87**

- the Douglas Skybolt, air-to-surface ballistic missile, later re-designated AGM-48.

## Ju-87

- the famous German "Stuka" dive-bomber, produced by Junkers, with sirens fitted into the wheel covers to quash the morale of the enemy.

## 87 ½

- lunch-counter code for "There's a pretty girl out front."

## 088

- Decimal ASCII code for [capital letter X].

## 88

- atomic number of the chemical element radium, symbol Ra.
- a high velocity gun with a caliber of 88mm, used by German forces as an antitank and antiaircraft weapon in World War II.
- in the slang of bingo players, before PC, 88 was known as "two fat ladies."
- ISBN group identifier for books published in Italy and Italian-speaking Switzerland.
- love and kisses (CB slang).
- most career Official Tournament wins in LPGA matches (golf record): by Kathy Whitworth, 1962–85.
- number of constellations in the sky, as recorded by the International Astronomical Union.
- number of counties in the state of Ohio.
- number of Earth days in Mercury's circuit of the sun.
- number of keys on a piano: 36 black and 52 white.
- number of the French Département of Vosges.
- retired football jersey number of:
  ෨ J. V. Cain, retired by the Arizona Cardinals.
  ⇨ Alan Page, retired by the Minnesota Vikings.
- sum of the first ten Fibonacci numbers (0 + 1 + 1 + 2 + 3 + 5 + 8 + 13 + 21 + 34).

- used as a code for "Heil, Hitler" in neo-Nazi circles, H being the eighth letter of the alphabet.

## Eighty Eight
- a small town in Barren County in south-central Kentucky

### *88 Men and 2 Women*
- Clinton Duffy's recounting of ninety executions he presided over as warden of San Quentin prison in the 1940s.

## FM 88
- Australia's traveler information radio station, offering "Australian history, music, tourist attractions, and goods."

## longest consecutive winning streak in NCAA basketball—eighty-eight games
- by UCLA Bruins under Coach John R. Wooden, "The Wizard of Westwood": fifteen at the end of the 1970–71 season, thirty in both 1971–72 and 1972–73, and then thirteen at the beginning of 1973–74. In his twenty-seven years of coaching at UCLA, Wooden's record was 620-147, including an unprecedented thirty-eight straight NCAA tournament victories.

## M-88
- designation for a series of 7.92 mm rifles and carbines made for the German Army in World War II, officially classified as GEW M1888.

## Olds 88
- full-size car, powered by a V-8 engine, sold by the Oldsmobile division of General Motors from 1949 to 1999. The longest continuous American car brand, over 10 million 88s were built, the last coming off GM's assembly plant in Orion Township, Michigan, on January 9, 1999. Car buffs called it "the 88."

## *089*

- Decimal ASCII code for [capital letter Y].

## *89*

- article of the New York State Education Law covering children with handicapping conditions.
- atomic number of the chemical element actinium, symbol Ac.
- 89 cm: maximum length of the blade in modern fencing, whether epee, foil, or sabre.
- ISBN group identifier for books published in Korea.
- number of the French Département of Yonne.
- retired football jersey number of:
  ⇨ Bob Dee, retired by the New England Patriots.
  ⇨ Gino Marchetti, retired by the Indianapolis Colts.

### Eighty-niners

- persons who participated in the Land Run of April 22, 1889, in Oklahoma, when the federal government opened unassigned lands to settlers. Under provisions of the Homestead Act of 1862, a legal settler could claim 160 acres and receive title after five years if they lived on the land and improved it.

### 89er's Day

- aka, Oklahoma Day, a holiday in that state; the anniversary of opening the Oklahoma Territory for settlement.

### Ford XRM-89

- Blue Scout 1, a three-stage missile based on NASA's Scout.

## *090*

- Decimal ASCII code for [capital letter Z].
- Dewey Decimal System designation for manuscripts and rare books.
- medical classification code for congenital syphilis.

# 90

- atomic number of the chemical element thorium, symbol Th.
- degrees in a right angle.
- in Hebrew, the numerical value of the letter *sadhe.*
- international telephone calling code for Turkey.
- ISBN group identifier for books published in the Netherlands and Flemish Belgium.
- 90 degrees: size of each interior or exterior angle of a square.
- 90 feet: distance between the bases on a regulation baseball field.
- 90 kilograms: maximum body weight for men weightlifters in category 7, Olympic competition.
- 90 yards: longest punt return in a Pro Bowl game (football record): by Billy Johnson for the AFC, 1976.
- number of the French Département of Territoire de Belfort.
- value of medieval Roman numeral N.
- value of Roman numeral XC.

### *Belle of the Nineties*
- Mae West sings "My Old Flame" and struts Western in this 1934 film directed by Leo McCarey.

### Ford XRM-90
- Blue Scout 2, a four-stage missile, used in orbital and suborbital tests.

### The Gay Nineties
- refers to the decade of the 1890s.

### I-90
- major east-west interstate highway from Seattle to Billings, Rapid City, Sioux Falls, Madison, Chicago, Cleveland, Buffalo, Syracuse, Albany, and Boston.

### MD-90
- mid-size, medium-range, fuel-efficient twinjet aircraft from Boeing, in service since 1995. Its high takeoff thrust makes it especially useful for short runways.

## 90-day embargo

- Congressional legislation passed in April 1812 to prepare American shipping for the possible forthcoming hostilities. An embargo was placed on all vessels in U.S. harbors.

## ninety-day wonder

- a U.S. Army second lieutenant commissioned after only 90 days of officer training.

## Ninety-East Ridge

- a ridge between the Mid-Indian Basin and the West Australian Basin, in the Indian Ocean.

## Ninety Mile Beach

- (a) aa stretch of seashore on the northernmost end of North Island, New Zealand.
- (b) beach area in the southeast corner of Victoria State, Australia.

## *Playhouse 90*

- a dramatic anthology on CBS-TV from October 1956 to September 1961, the quality standard for TV drama of the period featuring name actors in classic teleplays by the best writers, directed by well-known TV and film directors.

## Nonagenarian

- a person between ninety and ninety-nine years of age.

## T-90

- the most modern Russian tank, the main battle tank of the Russian Army. A derivative of the earlier T-72. In production since 1992.

## *091*

- Decimal ASCII code for [left bracket ( [ )].

# 91

- atomic number of the chemical element proactinium, symbol Pa.
- international telephone calling code for India.
- ISBN group identifier for books published in Sweden.
- number of the French Département of Essonne.
- sum of the first six square numbers (1 + 4 + 9 + 16 + 25 + 36).
- sum of the first thirteen integers.
- value of Roman numeral XCI.

## I-91

- north-south interstate highway from New Haven to Hartford, Springfield, Brattleboro (Vermont), and to Rock Island, Quebec.

## longest winning streak in boxing history—ninety-one bouts

- by Sugar Ray Robinson, begun by defeating Jackie Wilson on February 19, 1943 and ended by a loss to Randy Turpin in London on July 10, 1951.

## ninety-one cents

- the amount of money a person has if he holds one each of U.S. coins less than one dollar: (1¢ + 5¢ + 10¢ + 25¢ + 50¢).

## Rule 91

- National Hockey League Rulebook item that deals with tripping.

# 092

- Decimal ASCII code for [back slash (\)].

# 92

- Article of the Uniform Code of Military Justice (UCMJ) dealing with failure to obey an order or regulation.

- atomic number of the chemical element uranium, symbol U.
- international telephone calling code for Pakistan.
- ISBN group identifier for books published by international publishers, such as UNESCO and the EU, and by European Community Organizations.
- most individual goals scored in an NHL eighty-game season (hockey record): by Wayne Gretzky for the Edmonton Oilers, 1981–82.
- number of counties in the state of Indiana.
- number of known natural chemical elements; all others are man-made.
- number of nations represented at President Kennedy's funeral at Arlington National Cemetery on November 25, 1963.
- number of people killed when a Russian R-16 rocket exploded while it was being fueled at the Baikonur Space Center in Kazakhstan on October 24, 1960; seen as the single greatest disaster in the history of rocketry. Official toll is ninety-two (seventy-four military, eighteen civilians), but true estimates run as high as 165. Among those killed were Marshall Mitrofan Nedelin, commander of Soviet Strategic Rocket Forces, and Colonel A. Nosov, chief of Baikonus launch command, who had personally launched Sputnik in 1957.
- number of short incidents portrayed in Peter Greenaway's 1980 pseudo-documentary film, *The Falls*, which deals with the survivors—all of whose surnames begin with the letters FALL—of some VUE (Violent Unknown Event) that has befallen the world.
- number of the French Département of Hauts-de-Seine.
- record number of career hat tricks for a soccer player: by Pelé, 1955–77.
- retired football jersey number of Reggie White, retired by the Philadelphia Eagles.

**Beretta 92**

- semi-automatic pistol designed and manufactured by Beretta of Italy. First produced in 1975, the gun is still in widespread use in several different variations. It has been used by the military in the U.S., France, Italy, and South Africa.

**FIM-92**

- military designation of the American anti-aircraft Stinger missile, used effectively by Afghan forces against Soviet helicopters in the Soviet-Afghan war in the early 1980s.

**Haydn's *Symphony No. 92***

- "Oxford" symphony (in G major).

***The House on 92nd Street***

- 1945 espionage thriller directed by Henry Hathway in semi-documentary style that adds to its realism. Shot in New York City in actual locations, and based partially on fact. Cast includes William Eythe, Lloyd Nolan, and Signe Hasso. The original story won an Oscar for Charles G. Booth.

**the 92nd**

- the Buffalo Soldiers, an all-black infantry division that distinguished itself in northern Italy in World War II.

***92 in the Shade***

- 1975 film directed by Thomas McGuane that portrays the highly competitive life of fishing guides in Key West, Florida. The recipient of some terrible reviews, the film features Peter Fonda, Warren Oates, Margot Kidder, and Burgess Meredith.

**TI-92**

- model number of the once widely used Texas Instruments calculator.

## *92.5*

- percentage of sterling silver that is silver; the other 7.5 percent is usually copper.

## *093*

- Decimal ASCII code for [right bracket ( ] )].
- Dewey Decimal System designation for incunabula.

# 93

- age of President Ronald Reagan at his death on June 5, 2004, the longest-lived president in the history of the country.
- atomic number of the chemical element neptunium, symbol Np.
- ISBN group identifier for books published in India (shared with 81).
- international telephone calling code for Afghanistan.
- 93 miles: length of the Strait of Gibraltar.
- 93 million miles: average distance of Earth from the sun, called an astronomical unit (AU).
- 93 pounds: record IGFA confirmed weight for a king mackerel, caught by Steve Perez Graulau off San Juan, Puerto Rico, on April 18, 1939.
- 93 yards: longest completed pass in an NFL Pro Bowl game (football record): in 1996, to Yancey Thigpen from Jeff Blake for the AFC, scoring a TD.
- number of counties in the state of Nebraska.
- number of the French Département of Seine-Saint-Denis.
- number on the last car of the escape train in the Frank Sinatra film *Von Ryan's Express*.

**93**

- (*Quatre-vingt-treize*): a melodramatic historical novel by Victor Hugo (1879), his last, set in 1793 France and including many major historical figures of the French Revolution.

**United Flight 93**

- designator of the flight that was the fourth intended attack plane on September 11, 2001. Because takeoff was delayed, the passengers learned about the attacks on the World Trade Center and the Pentagon, and they realized that the terrorists controlling their plane were not mere hijackers. The passengers fought back and managed to crash the plane near Shanksville, Pennsylvania, sacrificing their lives to prevent the hijackers from completing their mission to hit the White House.

## United 93

- 2006 film, written and directed by Paul Greengrass, recounting the most likely chain of events that transpired on the doomed flight on September 11, 2001, when the passengers overtook their Arab captors to prevent the plane from attacking the White House; the film feels like both drama and documentary.

## 094

- Decimal ASCII code for [caret/circumflex (^)].

## 94

- Article of the New York State Civil Law dealing with admission to practice.
- Article of the Uniform Code of Military Justice (UCMJ) dealing with mutiny or sedition.
- atomic number of the chemical element plutonium, symbol Pu.
- international telephone calling code for Sri Lanka.
- most strikeouts pitched in World Series games (baseball record): by Whitey Ford, New York Yankees, 1950–64. Ford pitched in twenty-two games in eleven Series.
- number of the French Département of Val-de-Marne.
- 94 decibels: maximum noise level the human ear can sustain without damage for any period of time.
- 94 feet: length of a regulation basketball court.
- 94 inches: the width of the main mirror in the Hubble Space Telescope.
- ninety-four years: the life of the original Orient Express, from its first run on October 4, 1883, Paris to Giurgi (on the Danube in Romania), via Strasbourg, Vienna, Budapest, and Bucharest, through its 1921 route extension to Istanbul (as the Simplon-Orient-Express), to its last trip in May, 1977. Five years later, in May 1982, the line was reborn as the Venice-Simplon-Orient-Express, running from London to Venice.

## AN-94

- the Abakan, a Russian assault rifle designed to replace the standard AK-74. Adopted by the Russian military in 1994, it proved to be too complex for the ordinary soldier, so it is used only by some elite forces. AN stands for Avtomat Nikonova, the latter the name of its designer.

## Haydn's *Symphony No. 94*

- the "Surprise" symphony (in G major).

## I-94

- major east-west interstate highway from Billings to Bismarck, Minneapolis/St. Paul, Milwaukee, Chicago, Detroit, and to Port Huron (Michigan).

## *095*

- Decimal ASCII code for [underscore (_)].
- Dewey Decimal System designation for books notable for special bindings.

## *95*

- atomic number of the chemical element americium, symbol Am.
- international telephone calling code for Myanmar (formerly Burma).
- lunch-counter code for a customer leaving without paying, meaning, "Stop him!"
- 95 feet: length of the Thomas Hart Benton mural, *America Today,* along the wall of the Equitable Center building's north corridor in New York City.
- number of counties in each of the states of Tennessee and Virginia.
- number of the French Département of Val-d'Oise.
- a U.S. penny is 95 percent copper and 5 percent zinc.

## French 95

- a cocktail made with champagne, bourbon whiskey, lemon juice, and powdered sugar.

## I-95

- major north-south interstate highway from Miami to Jacksonville, Savannah, Fayetteville, Richmond, D.C., Baltimore, Philadelphia, New York, New Haven, Boston, Portland, Bangor, and across the Canadian border into Woodstock, New Brunswick.

## 95 percent

- confidence level sufficient in most research to reject the null hypothesis.

## 95 Theses

- "The Ninety-Five Theses on the Power and Efficacy of Indulgences," a document that Martin Luther posted on the door of the castle church in Wittenberg on October 31, 1517, in protest of the dispensation of indulgences granted to Johann Tetzel, a German Dominican preacher. Luther's document is credited with sparking the Protestant Reformation.

## Tu-95

- Tupolev Tu-95, a strategic bomber and missile carrier from the period of the Soviet Union. First flown in 1952, its turboprop engines provided for intercontinental missions.

## *96*

- age of screen legend Katharine Hepburn at her death on June 29, 2003. She won a record four Best Actress Oscars.
- atomic number of the chemical element curium, symbol Cm.
- drams in a pound (troy or apothecary).
- 96 miles per hour: minimal wind speed of a category 2 hurricane (up to 110 mph), as per the Saffir-Simpson Hurricane Scale.
- number of countries signing the nuclear test ban treaty in 1963; the country most obviously absent was France.
- a number of special significance in Hungary, symbolic of the year 1896, the millennium of the first settlement of the

Magyars, the forebears of the present Hungarian people. The Hungarian Parliament, designed around the number ninety-six, has ninety-six steps from the front door in the center up to the central dome, which is ninety-six meters high, sharing the honor of highest structure with the dome of St. Stephen's Basilica, also ninety-six meters high.

- number of the French Département of Val-d'Oise.

## Haydn's *Symphony No. 96*

- the "Miracle" symphony (in D major).

## Justice for the 96

- slogan by bereaved families and survivors who have been pressuring the British government to investigate and explain what went wrong on April 15, 1989, when ninety-six people were crushed to death by over exuberant fans crowding into the Hillsborough field of Sheffield Wednesday Football Club for a soccer game between Liverpool and Nottingham Forest. Called "The Hillsborough Disaster," the event produced a passionate coalition demanding a response from the authorities.

## *Mars 96 Orbiter*

- a spacecraft launched by Russia on November 16, 1996, intended to land small surface probes on the Martian surface to study the physics and astrophysics of the planet. But the spacecraft failed to achieve insertion into Mars cruise trajectory and crashed about four hours after launch into the Pacific Ocean in the vicinity of northern Chile.

## Ninety Six

- a British post in South Carolina during the American Revolution, to which American general Nathanael Greene laid siege in May and June of 1781 but failed to capture, in the first land battle of the American Revolution south of New England.

# *097*

- Decimal ASCII code for [lower-case (a)].

# 97

- Article of the Uniform Code of Military Justice (UCMJ) dealing with unlawful detention.
- atomic number of the chemical element berkelium, symbol Bk.
- largest prime number below 100.
- most consecutive free throws scored in an NBA season (basketball record): by Mike Williams for the Minnesota Timberwolves, between March 24 and November 9, 1993.
- most combined points scored in a Pro Bowl game (football record): NFC (fifty-five) against AFC (fifty-two), 2004.
- number of people aboard the *Hindenburg* when it crashed at Lakehurst, New Jersey at 7:25 PM, on May 6, 1937; thirty-six of those aboard died.
- 97 percent of the world's water is saline; the rest is fresh water.
- 97 pounds: the weight of the weakling who gets sand kicked in his face in the adverts for the Charles Atlas body-building course.
- total number of pitches thrown by New York Yankee Don Larsen in the only perfect game in World Series baseball history. The game, at Yankee Stadium on October 8, 1956, played against the the Brooklyn Dodgers, took two hours and six minutes, ending in a 2–0 victory.

### "The Wreck of the Old 97"

- recorded by Vernon Dalhart in 1924, the first American record (a Victor 78) to sell a million copies. The Old 97 was a mail train on the Southern Line that plunged off the 75-foot Stillhouse Trestle near Danville, Virginia, on September 27, 1903, killing eleven people.

# 098

- Decimal ASCII code for [lower-case (b)].
- Dewey Decimal System designation for printed forgeries and hoaxes.
- medical classification code for gonococcal infections.

# 98

- atomic number of the chemical element californium, symbol Cf.
- international telephone calling code for Iran.
- lunch-counter code for assistant manager.
- most men's U.S. Open singles tennis matches won: by Jimmy Connors (against seventeen losses), for a winning percentage of 85.2, 1970–89, 1991–92.
- 98 yards: longest punt in NFL history (football record): by Steve O'Neal for the New York Jets against the Denver Broncos, September 21, 1969.

## Generation of '98

- the Spanish writers, poets, and intellectuals who, after Spain's defeat in the Spanish-American War, worked to reenergize and revive her place in the literary world. Spain's loss of prestige and the last shreds of her empire motivated many writers and thinkers to reexamine her heritage and her destiny, and to strive to reestablish a sense of national self-respect.

## KAR 98

- basic infantry rifle of the German Army in World War II, "KAR" indicating it was a carbine.

## Olds 98

- top-of-the-line Oldsmobile car from General Motors, manufactured in 1946 to a model which, according to the *Rap Dictionary*, was considered in the 'hood to be the best make of car, year after year.

## 98.6 degrees Fahrenheit

- commonly accepted average normal body temperature (= 37 degrees Celsius).

# 099

- Decimal ASCII code for [lower-case (c)].

# 99

- age of Abraham when he was circumcised (Genesis 17:24).
- atomic number of the chemical element einsteinium, symbol Es.
- lunch-counter code for the manager, or "See the manager," or "Report to the boss."
- 99 yards:
  - ⇨ longest kickoff return in a Super Bowl game (football record): by Desmond Howard for the Green Bay Packers against the New England Patriots, SB-XXXI, 1997.
  - ⇨ longest pass completion in an NFL game (football record): several times, most recent, Gus Frerotte to Bernard Berrian for the Minnesota Vikings against the Chicago Bears, November 30, 2008.
  - ⇨ longest run from scrimmage in an NFL game (football record): by Tony Dorsett for the Dallas Cowboys against the Minnesota Vikings, January 3, 1983.

- ninety-nine years: the standard length of some land leases.
- number of beads on a Muslim rosary, called the *tasbih.*
- number of counties in the state of Iowa.
- number of Heavenly Tengri, the highest layer of dieties among the ancient Mongols, divided into 55 Western Tengri and 44 Eastern Tengri.
- number of Hershey's Kisses in one pound, according to the official Hershey website.
- retired football jersey number of:
  - ⇨ Jerome Brown, retired by the Philadelphia Eagles.
  - ⇨ Marshall Goldberg, retired by the Arizona Cardinals.

- retired hockey shirt number of Wayne Gretzky, retired by both the Edmonton Oilers and the Los Angeles Kings. On April 18, 1999, following Gretzky's retirement, it became the only shirt number retired by the entire National Hockey League. Gretzky played his last game on January 24, 1999, at the Ice Palace in Tampa. The number had also been worn by Leo Bourgeault with the Montreal Canadiens during the 1934–35 season, Rick Dudley with the Winnipeg Jets in 1981, Joe Lamb with the Montreal Canadiens during the 1934–35 season, and Wilf Paiement with the Toronto Maple Leafs, 1979–82.

- uniform number of Cleveland Indians' pitcher Rick "Wild Thing" Vaughn (played by Charlie Sheen) in the David S. Ward film *Major League.*

## Agent 99

- Agent Susan Hilton, played by Barbara Felton, on the TV show *Get Smart.* On NBC, 1965–69, and CBS, 1969–70.

## "99 Bottles of Beer on the Wall"

- drinking song in which each successive verse counts one less bottle.

## 99 percent

- said of an opinion or a piece of information that is not totally certain, as in "I'm 99 percent sure."

## The Ninety-Nines

- The International Organization of Women Pilots, founded in 1929.

## 99 to 0

- the Senate confirmation vote on September 21, 1981, to approve Sandra Day O'Connor, the first woman appointed to the Supreme Court.

## *100*

- atomic number of the chemical element fermium, symbol Fm.
- cents in a dollar; in the UK, 100 pence in a pound.
- Decimal ASCII code for [lower-case (d)].
- in Hebrew, the numerical value of the letter *koph.*
- in India, 100 is the telephone number of the police.
- lepta in a drachma, an aluminum coin of modern Greece.
- links in a chain (surveyor's measure).
- most pass completions in Super Bowl career (football record): by Tom Brady for the New England Patriots, SB-XXXVI, 2002; SB-XXXVIII, 2004; SB-XXXIX, 2005; and SB-XLII, 2008.
- number of counties in the state of North Carolina.
- number of eyes on the all-seeing Argus in Greek mythology.

- number of senators in U.S. Congress (two from each state).
- number of tiles in a Scrabble set.
- 100 feet:
  - ⇨ width of a regulation ice hockey rink.
  - ⇨ longest putt in golf Masters Tournament history: sunk by Nick Faldo for a birdie on the second hole at Augusta National in April, 1989.

- 100 kilograms: maximum body weight for men weightlifters in category 8, Olympic competition.
- 100 meters: shortest track event recognized by both the IAAF and the Olympics. Current ratified record for men is 9.58 seconds, made by Jamaican Usain Bolt at the IAAF World Athletics Championships in Berlin on August 16, 2009; and for women, 10.49 seconds by American Florence Griffith-Joyner in Indianapolis, Indiana, on July 16, 1988.
- 100 yards:
  - ⇨ length of a football field between the goalposts.
  - ⇨ length of a regulation field hockey field.
  - ⇨ longest interception return in a Super Bowl game (football record): by James Harrison for the Pittsburgh Steelers, SB-XLIII, 2009.

- hundred years:
  - ⇨ age of comedian Bob Hope when he died on July 27, 2003.
  - ⇨ length of Sleeping Beauty's nap before being awakened by a prince.

- operator telephone number in the UK.
- pounds in a short hundredweight.
- record number of goals scored in one NHL season, including playoffs (hockey record): by Wayne Gretzky for the Edmonton Oilers, 1983–84 (87 goals in seventy-four regular games, and 13 goals in nineteen playoff games).
- record number of individual points scored in a professional basketball game, by Wilt Chamberlain for the Philadelphia Warriors against the New York Knicks on March 2, 1962 (thirty-six field goals, twenty-eight free throws).
- the square of 10.
- sum of the first four cubic numbers (1 + 8 + 27 + 64).

- sum of the first nine prime numbers $(2 + 3 + 5 + 7 + 11 + 13 + 17 + 19 + 23)$.
- value of Roman numeral C.
- yards in a bolt (cloth measure).
- years in a century.

## F-100

- U.S. Air Force Super Sabre, nicknamed "The Hun," produced by North American Aviation, the first fighter capable of supersonic speed in level flight. The first prototype, YF-100, bettered the speed of sound on its first flight in May 1953. Later models achieved speeds over 1,000 miles per hour.

## Haydn's *Symphony No. 100*

- "Military" symphony (in G major).

## Hecatoncheires

- hundred-handed beings, sons of Uranus and Gaea. According to Hesiod, they were three in number: Briareus, Cottus, and Gyges.

## hundred aces

- in the card game pinochle, a meld of four aces, each of a different suit, is worth hundred points.

## The Hundred Days

- (a) the short rule of Napoleon (March 20 to June 22, 1815) in his attempt to regain the throne of France, concluding with his abdication on June 22 after his severe defeat at Waterloo, June 18. The period is measured up to the day of the second restoration of Louis XVIII.
- (b) the first days of the Roosevelt administration in 1933, during which the new president pushed through legislation on banking, agriculture, employment, conservation, and other components of his New Deal.

## The Hundred Days of Reform

- following China's defeat by Japan in 1895, pressures increased to westernize China and to abandon the monarchy in favor of a constitutional state. Under a government led by reformer K'ang Yu-wei, several edicts emerged from the imperial court, intended to modernize Chinese society. Included were the adoption of a public school system, the inclusion of Western subjects in Chinese education, popular election of local officials, remodeling of the Chinese bureaucracy, and eventual creation of a national parliamentary government. Most of China ignored the edicts of the reform government, and after barely three months (June 11 to September 21, 1898), a coup d'état returned power to the Empress Dowager, Ci Xi.

## 100 degrees Celsius

- temperature at which water boils (= 212 degrees Fahrenheit).

## 100 degrees Fahrenheit

- = 37.8 degrees Celsius.

## Hundred Flowers Campaign

- movement started in 1956 within the Chinese Communist government to ease restrictions on Chinese intellectuals and to allow some freedom of thought and speech. Sparked by Nikita Khrushchev's disparagement of Stalin, Mao invited criticism of the Communist Party's policies with the slogan, "Let a hundred flowers bloom, and a hundred schools of thought contend." But the response went too far, and by the summer of 1957, the government was cracking down on critics of the regime.

## 100 Mile House

- a town on Cariboo Highway in south-central British Columbia, Canada.

## hundredweight

- unit of weight equal to 100 pounds in the U.S., called a short hundredweight; in the UK, equal to 112 pounds, known as a long hundredweight. As a unit of mass, the short hundredweight equals 45.4 kilograms, and the long hundredweight equals 50.8 kilograms.

## Hundred Years' War

- a long series of wars between England and France (1337–1453), beginning with Edward III's claim to the French crown, ending with England losing all her French possessions except Calais. Notable were the battles of Cressy (1346), Poitiers (1356), and Agincourt (1415).

## one hundred dollars

- (a) denomination of U.S. paper money that bears the portrait of Benjamin Franklin.
- (b) denomination of U.S. savings bond that bears the portrait of Thomas Jefferson.
- (c) denomination of U.S. treasury bond that bears the portrait of Andrew Jackson.

## *One Hundred Men and a Girl*

- charming mixture of music and humor, featuring Deanna Durbin singing while Leopold Stokowski conducts. This 1937 film earned an Oscar for Charles Previn's musical supervision.

## 100-point card

- in the game of canasta, a 3 of diamonds or 3 of hearts.

## 100 proof

- (a) describes an alcoholic drink having a 50 percent alcohol content, as many vodkas do.
- (b) genuine; strongest; top-of-the-line (from 100-proof whiskey, the highest alcohol content allowed in the U.S.).

## *One Hundred Years of Solitude*

- Gabriel García Márquez's classic novel (first published in Spanish in 1967 as *Cien Años de Soledad*), which chronicles a century of life in the mythic Colombian village of Macondo. Considered a masterpiece for its "astounding vision of human life and death, a full measure of humankind's inescapable potential and reality."

**Second Hundred Years' War**

- between England and France (1689–1763), which virtually eliminated France as a power in North America and thus reduced the colonists' need to look to England for protection, thereby planting the seeds of future independence.

**sweet one hundreds**

- cherry tomatoes, so called because one hundred make a pound.

# ACKNOWLEDGMENTS

Writing is known to be a solitary avocation, but converting the written word into a medium that can reach others is a group effort. With that in mind, I must express my gratitude, in varying degrees, to several folks who have contributed to this insanely detailed enterprise. To David Fischer, as always, for being a bottomless font of useful information on all aspects of sports; to David Vogel for his knowledge of matters scientific and technical, and for his ability to transmit that knowledge comprehensibly; to Ted Sennett, for his broad familiarity with movies, and a memory providing facile accessibility to his store of such data; to Joyce Altman for her expert editorial advice on some sticky issues that I couldn't comfortably resolve on my own; to my daughter Liza, for helping me distinguish the significant from the inconsequential; to Eric Schaper, for keeping my computer functioning after several potentially disastrous events; and to Marvin Duckler, just for being Marvin Duckler. And to Mark Weinstein, along with his colleagues at Skyhorse Publishing, for putting up with the vagaries of an author who had to have things a particular way, and for suffering such imperatives with notable tolerance.

Thank you every one, and to all a good night.

# ABOUT THE AUTHOR

Herb Reich spent fifteen years as a senior acquisitions editor at John Wiley & Sons. Before that, he administered the behavioral science publishing program at Basic Books, and served as editorial director of the Macmillan Book Clubs. Along the way, he spent two post-grad years affiliated with the Research Foundation of the State University of New York; edited *The Odyssey Scientific Library;* and contributed to *The Random House Dictionary of the English Language* and the Corsini *Encyclopedia of Psychology.* The author of *Don't You Believe It!,* he lives in Hastings-On-Hudson, New York.